Simon Gogl

Laying the Foundations of Occupation

Jahrbuch für Wirtschaftsgeschichte

Im Auftrag der Herausgeber des Jahrbuchs
für Wirtschaftsgeschichte herausgegeben von
Alexander Nützenadel und Jochen Streb

Beiheft 27

Simon Gogl

Laying the Foundations of Occupation

Organisation Todt and the German Construction Industry in Occupied Norway

DE GRUYTER
OLDENBOURG

This thesis has been submitted for the degree of Philosophiae Doctor (PhD) to the Faculty of Humanities at the Norwegian University of Science and Technology in June 2019. It has been defended in Trondheim on 25 October 2019.

ISBN 978-3-11-069410-9
e-ISBN (PDF) 978-3-11-069429-1
e-ISBN (EPUB) 978-3-11-069433-8

Library of Congress Control Number: 2020935875

Bibliografische Information der Deutschen Nationalbibliothek
Die Deutsche Nationalbibliothek verzeichnet diese Publikation in der Deutschen Nationalbibliografie; detaillierte bibliografische Daten sind im Internet über http://dnb.dnb.de abrufbar.

Coverabbildung: Organisation Todt construction site in Aukra (Møre og Romsdal).
Source: Romsdalsmuseet, Molde.
Satz: Integra Software Services Pvtl. Ltd.
Druck und Bindung: CPI books GmbH, Leck

www.degruyter.com

Acknowledgements

The present publication is a slightly revised version of my PhD thesis that I submitted to the Faculty of Humanities at the Norwegian University of Science and Technology, Trondheim, in June 2019. It could not have been written without the ceaseless support and advice of many people. First and foremost, I want to thank my supervisors Professor Jonas Scherner and Professor Hans Otto Frøland at the Norwegian University of Science and Technology who encouraged me to write this thesis. They have not only patiently answered, but also posed the right questions to open up ever new perspectives on my research topic to me. They have commented on my drafts more times than I can count and have supported my work in every imaginable way.

I have written this thesis in close association with the Trondheim-based research project *The Political Economy of Forced Labour: Organisation Todt in Norway during World War II.* Starting almost from nothing, this project has investigated the history of the *Organisation Todt* in occupied Norway. I owe a debt of gratitude to the members of the project: Ketil Gjølme Andersen, Mats Ingulstad, Gunnar D. Hatlehol and Torgeir E. Sæveraas. Thank you for your comments, ideas and pioneering archival work in the Organisation's archives at the National Archives in Oslo.

Many others have given me the opportunity to discuss my work with them on various occasions, answered my questions and provided me with valuable information and sources: Professor Harald Espeli, Professor Maria Fritsche, Professor Rüdiger Hachtmann, Eirik Holmen, Mari Olafson Lundemo, Pål Nygaard, Professor Pål Thonstad Sandvik, Professor Mark Spoerer, Björn Stemler and Michael Stokke. I also want to thank all my colleagues at NTNU's Department for Historical Studies and the members of the research project *In a World of Total War: Norway 1939–1945.* All participants of seminars, conferences and workshops in Oslo, Berlin, Trondheim, Mannheim, Narvik and Dortmund deserve thanks for their suggestions and critical comments, which were crucial in sharpening my arguments.

This thesis was partly written and completed in Bergen, Norway. I was welcomed with open arms at the University of Bergen's Department of Archaeology, History, Cultural Studies and Religion, and was generously allowed to use their research infrastructure. In particular, I want to thank Professor Camilla Brautaset for her invaluable personal and professional support. Furthermore, I want to thank Professor Christhard Hoffmann and UiB's research group on transnational history for the great opportunity to present and defend the draft of my thesis in a *sluttseminar*. It goes without saying that all inaccuracies and mistakes remain my own.

I also owe thanks to the staff of the archives and libraries that I have visited and used for my work: The staff at the Federal Archives in Berlin and Freiburg im Breisgau, the National Archives in Oslo, the *Romsdalsmuseet* in Molde, the (Main) State Archives in Weimar, Wiesbaden, Berlin, Merseburg and Munich, the archives

https://doi.org/10.1515/9783110694291-202

of the *Institut für Zeitgeschichte* and *Deutsches Museum* in Munich, and of the *Institut für Stadtgeschichte* in Frankfurt am Main. Many thanks to Dr. Martin Krauß at the *Unternehmensarchiv der Bilfinger SE* in Mannheim, and the staff at the university libraries in Trondheim, Bergen and Berlin, the *Staatsbibliothek zu Berlin* and the Bavarian State Library in Munich.

Last, but not least, I want to thank my friends and family for their support during the past years, particularly my wife Jana, who has made all this possible, and my son August, who is blessed with good sleep – which has probably been the most important factor in the successful completion of this PhD project.

Bergen, March 2020.
Simon Gogl

All translations from German and Norwegian are my own, unless stated otherwise.
Tonnes means metric tonnes throughout.

Contents

List of Abbreviations

ABL	Abschnittsbauleitung der Organisation Todt
Arge	Arbeitsgemeinschaft
A/S	Aksjeselskap
BL	Bauleitung der Organisation Todt
DAF	Deutsche Arbeitsfront
DAV	Verordnung zur Begrenzung von Gewinnausschüttungen
D.B.H.G.	Deutsche Bergwerks und Hüttenbau GmbH
ERKA	Reichs-Kredit-Gesellschaft
EW	Einsatzgruppe Wiking
GAV	Verordnung über die Gewinnabführung
GB Chem	Generalbevollmächtigter für Sonderfragen der chemischen Erz/eugung
GBA	Generalbevollmächtigter für den Arbeitseinsatz
GBI	Generalbauinspektor für die Reichshauptstadt
GB Bau	Generalbevollmächtigter für die Regelung der Bauwirtschaft
GdSt	Generalinspektor für das deutsche Straßenwesen
GmbH	Gesellschaft mit beschränkter Haftung
HAFRABA	Verein zur Vorbereitung der Autobahn Hansestädte – Frankfurt – Basel
IdLN	Inspekteur des Landesbefestigung Nord
InWest	Inspekteur der Westbefestigungen
KBN	Kriegsgefangenen-Bezirkskommandant Norwegen
KG	Kommanditgesellschaft
kr.	Norske kroner
KWVO	Kriegswirtschaftsverordnung
LAA	Landesarbeitsamt
LSBÖ	Leitsätze für die Preisermittlung auf Grund der Selbstkosten bei Bauleistungen für öffentliche Auftraggeber
MStGB	Militärstrafgesetzbuch
NACO	Norsk Aluminium Company
NATM	New Austrian Tunnelling Method
NS	Nasjonal Samling
NSB	Norges Statsbaner
NSDAP	Nationalsozialistische Deutsche Arbeiterpartei
NSKK	Nationalsozialistisches Kraftfahrkorps
OBL	Oberbauleitung der Organisation Todt
OBR	Oberste Bauleitung der Reichsautobahnen
OKM	Oberkommando der Marine
OKW	Oberkommando der Wehrmacht
OT	Organisation Todt
OTZ	Organisation Todt-Zentrale
PA	Principal-Agent
POW	Prisoner of war
RAD	Reichsarbeitsdienst
RAM	Reichsarbeitsministerium
RdI	Reichsverband der deutschen Industrie
RfAA	Reichsanstalt für Arbeitsvermittlung und Arbeitslosenversicherung
RFM	Reichsfinanzministerium
RK	Reichskommissariat für die besetzten norwegischen Gebiete

https://doi.org/10.1515/9783110694291-204

RK See	Reichskommissar für die Seeschiffahrt
RM	Reichsmark
RoE	Return on Equity
RoS	Return on Sales
Rpf.	Reichspfennig
RTOLL	Reichstarifordnung über den Leistungslohn im Baugewerbe
RVM	Reichsverkehrsministerium
RWM	Reichswirtschaftsministerium
SA	Sturmabteilung der NSDAP
SRA	Statistisches Reichsamt
Strabag	Straßenbau Aktiengesellschaft
SS	Schutzstaffel der NSDAP
VOB	Verdingungsordnung für Bauleistungen
WBN	Wehrmachtbefehlshaber Norwegen
WDStO	Wehrmachtsdisziplinarstrafordnung
Wigru Bau	Wirtschaftsgruppe Bauindustrie

I Introduction: Rebuilding a Continent

Between 1933 and 1945, under the aegis of National Socialist economic policy, German industry overcame a disastrous economic crisis and experienced a boom fuelled by rearmament, before ultimately placing itself in the service of an all-embracing war machinery. No aspect of economic life in Germany was able to elude this turn of events. But few sectors of the industry mirrored this development as obviously as construction. The turnovers and profits of German construction companies soared in the course of the 1930s. Industrial and military building programmes as well as the construction of the *Autobahn* and the *Westwall* (Siegfried Line) helped to reduce unemployment and prepared the Reich for the war to come. In June 1938, almost 10 percent of the German workforce was employed in construction alone, making it the largest industrial sector in the country's economy.[1]

After the outbreak of the war, this massive German construction activity spread all over occupied Europe. The bunkers and gun emplacements of the Atlantic Wall from the Bay of Biscay to the Arctic Sea, airfields, roads and bridges, railways and dams were meant to secure German rule and bind the occupied territories to the new hegemon. More and more often, public authorities like the Armaments Ministry, the Office of the Four Year Plan (*Vierjahresplanbehörde*) and the *Wehrmacht* entrusted the planning and administration of their projects to an organisation that had proven its effectiveness during the construction boom of the 1930s: *Organisation Todt* (OT).

The Organisation administered its huge projects with a comparatively small administrative staff consisting of architects and engineers. It provided the construction sites with labour and building materials, while the execution of the projects lay in the hands of private companies. The latter brought along skilled workers, administrative personnel, machines and their technical expertise. During the war, the Organisation had the legal status of a *Wehrmacht* auxiliary force (*Wehrmachtgefolge*).

This thesis tells the story of the relation between the *Organisation Todt* and the German construction industry. In particular, it will focus on *Einsatzgruppe Wiking* (EW), the Organisation's subsidiary in occupied Norway, and its contractors.[2] The thesis rests on the assumption that the OT's German contractors were so vital to its *modus operandi* that we can only understand the Organisation if we enquire into private industry's role in it. By investigating this relation, I will address three closely related questions. First, I ask what kind of organisation the OT actually was. Only an analysis of the relation between the OT and the construction industry will enable us to situate the OT more precisely in the field between National Socialist

1 *Statistisches Handbuch von Deutschland 1928–1944* (München: Franz Ehrenwirth-Verlag, 1949), 478–79.

2 *Einsatzgruppe Wiking* also carried out projects in Denmark and, from summer 1944 onwards, Finland. This thesis, however, focuses on Norway, which was EW's most important area of activity.

https://doi.org/10.1515/9783110694291-001

state agencies, the military and private business. Second, I ask to what degree and in which fields of business decisions German construction firms were able to maintain their entrepreneurial freedom of action both under the OT and in the face of other institutions' attempts to steer the construction sector. While construction was a highly regulated sector in the Third Reich, the question will be to what degree the state was able to effectively implement regulations and monitor their observance. Third, I put the focus on Norway, in order to answer the question of how the relation between the OT and its contractors and state regulation of German construction activities worked in an occupied territory. Norway is a particularly promising object of research as there is probably no other occupied country where construction took on such a central role during the Second World War and where such a large amount of sources from an OT mission has survived in the archives. The focus on Norway will also contribute to a better and partly novel understanding of how German construction activities affected the economy of occupied Norway.

The OT was led by and named after the road engineer Fritz Todt, who had been appointed General Inspector for German Roads (*Generalinspektor für das deutsche Straßenwesen*, GdSt) in 1933 to supervise the construction of the German *Autobahn*. Within a decade, he became one of the most powerful figures of the Third Reich by accumulating more than half a dozen offices. He was commissioned with the construction of the *Westwall*, the birthplace of the OT, in 1938, became General Commissioner for the Regulation of the Construction Industry (*Generalbevollmächtigter für die Regelung der Bauwirtschaft*, GB Bau) in December of the same year, before becoming Reich Minister for Armaments and Munitions in 1940.[3] After Todt's fatal accident in 1942, Albert Speer succeeded him in all his positions, managing both the armaments and the construction sector.[4] However, keeping the Reich's armaments production running occupied Speer more and more, and thus the head of the OT headquarters, Xaver Dorsch, was *de facto* leading the Organisation during the last years of the war. Although he was one of the less prominent men within the Nazi leadership circles, the ambitious Dorsch soon became the most influential person in the German construction sector.[5] What Todt, Speer and Dorsch had in common was their absolute loyalty to Adolf Hitler and unconditional devotion to the cause of National Socialism. Something else they had in common was that they were remembered as unpolitical technocrats – not only in post-war Germany, but also in the English-speaking world. Their memory was perpetuated by apologetic historians in the case of Todt and Dorsch,[6] and

3 On Fritz Todt, see chapter II.2.a.
4 On Albert Speer, see chapter IV.1.c.
5 On Xaver Dorsch, see chapter IV.1.a.
6 For some particularly apologetic views on Todt, see for example: Franz W Seidler, *Fritz Todt. Baumeister des Dritten Reiches* (1986; repr., München: F. A. Herbig, 1988), 334, 356; Karl-Heinz Ludwig, *Technik und Ingenieure im Dritten Reich* (Düsseldorf: Droste Verlag, 1974), 382–93.

personal testimonies full of half-truths and clever fabrications in the case of Speer.[7]

As leaders of the OT, the three men shaped the way in which public construction projects in the German Reich were carried out. During the construction of the *Autobahn*, Todt, Fritz had already avoided establishing a large central administration, instead setting up smaller units – *Oberste Bauleitungen der Reichsautobahnen* (OBRs) – that supervised construction works locally. Building on the experiences from the mid-1930s, OT's administrative apparatus remained lean and flexible by combining the power of a state agency with the flexibility of private industry in a unique way. It was an entirely new and powerful protagonist that transcended the boundaries of established institutions both within the Reich and the occupied territories.

During the Second World War, the centre of OT's activities was occupied Europe, where the Organisation became the most important German construction customer besides the *Wehrmacht*. From 1942 onwards, the Organisation established regional task forces, *Einsatzgruppen*, to carry out construction projects with a high degree of autonomy. In occupied Norway, *Einsatzgruppe Wiking* was established on 1 April 1942. The EW administered most of the construction activities in Norway, including the massive coastal fortifications that were part of the Atlantic Wall, roads, railways, power stations and submarine bases. Before long, EW's projects took a central place in the country's economy and the occupants brought approximately 130,000 foreign labourers to Norway, making it the only occupied territory that became a net importer of forced labour during the war.[8] The majority of these forced labourers worked for the EW at some point in time between 1942 and 1945. EW distributed them between the construction sites that were operated by the companies. Around 500 German firms followed the call northwards as contractors of the EW.

1 Literature

To start with, the following literature review assesses the research on the OT and the construction industry in National Socialist Germany in general. To what extent did it allow for the close and unique integration of private business and state administration? Likewise, I discuss whether the literature concerned with the occupation of

7 On Speer's self-staging in post-war Germany, see most recently: Isabell Trommer, *Rechtfertigung und Entlastung: Albert Speer in der Bundesrepublik* (Frankfurt am Main: Campus Verlag, 2016); Magnus Brechtken, *Albert Speer. Eine deutsche Karriere* (München: Siedler, 2017).

8 Hans Otto Frøland, Gunnar D. Hatlehol, and Mats Ingulstad, 'Regimenting Labour in Norway during Nazi Germany's Occupation', *Working Papers of the Independent Commission of Historians Investigating the History of the Reich Ministry of Labour (Reichsarbeitsministerium) in the National Socialist Period No. 12*, 2017, 12.

Norway has gained some understanding of the massive transformation that the EW's engagement implied for the Norwegian economy. In part b, I will then review the extensive literature on state-business relations in the Third Reich that has generated important insights into how the state attempted to steer and regulate the industry and how companies operated within this environment. The aim is to identify fruitful perspectives that can guide an analysis of the relation between the OT and its contractors. Finally, I will take a closer look on the research that has highlighted the importance of contracts in state-business relations under the National Socialist rule. I will also present the assumptions of Principal-Agent (PA) theory, which earlier research has productively used to understand contractual relations between awarding authorities and contractors in construction.

a Organisation Todt and the German Construction Sector

In late November 1943, the city of Berlin experienced some of heaviest air raids of the war. After having hit the buildings of the Ministry of Armaments and War Production the previous night, the bombers of the Royal Air Force headed to the western districts of the city once more on 23 and 24 November. This time, their bombs destroyed the *OT-Zentrale* (OTZ), the headquarters of *Organisation Todt*, and with it, tonnes of documents on the history, structure and projects of the Organisation.[9]

Due to the loss of most of its records, the OT is still one of the Third Reich's organisations we know least about. Until the late 1990s, research was based mainly on the fragmentary records at the German Federal Archives. Hedwig Singer has edited a compilation of key documents, together with a brief overview over the history of the OT.[10] The volume also contains accounts by Xaver Dorsch and Rudolf Dittrich, a former staff member of the Armaments Ministry.[11] German historian Franz W. Seidler has written the only extensive history of the OT and the only biography on Fritz Todt, the latter also containing a comprehensive chapter on the Organisation.[12] Seidler

9 Hedwig Singer, *Entwicklung und Einsatz der Organisation Todt*, Quellen zur Geschichte der Organisation Todt, 1/2 (Osnabrück: Biblio Verlag, 1998), 6; Martin Middlebrook and Chris Everitt, *The Bomber Command War Diaries. An Operational Reference Book 1939–1945* (Barnsley: Pen & Sword Aviation, 2014), 454.

10 Singer, *Organisation Todt.*

11 Franz Xaver Dorsch, *Die Organisation Todt*, Quellen zur Geschichte der Organisation Todt, 1/2 (1950; repr., Osnabrück: Biblio Verlag, 1998); Rudolf Dittrich, *Vom Werden, Wesen und Wirken der Organisation Todt. Ausarbeitung für die Historical Division / US Army in Europe*, Quellen zur Geschichte der Organisation Todt, 1/2 (Osnabrück: Biblio Verlag, 1998).

12 Franz W Seidler, *Die Organisation Todt. Bauen für Staat und Wehrmacht 1938–1945*, 2nd ed. (1987; repr., Koblenz: Bernard & Graefe Verlag, 1998); Seidler, *Fritz Todt.* Two more articles on the Organisation by Seidler are: Franz W Seidler, 'Das Nationalsozialistische Kraftfahrkorps und die

provides a broad and helpful overview over the OT's organisational structure and mode of operation. As he based his research exclusively on sources from a handful of German archives, however, his works lack the details and fruitful perspectives that can be obtained from non-German or company archives. When it comes to forced labour, the involvement of OT members in war crimes or his interpretation of Fritz Todt's role in the Third Reich, Seidler's research is clearly apologetic and outdated. Nevertheless, it remains a point of departure for every research on the OT. Another source is a report on the OT of March 1945 by Allied military intelligence, which draws on interrogations of OT members and documents that fell into the hands of Allied troops during their advance in the last months of the war.[13] Also in 1945, the International Labour Office in Montreal published a short overview of the Organisation that was mostly based on newspaper articles.[14] Taken together, however, these works have not allowed us to gain more than a rough idea of the structure and *modus operandi* of the OT.

Another explanation for the noticeable lack of research on the OT might be that academic works on the Nazi economic system or even on a genuinely "European" subject like forced labour for a long time focused on the situation within the German Reich. The OT, however, was active mostly in the occupied territories. This might be one reason why the construction firms working under the OT, too, have received little attention. Articles often are based on records that *Einsatzgruppen* and OBLs left behind during their retreat in 1944–45 and focus on the Organisation's activities in a certain region of Germany or Europe. Other historians have taken the booming research on National Socialist forced labour as a starting point and laid the focus on certain groups of labourers in the OT's construction sites.[15] Fabian Lemmes' study of the OT in Italy and France is certainly the most detailed and comprehensive work on

Organisation Todt im Zweiten Weltkrieg. Die Entwicklung des NSKK bis 1939', *Vierteljahreshefte für Zeitgeschichte* 32, no. 4 (1984): 625–36; Franz W Seidler, 'L'Organisation Todt', *Revue d'histoire de la Deuxième Guerre mondiale et des conflits contemporains* 34, no. 134 (1984): 33–58.

13 *Handbook of the Organisation Todt by the Supreme Headquarters Allied Expeditionary Force Counter-Intelligence Sub-Division MIRS/MR-OT/5/45*, Quellen Zur Geschichte Der Organisation Todt 4 (Osnabrück: Biblio Verlag, 1992). The report has been republished recently (with minor changes): John Christopher, ed., *Organisation Todt: From Autobahns to the Atlantic Wall* (Stroud: Amberley, 2014).

14 John H. E. Fried, *The Exploitation of Foreign Labour by Germany* (Montreal: International Labour Office, 1945), chap. VII.

15 Edith Raim, *Die Dachauer KZ-Außenkommandos Kaufering und Mühldorf. Rüstungsbauten im letzten Kriegsjahr 1944/45* (Landsberg am Lech: Landsberger Verlagsanstalt Martin Neumeyer, 1992); Madeleine Bunting, *The Model Occupation: The Channel Islands under German Rule 1940–1945* (London: HarperCollins, 1995); Johannes Grötecke, *Edertalsperre, Wiederaufbau nach der Zerstörung: 1943–1945* (Marburg: Jonas Verlag, 1996); Stefan Laube, 'Technik, Arbeit und Zerstörung. Die Organisation Todt in Prag (1944–1945)', *Bohemia* 40, no. 2 (1999): 387–416; Bella Guttermann, 'Jews in the Service of Organisation Todt in the Occupied Soviet Territories, October 1941–March 1942', *Yad Vashem Studies* 29 (2001): 65–107; Louise Willmot, 'The Goodness

regional OT missions in recent historiography.[16] However, the research on the Organisation remains patchy and many questions are still unanswered.

This is particularly true of the relation between the OT and the construction companies, which formed the backbone of the Organisation. Many questions have not yet been analysed systematically, such as why German construction companies followed the OT even to the remotest regions of the European continent. Did the OT use threats and coercion or contractual incentives to integrate the private construction industry into its organisation? How big was the companies' entrepreneurial room for manoeuvre?

Existing literature seldom addresses such questions and hardly differentiates at all between the OT's administrative apparatus and the construction companies.[17] It often completely omits the phase from the first contacts between the OT and the

---

of Strangers: Help to Escaped Russian Slave Labourers in Occupied Jersey, 1942–1945', *Contemporary European History* 11, no. 2 (2002): 211–27; Piotr Kruszyński, 'Die Ausbeutung der Häftlingsarbeit im Komplex Riese im Eulengebirge durch die Organisation TODT und mitarbeitende Firmen', in *Die Ausnutzung der Zwangsarbeit der Häftlinge des KL Groß-Rosen durch das Dritte Reich* (1999; repr., Wałbrzych: Muzeum Gross-Rosen, 2004), 40–54; Sabine Rutar, 'Arbeit und Überleben in Serbien. Das Kupfererzbergwerk Bor im Zweiten Weltkrieg', *Geschichte und Gesellschaft* 31, no. 1 (2005): 101–34; Christine Glauning, *Entgrenzung und KZ-System. Das Unternehmen „Wüste" und das Konzentrationslager in Bisingen 1944/45*, Geschichte der Konzentrationslager 1933–1945 7 (Berlin: Metropol Verlag, 2006); Scott Soo, 'Ambiguities at Work: Spanish Republican Exiles and the Organisation Todt in Occupied Bordeaux', *Modern & Contemporary France* 15, no. 4 (2007): 457–77; Sabine Schalm, *Überleben durch Arbeit? Außenkommandos und Außenlager des KZ Dachau 1933–1945*, Geschichte der Konzentrationslager 1933–1945 10 (Berlin: Metropol Verlag, 2009). For Italian and French historiography on the OT, see Fabian Lemmes, 'Arbeiten für das Reich: die Organisation Todt in Frankreich und Italien, 1940–1945' (European University Institute Firenze, 2009), 6–30.

16 Lemmes, 'Arbeiten für das Reich'. The publication of Lemmes' study has been announced for mid-2020. See also: Fabian Lemmes, 'Zwangsarbeit im besetzten Europa. Die Organisation Todt in Frankreich und Italien, 1940–1945', in *Rüstung, Kriegswirtschaft und Zwangsarbeit im 'Dritten Reich'*, ed. Andreas Heusler, Mark Spoerer, and Helmuth Trischler (München: Oldenbourg Verlag, 2010); Fabian Lemmes, 'Arbeiten für den Besatzer: Lockung und Zwang bei der Organisation Todt in Frankreich und Italien 1940–1945', in *Zwangsarbeit in Hitlers Europa. Besatzung – Arbeit – Folgen*, ed. Dieter Pohl and Tanja Sebta (Berlin: Metropol Verlag, 2013).

17 Symptomatically, Laurie R. Cohen writes of "the building company Organisation Todt": Laurie R. Cohen, *Smolensk under the Nazis: Everyday Life in Occupied Russia* (Rochester: University of Rochester Press, 2013), 74. Armin Bergmann's short entry on the OT in the *Enzyklopädie des Nationalsozialismus* does not mention the integration of private construction companies at all: Armin Bergmann, 'Organisation Todt (OT)', in *Enzyklopädie des Nationalsozialismus*, ed. Wolfgang Benz, Hermann Graml, and Hermann Weiß (München: Deutscher Taschenbuch Verlag, 1997), 629. The private construction industry is also ignored in the chapter on the OT in: Rudolf Absolon, *Die Wehrmacht im Dritten Reich*, vol. VI (Boppard am Rhein: Harald Boldt Verlag, 1995), 54–69. Some firms working under the OT on the railway construction sites in Northern Norway are mentioned cursorily in: Joachim Petersen, *Hitlers Polareisenbahnpläne 1940 bis 1945 in Nordnorwegen* (Friedberg: Podzun-Pallas-Verlag, 1992).

companies to the contract negotiations and the signing of the contracts, or deals with the topic in a single sentence: first, companies were simply conscripted into the OT, and then forced labour was assigned to the building sites.[18]

There are some studies, however, that stand out when it comes to the relation between companies and the OT. Edith Raim's work on the construction projects in the Bavarian towns of Kaufering and Mühldorf in 1944–45 provides the most detailed study on contract negotiations and financial aspects of the OT's cooperation with private construction firms.[19] The study by Fabian Lemmes focuses on German and French enterprises under OT's *Einsatzgruppe West*. The research of Marc Buggeln on the OT's construction of the naval bunker *Valentin* in Bremen deals with the dispute between construction firms and the OT on the distribution of profits.[20] Steen Andersen has researched Danish multinational construction firms and their contacts with the OT, especially Christiani & Nielsen's engagement in France and Norway.[21] As these studies have shown, the German construction projects opened up huge opportunities for private industry. The question remains, however, whether the latter had to pay a price in the form of lost autonomy.

The historiography on the construction sector in general is likewise limited. While most works on the National Socialist economy do in fact contain some remarks on the sector's development in the course of the 1930s, we are still lacking a comprehensive history of the construction industry in the Third Reich. Although there is a trend in the field of economic history in recent years to investigate the role of private business under Nazi rule more closely, research so far has focused mostly on the armaments and autarky sector, and thus we still know surprisingly little about the German construction industry and its relation to the National Socialist state. As a matter of fact, the best-known construction projects of Nazi Germany's peace years, the *Autobahn* and the *Westwall*, have gained some attention – the former often

18 Seidler, *Die Organisation Todt*, 140–46, 238; Erhard F. Knechtel, *Arbeitseinsatz im Baugewerbe im Deutschen Reich 1939–1945* (Berlin: Hauptverband der Deutschen Bauindustrie, 2001), 15; Grötecke, *Edertalsperre*, 57.

19 Raim, *Kaufering und Mühldorf.*

20 Marc Buggeln, '"Menschenhandel" als Vorwurf im Nationalsozialismus. Der Streit um den Gewinn aus den militärischen Großbaustellen am Kriegsende (1944/45)', in *Rüstung, Kriegswirtschaft und Zwangsarbeit im 'Dritten Reich'*, ed. Andreas Heusler, Mark Spoerer, and Helmuth Trischler (München: Oldenbourg Verlag, 2010), 199–218; Marc Buggeln, *Der U-Boot-Bunker 'Valentin': Marinerüstung, Zwangsarbeit und Erinnerung* (Bremen: Edition Temmen, 2017).

21 Steen Andersen, *De gjorde Danmark større . . . De multinationale danske entreprenørfirmaer i krise og krig 1919–1947* (København: Lindhardt og Ringhof, 2005); Steen Andersen, 'Between Imperative and Risk. The Case of Christiani & Nielsen's Market Entry in Norway 1941–45', *Scandinavian Economic History Review* 59, no. 1 (2011): 3–28; Steen Andersen, 'The French Opportunity. A Danish Construction Company Working for the Germans in France, 1940–1944', in *Economies under Occupation. The Hegemony of Nazi Germany and Imperial Japan in World War II*, ed. Marcel Boldorf and Tetsuji Okazaki (London, New York: Routledge, 2015).

among historians interested in its impact on the reduction of unemployment and the surrounding propaganda, the latter mostly among military historians.[22] Still, the companies that carried out those projects are seldom analysed.[23] Studies such as those by Mark Spoerer on the profits of German companies and by Jonas Scherner on investment contracts, which have enhanced our knowledge by presenting broad empirical evidence on German businesses under the Nazi rule, do not cover construction companies either.[24] However, to understand the construction industry's bargaining position and strategies during the negotiations with the OT, we need to know more on the economic conditions in different sub-sectors of the industry at different points in time. Similarly, there is still a research deficit regarding the rationing and allocation of building materials and machines, which were two of the most important administrative means to steer the sector.[25]

---

22 On the *Autobahn*: Hansjoachim Henning, 'Kraftfahrzeugindustrie und Autobahnbau in der Wirtschaftspolitik des Nationalsozialismus 1933 bis 1936', *Vierteljahrschrift für Sozial- und Wirtschaftsgeschichte* 65, no. 2 (1978): 217–42; Rainer Stommer, ed., *Reichsautobahn: Pyramiden des Dritten Reiches. Analysen zur Ästhetik eines unbewältigten Mythos* (Marburg: Jonas Verlag, 1982); James D. Shand, 'The Reichsautobahn: Symbol for the Third Reich', *Journal of Contemporary History* 19, no. 2 (1984): 189–200; Erhard Schütz and Eckhard Gruber, *Mythos Reichsautobahn. Bau und Inszenierung der 'Straßen des Führers' 1933–1941* (Berlin: Christoph Links Verlag, 1996); Friedrich Hartmannsgruber, '". . . ungeachtet der noch ungeklärten Finanzierung". Finanzplanung und Kapitalbeschaffung für den Bau der Reichsautobahnen 1933–1945', *Historische Zeitschrift* 278, no. 3 (2004): 625–81. On the *Westwall*: Walter Frederik Renn, 'Hitler's West Wall: Strategy in Concrete and Steel 1938–1945' (The Florida State University, 1970); Manfred Gross, *Der Westwall zwischen Niederrhein und Schnee-Eifel* (Köln: Rheinland-Verlag, 1982); Dieter Bettinger and Martin Büren, *Der Westwall. Die Geschichte Der Deutschen Westbefestigungen Im Dritten Reich*, vol. 1, 2 vols (Osnabrück: Biblio Verlag, 1990).

23 Two exceptions are the studies of Lärmer and Silverman, which investigate the motives of companies participating in the construction of the *Autobahn*: Karl Lärmer, *Autobahnbau in Deutschland 1933 bis 1945. Zu den Hintergründen* (Berlin (Ost): Akademie Verlag, 1975); Dan P. Silverman, *Hitler's Economy: Nazi Work Creation Programs, 1933–1936* (Cambridge/Mass., London: Harvard University Press, 1998). A short discussion of the role of entrepreneurs at the *Westwall* is provided by Gross, *Westwall*, 248–51.

24 Mark Spoerer, *Von Scheingewinnen zum Rüstungsboom: Die Eigenkapitalrentabilität der deutschen Industrieaktiengesellschaften 1925–1941* (Stuttgart: Franz Steiner Verlag, 1996); Jonas Scherner, *Die Logik der Industriepolitik im Dritten Reich – Die Investitionen in die Autarkie- und Rüstungsindustrie und ihre staatliche Förderung* (Stuttgart: Franz Steiner Verlag, 2008).

25 There is some research on the steel rationing system, though: Thomas Sarholz, 'Die Auswirkungen der Kontingentierung von Eisen und Stahl auf die Aufrüstung der Wehrmacht von 1936 bis 1939' (TU Darmstadt, 1983); Ulrich Hensler, *Die Stahlkontingentierung im Dritten Reich* (Stuttgart: Franz Steiner Verlag, 2008); Ulrich Hensler, 'Iron and Steel Rationing During the Third Reich', in *German Industry in the Nazi Period*, ed. Christoph Buchheim (Stuttgart: Franz Steiner Verlag, 2008). Jonas Scherner has estimated the production of construction equipment and machinery up to 1944, which allows conclusions to be drawn with respect to the performance of the construction industry during the war: Jonas Scherner, 'Nazi Germany's Preparation for War: Evidence from Revised Industrial Investment Series', *European Review of Economic History* 14 (2010): 433–68.

The extraordinary profits that German construction companies realised over the course of the 1930s have attracted the interest of economic historians concerned with the Nazi state's struggle to control prices and profits.[26] Most notably, Jochen Streb has analysed the contract types and negotiations between procurement agencies and construction companies, as well as documented the unsuccessful attempts to get a grip on the problem of excess profits in the sector.[27] In particular, Streb has pointed out the close contacts and shared interests between state officials and businessmen, as well as corruption. These phenomena could help to explain why construction companies continued to realise high profits throughout the war.[28] It is one aim of this thesis to ascertain whether we can observe such mergings of interests also in the case of the OT.

Only little information can be gleaned from business histories and anniversary *festschriften* published by the construction companies themselves. Until the 1990s, the German construction industry barely mentioned their wartime activities or cooperation with the OT. While, as Sebastian Brünger has recently shown, the successor companies of the IG Farben corporation actively sought to shape the public image of their role in the Third Reich as victims of a command economy, nothing similar can be said about construction firms.[29] Unlike the representatives of IG Farben at Nuremberg, German construction companies were not put in the dock. For the latter, there was no immediate need to elaborate a defence narrative, as the focus of the Allied prosecutors lay elsewhere. If at all, the years 1933 to 1945 were only mentioned cursorily.[30] The wish to let sleeping dogs lie maybe becomes most evident in the case of Philipp Holzmann's 1949 anniversary *festschrift*. Only a pared-down version of the chapter on the Nazi period made it into the

26 Timothy W. Mason, *Arbeiterklasse und Volksgemeinschaft. Dokumente und Materialien zur deutschen Arbeiterpolitik 1936–1939*, Schriften des Zentralinstituts für sozialwissenschaftliche Forschung der Freien Universität Berlin 22 (Opladen: Westdeutscher Verlag, 1975); Cornelia Rauh-Kühne, 'Hitlers Hehler? Unternehmerprofite und Zwangsarbeiterlöhne', *Historische Zeitschrift* 275, no. 1 (2002): 1–55; Dietrich Eichholtz, *Geschichte der deutschen Kriegswirtschaft 1939–1945. Band II/2: 1941–1943* (1985; repr., München: K. G. Saur, 2003), 513.

27 Jochen Streb, 'Das Scheitern der staatlichen Preisregulierung in der nationalsozialistischen Bauwirtschaft', *Jahrbuch für Wirtschaftsgeschichte*, no. 1 (2003): 27–48; Jochen Streb, 'Negotiating Contract Types and Contract Clauses in the German Construction Industry during the Third Reich', *The RAND Journal of Economics* 40, no. 2 (2009): 364–79.

28 Streb, 'Preisregulierung', 45–47.

29 Sebastian Brünger, *Geschichte und Gewinn. Der Umgang deutscher Konzerne mit ihrer NS-Vergangenheit* (Göttingen: Wallstein Verlag, 2017).

30 Thormann-und-Stiefel-AG Augsburg, *75 Jahre 'Thosti', Thormann & Stiefel Aktiengesellschaft Augsburg, 1876–1951* (Darmstadt: Hoppenstedt, 1951); Beton- und Tiefbau Mast Aktiengesellschaft, *Von einem, der auszog das Bauen zu lernen: 50 Jahre Beton- und Tiefbau Mast Aktiengesellschaft* (Berlin: Mast, 1955); Gerd v. Klass, *Weit spannt sich der Bogen. 1865–1965. Die Geschichte der Bauunternehmung Dyckerhoff & Widmann* (Wiesbaden: Verlag für Wirtschaftspublizistik, 1965); Ed. Züblin AG, ed., *75 Jahre Züblin-Bau, 1898–1973* (Stuttgart: Karl Krämer Verlag, 1973).

finally published book. The original draft had still contained detailed information on Philipp Holzmann's works for the OT, the note that members of the firm had been delegated to all public authorities concerned with construction, as well as statistics showing that 25 percent of the firm's turnover between 1933 and 1939 came from military construction sites (including the *Westwall*) and that the firm's market share had grown significantly after 1933. Obviously, the Philipp Holzmann AG did not deem it wise to publish these facts in 1949.[31] And even in 1969, in an act of blatant self-victimisation, one of the leading German construction firms Sager & Woerner dealt with its engagement in occupied Europe in only one short paragraph:

> Today, 25 years after the end of the war, a presentation of the wartime construction projects outside of the Reich does not seem appropriate. Remembering the bitter sacrifices that they have demanded from the German construction industry in the form of blood and property would only open old wounds.[32]

The situation only changed in the 1990s with the public debate on compensation payments for former forced labourers. More recent business histories still emphasise elements of coercion and the construction firms' lacking scope of action and do not investigate how their cooperation with the OT actually came about. However, they already draw a much more nuanced picture than earlier contributions in this field.[33]

There is far less to follow up in the case of Norwegian historiography. Despite the EW's crucial role in the economy of the occupied country, it did not leave any

31 Hans Meyer-Heinrich, *Philipp Holzmann Aktiengesellschaft im Wandel von hundert Jahren* (Frankfurt am Main: Umschau Verlag, 1949). The draft of the book can be found at the *Archiv des Instituts für Stadtgeschichte, Frankfurt am Main* (hereafter: IfS), W1-2/298.

32 Sager & Woerner AG, *Sager & Woerner 1864–1964*, vol. 2: Sammlung-Einsatz-Krise-Erstarkung, 1920–1945 (München: Sager & Woerner AG, 1969), 174.

33 Karlheinz Fuchs and Senta Everts-Grigat, *Züblin. 100 Jahre Bautechnik, 1898–1998* (Stuttgart: Ed. Züblin AG, 1998); Manfred Pohl, *Die Strabag: 1923 bis 1998* (München: Piper, 1998); Manfred Pohl, *Philipp Holzmann: Geschichte eines Bauunternehmens 1849–1999* (München: C.H. Beck, 1999); Manfred Pohl and Birgit Siekmann, *HOCHTIEF und seine Geschichte: Von den Brüdern Helfmann bis ins 21. Jahrhundert* (München, Zürich: Piper, 2000); Bernhard Stier and Martin Krauß, *Drei Wurzeln – ein Unternehmen. 125 Jahre Bilfinger Berger AG* (Heidelberg, Ubstadt-Weiher, Basel: Ifu im Verl. Regionalkultur, 2005). Two studies on Austrian construction companies are: Herbert Matis and Dieter Stiefel, *'Mit der vereinigten Kraft des Capitals, des Credits und der Technik . . . ': Die Geschichte der österreichischen Bauwesens am Beispiel der Allgemeinen Baugesellschaft – A. Porr Aktiengesellschaft* (Wien, Köln, Weimar: Böhlau, 1994); Stefan Lütgenau and Alexander Schröck, *Zwangsarbeit in der österreichischen Bauindustrie. Die Teerag-Asdag AG 1938–1945* (Innsbruck, Wien, München: Studien-Verlag, 2001). To date, the study of Lütgenau and Schröck is the most thorough business history, when it comes to the topics forced labour and state-business relations in construction.

traces in Norwegian historiography until recently. One reason for this is that economic aspects in general, and economic collaboration in particular, have scarcely been a part of the narrative on occupation.[34] Pioneering studies were only provided by historians from outside Norway: *The Fascist Economy in Norway* by Alan S. Milward, and the study on the *Reichskommissariat Norwegen* by Robert Bohn.[35] While the OT does not take a central position in Milward's book, Bohn has devoted one chapter to the construction plans of the *Wehrmacht* and the OT, showing how the lack of building materials, labour and transport capacities placed the projects in constant jeopardy.[36] Some works by Norwegian historians, on the other hand, have shed some light on the collaboration of Norwegian building entrepreneurs, the so-called barrack barons (*brakkebaroner),* with the German occupants.[37] *Festschriften* and business histories on Norwegian construction firms have largely ignored the occupation years, or established a rather biased narrative of German threats and Norwegian resistance.[38]

However, there are several trends in Norwegian historiography, too, that have contributed to the development that OT's *Einsatzgruppe Wiking* has become an

---

34 Ole Kristian Grimnes, 'Kollaborasjon og oppgjør', in *I krigens kjølvann. Nye sider ved norsk krigshistorie og etterkrigstid*, ed. Stein Ugelvik Larsen (Oslo: Universitetsforlaget, 1999), 47–57; Harald Espeli, 'Det økonomiske forholdet mellom Tyskland og Norge 1940–45', in *Danske tilstander – norske tilstander. Forskjeller og likheter under tysk okkupasjon 1940–45*, ed. Hans Fredrik Dahl et al. (Oslo: Forlaget Press, 2010), 136–37; Gaute Fredriksen, '" . . . og det er en annen historie": Den økonomiske historieskrivingen om årene 1940–1945 i Norges historie. En historiografi.' (Universitetet i Oslo, 2011), 30. The OT is only cursorily, if at all, mentioned in books on the occupation and the economic history of Norway: Erling Petersen, 'Økonomiske forhold', in *Norges krig 1940–1945: 3*, by Sverre Steen (Oslo: Gyldendal, 1950), 495–568; Fritz Hodne, *Norges økonomiske historie 1815–1970* (Oslo: J. W. Cappelens Forlag, 1981); Fritz Hodne, *The Norwegian Economy 1920–1980* (New York: St. Martin's Press, 1983); Guri Hjeltnes, *Hverdagsliv*, Norge i krig. Fremmedåk og frihetskamp 1940–1945 5 (Oslo: Aschehoug, 1986); Magne Skodvin, *Norsk historie 1939–1945: Krig og okkupasjon* (Oslo: Det Norske Samlaget, 1991); Fritz Hodne and Ola Grytten, *Norsk økonomi 1900–1990* (Oslo: Tano, 1992); Hans Fredrik Dahl et al., eds., *Norsk krigsleksikon 1940–1945* (Oslo: J. W. Cappelens Forlag, 1995); Fritz Hodne and Ola Grytten, *Norsk økonomi i det tyvende århundre* (Bergen: Fagbokforlaget, 2002); Berit Nøkleby, *Hitlers Norge: Okkupasjonsmakten 1940–1945* (Oslo: Cappelen Damm, 2016).

35 Alan S Milward, *The Fascist Economy in Norway* (Oxford: Clarendon Press, 1972); Robert Bohn, *Reichskommissariat Norwegen. 'Nationalsozialistische Neuordnung' und Kriegswirtschaft* (München: Oldenbourg, 2000).

36 Bohn, *Reichskommissariat Norwegen*, 355–82.

37 Terje Valen, *De tjente på krigen: hjemmefronten og kapitalen* (Oslo: Oktober, 1978); Dag Ellingsen, *Krigsprofitørene og rettsoppgjøret* (Oslo: Gyldendal, 1993); Ingeborg Solbrekken, *Galskap og rettergang: landssviksaken mot Kirsten Flagstad og Henry Johansen* (Oslo: Transit, 2007); Ola Karlsen, *Profitørene: De ukjente landssvikerne* (Oslo: Gyldendal Norsk Forlag, 2017).

38 Øyvind Steen, *I første rekke. Ingeniør F. Selmer A/S 1906–1981* (Oslo: Ingeniør F. Selmer A/S, 1981); Gun Helena Svendsen, *Anleggsprat. Jubileumsnummer 1990: 50 år* (Trondheim: A/S Anlegg, 1991).

object of research in recent years. First, especially from the 1980s onwards, several Norwegian historians have increasingly questioned the dominant Norwegian post-war narrative of the occupation as a period exclusively characterised by economic suffering and destruction. Instead, these historians have shown that the five years under German rule were also the time of massive investments in the country's industry and infrastructure.[39] Although not yet focusing on the EW, their perspective sooner or later begged the question of who actually ordered and carried out these investments. Second, Norwegian historiography has in the last two decades shown a broader interest in the economic history of the occupation years, especially in what Norwegian companies contributed to the German war economy.[40] Third, also the booming research on National Socialist forced labour left its traces in Norway. Since the EW was one of the most important employers of forced labour in the country, studies on forced labour in occupied Norway have increased our knowledge on

39 Gunnar Wasberg and Arnljot Svendsen, *Industriens historie i Norge* (Oslo: Norges Industriforbund, 1969), 219–31; Olav Wicken, 'Industrial Change in Norway during the Second World War: Electrification and Electrical Engineering', *Scandinavian Journal of History* 8 (1983): 119–50; Tore Jørgen Hanisch and Even Lange, *Veien til velstand: industriens utvikling i Norge gjennom 50 år* (Oslo: Universitetsforlaget, 1986), 44–46; Hjeltnes, *Hverdagsliv*, 67–95; Tjelmeland, Hallvard, 'Kva gjorde krigen med Norge?', *Nytt Norsk Tidsskrift* 13, no. 1 (1996): 28–42; Even Lange, *Samling om felles mål: 1935–1970*, Aschehougs norgeshistorie 11 (Oslo: Aschehoug, 1998), 107–11.

40 Jan Didriksen, *Industrien under hakekorset* (Oslo: Universitetsforlaget, 1987); Trond Bergh, Harald Espeli, and Knut Sogner, *Brytningstider: storselskapet Orkla 1654–2004* (Oslo: Orion, 2004); Pål Thonstad Sandvik, *Falconbridge Nikkelverk 1910 – 1929 – 2004: Et internasjonalt selskap i Norge* (Kristiansand: Falconbridge Nikkelverk, 2004); Ketil G. Andersen, *Flaggskip i fremmed eie: Hydro 1905–1945* (Oslo: Pax Forlag, 2005); Sverre A. Christensen and Harald Rinde, *Nasjonale utlendinger: ABB i Norge 1880–2010* (Oslo: Gyldendal akademisk, 2009); Hans Otto Frøland and Jan Thomas Kobberrød, 'The Norwegian Contribution to Göring's Megalomania. Norway's Aluminium Industry during World War II', *Cahiers d'histoire de l'aluminium* 42–43 (2009): 137–53; Hans Otto Frøland, 'Nazi Planning and the Aluminium Industry', in *Alan S. Milward and a Century of European Change*, ed. Fernando Guirao, Frances Lynch, and Sigfrido Ramirez Perez (New York: Routledge, 2012); Anette Storeide, *Norske krigsprofitører – Nazi-Tysklands velvillige medløpere* (Oslo: Gyldendal, 2014); Christine Myrvang, *Troskap og flid: Kongsberg Våpenfabrikks historie 1814–1945* (Oslo: Pax Forlag, 2014); Bjørn Westlie, *Fangene som forsvant: NSB og slavearbeiderne på Nordlandsbanen* (Oslo: Spartacus Forlag, 2015); Harald Rinde, *Det moderne fylket*, Nordlands historie 3 (Oslo: Fagbokforlaget, 2015), 180–95; Hans Otto Frøland, Mats Ingulstad, and Jonas Scherner, eds., *Industrial Collaboration in Nazi-Occupied Europe: Norway in Context* (London: Palgrave Macmillan, 2016); Bjørn Westlie, 'NSB, krigsfangene og andre verdenskrig' (Universitetet i Oslo, 2018). An early overview, however, was already provided in 1962 when the Norwegian Ministry for Justice and Police published a summary report on the Norwegian post-war trials (*landssvikoppgjøret*). The report also dealt with Norwegian firms accused for "economic treason": Justis- og politidepartementet, *Om landssvikoppgjøret. Innstilling fra et utvalg nedsatt for å skaffe tilveie materiale til en innberetning fra Justisdepartementet til Stortinget. Utvalget oppnevnt 22. desember 1955. Innstilling avgitt 11. januar 1962* (Gjøvik: Mariendals Boktrykkeri, 1962).

EW, too.[41] The most important impetus to the field, however, was the fact that the source material on *Einsatzgruppe Wiking* at the Norwegian National Archives was made publicly available in 2011. Since then, a research project at the Department of Historical Studies at the Norwegian University of Science and Technology has put great effort into investigating the history of the EW. This thesis was written in close association with the project and has benefitted greatly from the knowledge generated by its members.[42]

No studies, however, have hitherto focused on the presence of approximately 500 German construction firms in occupied Norway. Their motives, economic significance and role in the system of forced labour and economic exploitation remain unclear. They are either mentioned only cursorily or sometimes they vanish into the blurred category of "the Germans", which can refer to the *Wehrmacht*, private companies, officials of the Reich Commissariat and the OT alike. One notable exception are two chapters in Torgeir E. Sæveraas' doctoral thesis, which deal with the relation between the EW and both German and Norwegian construction companies.[43]

---

41 Birgit Koch, 'De sovjetiske, polske og jugoslaviske krigsfanger i tysk fangenskap i Norge 1941–1945' (Universitetet i Oslo, 1988); Emilia Denkiewicz-Szczepaniak, 'Polske OT-tvangsarbeidere og krigsfanger i Norge under annen verdenskrig', *Historisk Tidsskrift* 76, no. 2 (1997): 268–81; Marianne Neerland Soleim, *Sovjetiske krigsfanger i Norge 1941–1945. Antall, organisering og repatriering* (Tromsø: Universitetet i Tromsø, 2004); Einar Kristian Steffenak, *Russerfangene* (Oslo: Humanist Forlag, 2008); Michael Stokke, 'Sovjetiske og franske sivile tvangsarbeidere i Norge 1942–1945: en sammenligning av arbeids- og leveforhold' (Universitetet i Bergen, 2008); Marianne Neerland Soleim, *Sovjetiske krigsfanger i Norge 1941–1945. Antall, organisering og repatriering* (Oslo: Spartacus Forlag, 2009); Marianne Neerland Soleim, ed., *Prisoners of War and Forced Labour: Histories of War and Occupation* (Newcastle: Cambridge Scholars Publishing, 2010); Gunnar D. Hatlehol, '"Norwegeneinsatz" 1940–1945. Organisation Todts arbeidere i Norge og gradene av tvang' (NTNU Trondheim, 2015); Hans Otto Frøland and Anders Lervold, 'Forced Labour in Norway during the German Occupation: French and Soviet Workers in the Light Metals Program', *Revue d'histoire Nordique* 17 (2014): 71–100; Hans Otto Frøland and Gunnar D. Hatlehol, 'Organisation Todt and Forced Labour in Norway during the Nazi Occupation: Preliminary Remarks from an Ongoing Research Project', in *Painful Heritage. Studies in the Cultural Landscape of the Second World War*, ed. Marek E. Jasinski and Leiv Sem (Trondheim: DKNVS, 2015); Frøland, Hatlehol, and Ingulstad, 'Regimenting Labour'.

42 See the many references to the research of Ketil G. Andersen, Hans Otto Frøland, Gunnar D. Hatlehol, Mats Ingulstad and Torgeir E. Sæveraas throughout this thesis. Results of the research project have been brought together in a special issue. See *Historisk Tidsskrift* 97, no. 3 (2018).

43 Torgeir Ekerholt Sæveraas, '"Beton macht Geschichte" – Organisation Todt og utbyggingen av Festung Norwegen' (NTNU Trondheim, 2016), chaps 3, 4. Another exception is Dag Ellingsen's short account of the cooperation between the largest Norwegian construction firm Høyer-Ellefsen and the German firms Straßenbau AG and Sager & Woerner in 1940–41. They were working for the *Wehrmacht*, not the OT, though: Ellingsen, *Krigsprofitørene og rettsoppgjøret*, 37–49.

b The Debate on Entrepreneurial Freedom in the Third Reich

Discussions of the relation between the construction industry and the OT are able to draw upon extensive research on state-business relations in the Third Reich. Ever since historians tried to grasp the nature of the National Socialist state, questions of the relation between the economy and politics and, more specifically, the relation between private business and the German administration, have been a key subject of scholarly debate. Historians argued about who really had held the reins in the Third Reich: the National Socialists or the industry and banks? Did the Nazis dictate production and investment decisions, not stopping at nationalising companies and attacking unruly entrepreneurs? Or were they in fact just the marionettes of "big business"?[44]

Positions stressing the "primacy of politics" soon became consensus among Western historians. In Richard Overy's words, "the Third Reich [. . .] set about reducing the autonomy of the economic élite and subordinating it to the interests of the Nazi state."[45] Overy concedes "that there were areas where the interests of Nazism and big business appeared to coincide": after all, profits were rising. However, he assumes that these profits were "tightly controlled" and concludes that "under such a system businessmen were regarded as economic functionaries serving the interests of the nation rather than as independent and enterprising creators of wealth. The concept of the 'managed economy' suited the regime's ideological ambitions, but stifled enterprise."[46] Ernst Nolte wrote that "[. . .] industrialists were completely eliminated as a major political factor, since they had no voice in ultimate policy-making decisions",[47] while Karl Dietrich Bracher claimed an "[. . .] absolute primacy of the political goals. [. . .] The economy was controlled by the Government and subject to the subsidies, retrenchments, plans, and controls of the Nazi regime. The voice of special interests disappeared; assent and cooperation offered the only chance for success – a clearly lopsided alliance of economy and dictatorship [. . .]."[48]

By the 1980s, however, the overly theoretical debate had become increasingly bloodless. Apart from orthodox East German historiography and a handful of

44 Two exhaustive reviews of the debate are provided by: Ian Kershaw, *The Nazi Dictatorship: Problems and Perspectives of Interpretation*, 3rd ed. (1985; repr., London: Arnold, 1993), 40–58; Klaus Hildebrand, *The Third Reich* (London, New York: Routledge, 1984), 121–31. See also: Henry A. Turner, 'Unternehmen unter dem Hakenkreuz', in *Unternehmen im Nationalsozialismus*, ed. Lothar Gall and Manfred Pohl (München: C.H. Beck, 1998), 15–23.

45 Richard Overy, *The Nazi Economic Recovery 1932–1938* (London: Macmillan, 1982), 58.

46 Overy, 55–57.

47 Ernst Nolte, 'Big Business and German Politics: A Comment', *The American Historical Review* 75, no. 1 (1969): 76.

48 Karl Dietrich Bracher, *The German Dictatorship. The Origins, Structure, and Consequences of National Socialism* (1969; repr., London: Penguin Books, 1973), 416–17.

Western Marxist scholars, no one earnestly assumed a one-way causal connection between business interests and Hitler's seizure of power in 1933, the occupation of large parts of Europe or the Nazi's genocidal policy.[49] Moreover, scholars like Hans-Erich Volkmann and Ian Kershaw rejected the misleading and over-simplified idea of politics and economics as two separable spheres and thus refuted the basic premise of the debate.[50] In retrospect, the controversy might tell us more about the ideological battles of the 1960s and 70s in German historiography than about the relation between state and industry in the Nazi period.

Broader empirical evidence on the individual companies' strategies, motives and room for manoeuvre, however, was still lacking. This first changed over the course of the 1980s and 90s, with what has been called a "business history turn" in the research on National Socialism.[51] As Neil Gregor argues, the boom of business history had its roots both in the historiography of the previous decade and in political developments. With a generation of historians that had internalised the interpretation of National Socialism as a polycratic system and who had read works like Ulrich Herbert's seminal study on forced labour,[52] it seemed natural to look at individual companies as relevant, acting entities under the dictatorship. The negotiations on the compensation of former forced labourers and increasing public pressure on German companies to face up to their own role in the Third Reich gave further impetus to the field of research.[53]

Although now resting on broader empirical foundations, the more recent studies and business histories still draw conflicting conclusions when it comes to the question of entrepreneurial freedom of action. Some stress how limited the German companies' room for manoeuvre was. They continue to echo the predominant "primacy of politics" interpretation and focus on episodes where the Nazi state used force and threats to reach its goals against the will of private industry. In this perception, the National Socialist economy was neither "(partially) 'capitalist' nor a

49 Kershaw, *Nazi Dictatorship*, 44.

50 Kershaw, 48; Hans-Erich Volkmann, 'Politik, Wirtschaft und Aufrüstung unter dem Nationalsozialismus', in *Hitler, Deutschland und die Mächte: Materialien zur Außenpolitik des Dritten Reiches*, ed. Manfred Funke (Düsseldorf: Droste Verlag, 1976), 279–80.

51 Werner Abelshauser, Jan-Otmar Hesse, and Werner Plumpe, 'Wirtschaftsordnung und Unternehmen im Nationalsozialismus: Neuere Forschungen zur Wirtschaftsgeschichte des Nationalsozialismus', in *Wirtschaftsordnung, Staat und Unternehmen: Neue Forschungen zur Wirtschaftsgeschichte des Nationalsozialismus*, ed. Werner Abelshauser, Jan-Otmar Hesse, and Werner Plumpe (Essen: Klartext Verlag, 2003), 10–11.

52 Ulrich Herbert, *Fremdarbeiter: Politik und Praxis des 'Ausländer-Einsatzes' in der Kriegswirtschaft des Dritten Reiches* (1985; repr., Bonn: Dietz, 1999).

53 Neil Gregor, 'Wissenschaft, Politik, Hegemonie. Zum Boom der NS-Unternehmensgeschichte', in *Unternehmen im Nationalsozialismus: Zur Historisierung einer Forschungskonjunktur*, ed. Norbert Frei and Tim Schanetzky (Wallstein Verlag, 2010), 79–93. See also: Brünger, *Konzerne*, especially chapter 5.

'marked economy'".[54] Similarly, Harold James characterises the economy as a "command economy".[55] Peter Hayes, writing on Degussa, concludes that from 1937 onwards, the firm's leaders "largely lost control, without comprehending it yet, over the direction of their own enterprise." Meanwhile, "the Reich's cumulative success in creating a 'carrot and stick' economy, one in which a militantly expansionist government alternately could lure and lash corporations in the directions it preferred, began to get the better of Degussa's chairman."[56] Referring to IG Farben, several historians state concordantly that "the concern became a mere executor of government orders",[57] faced the threat of nationalisation in the case of Buna production[58] and had lost a considerable part of its entrepreneurial freedom by the end of the 1930s.[59] Most pointedly, Peter Temin has emphasised the similarities between the National Socialist and the Soviet economic systems. On investment contracts between the Nazi state and private businesses, he writes: "These contracts were nominally contracts expressing agreement between the two parties. But they were decidedly unequal. The Nazis viewed private property as conditional on its use – not as a fundamental right. If the property was not being used to further Nazi goals, it could be nationalized."[60]

The state certainly regulated the input of raw materials and labour, while the striving for rearmament and autarky, and ultimately war, necessitated the adjustment of companies' investment and production strategies. As Gerd Höschle has shown for the textile sector, the state successfully restricted the import of raw materials and negotiated bilateral trade agreements to ease Germany's balance of payments.[61] Some firms in the sector were no longer allowed to expand their production capacities and prices were regulated.[62] By making their use in armament production a precondition for the allocation of iron and steel, the state gained influence over the production programmes of private companies, as is documented

54 Michael von Prollius, *Das Wirtschaftssystem der Nationalsozialisten 1933–1939. Steuerung durch emergente Organisation und Politische Prozesse* (Paderborn: Ferdinand Schöningh, 2003), 244.

55 Harold James, *Die Deutsche Bank im Dritten Reich* (München: C.H. Beck, 2003), 208.

56 Peter Hayes, *From Cooperation to Complicity: Degussa in the Third Reich* (Cambridge: Cambridge University Press, 2004), 113–14.

57 Peter Hayes, *Industry and Ideology. IG Farben in the Nazi Era* (1987; repr., Cambridge: Cambridge University Press, 2001), 326.

58 Bernhard Lorentz and Paul Erker, *Chemie und Politik. Die Geschichte der Chemischen Werke Hüls 1938 bis 1979. Eine Studie zum Problem der Corporate Governance* (München: C.H. Beck, 2003), 9.

59 Bernd C. Wagner, *IG Auschwitz. Zwangsarbeit und Vernichtung von Häftlingen des Lagers Monowitz 1941–1945* (München: K. G. Saur, 2000), 36.

60 Peter Temin, 'Soviet and Nazi Economic Planning in the 1930s', *Economic History Review* 44, no. 4 (1991): 576.

61 Gerd Höschle, *Die deutsche Textilindustrie zwischen 1933 und 1939. Staatsinterventionismus und ökonomische Rationalität* (Stuttgart: Franz Steiner Verlag, 2004), 74–87.

62 Höschle, 57–58, 135–69.

in the case of the Flick corporation.[63] Writing on the Vereinigte Stahlwerke AG, Alexander Donges argues that it was in particular the combination of several regulations that steered the company in the direction preferred by the state.[64] Prices for raw materials and semi-finished goods were monitored, access to the capital market was restricted and iron and steel for the extension of plants rationed. At the same time, the company faced a liquidity crisis. Therefore it invested in processing and armament plants, for which the state provided capital and building materials and where higher prices could be realised.[65] The IG Farben AG and the Degussa were likewise vulnerable to capital market restrictions as they struggled with liquidity crises in the late 1930s.[66] The majority of recent studies follows this interpretation of an economic regime that "within an interventionist framework, wilfully exploited the profit seeking of private businesses through effective incentives [. . .]."[67] The design of this interventionist framework has been described in detail for different sectors, with measures like capital market controls, wage and price monitoring, investment restrictions, rationing of raw materials or the control of excess profits. The economic system under Nazi rule has thus been characterised as "deformed",[68] a "regulated market economy"[69] or a market economy that already resembled a wartime economy in an economic state of emergency before the war broke out.[70]

Nevertheless, when describing the countless measures that formed the regulatory framework for these state-industry encounters, there is a strong trend in recent

---

63 Norbert Frei et al., *Flick: Der Konzern, die Familie, die Macht* (München: Karl Blessing Verlag, 2009), 199–200.

64 Alexander Donges, *Die Vereinigte Stahlwerke AG im Nationalsozialismus. Konzernpolitik zwischen Marktwirtschaft und Staatswirtschaft* (Paderborn: Ferdinand Schöningh, 2014), 404.

65 Donges, 402–4.

66 Jonas Scherner, 'Keine primäre moralische Verantwortung? Zum Umfang und zu Grenzen unternehmerischer Handlungsspielräume in der NS-Zeit', in *Ökonomie und Ethik. Beiträge aus Wirtschaft und Geschichte*, ed. Gert Kollmer-von Oheimb-Loup, Sibylle Lehmann-Hasemeyer, and Stefanie van de Kerkhof (Ostfildern: Jan Thorbecke Verlag, 2017), 176–77.

67 Höschle, *Textilindustrie*, 320.

68 Christoph Buchheim, 'Die Wirtschaftsentwicklung im Dritten Reich. Mehr Desaster als Wunder. Eine Erwiderung auf Werner Abelshauser', *Vierteljahreshefte für Zeitgeschichte* 49, no. 4 (2001): 653–64; Michael C. Schneider, *Unternehmensstrategien zwischen Weltwirtschaftskrise und Kriegswirtschaft. Chemnitzer Maschinenbauindustrie in der NS-Zeit 1933–1945* (Essen: Klartext Verlag, 2005), 22.

69 Dieter Ziegler, '"A Regulated Market Economy": New Perspectives on the Nature of the Economic Order of the Third Reich, 1933–1939', in *Business in the Age of Extremes. Essays in Modern German and Austrian Economic History*, ed. Hartmut Berghoff, Jürgen Kocka, and Dieter Ziegler (Cambridge: Cambridge University Press, 2013), 139–52.

70 Jochen Streb, 'Das nationalsozialistische Wirtschaftssystem: Indirekter Sozialismus, gelenkte Marktwirtschaft oder vorgezogene Kriegswirtschaft?', in *Der Staat und die Ordnung der Wirtschaft: Vom Kaiserreich bis zur Berliner Republik*, ed. Werner Plumpe and Joachim Scholtyseck (Stuttgart: Franz Steiner Verlag, 2012), 83.

literature to highlight the deficits of these regulations, such as the impossibility of controlling them, their unintended effects and their ineffectiveness. This thesis is strongly informed by this perspective: we cannot readily infer from the existence of a multitude of regulations that the state was necessarily able to enforce them effectively.

Recent research on the National Socialist economy has increasingly focused on this aspect. Capital market controls, for example, remained a relatively weak tool for steering private industry's investment decisions, since most German companies were quite solvent in the 1930s, with the situation at Vereinigte Stahlwerke, IG Farben and Degussa being the exception rather than the rule.[71] Moreover, the determination of prices for various products and the control of the general price stop after 1936 remained incomplete.[72] While controlling prices was easier in the case of basic, standardised goods like coal and steel, it was almost impossible in sectors with a varied range of products, like the textile industry,[73] not to mention the construction industry, where every construction is more or less unique.[74] In particular, companies used the introduction of allegedly "new" products, that is, slightly changed versions of existing products, to circumvent regulations.[75] As Hans Dichgans, a leading official at the Reich Price Commissioner (*Reichskommissar für die Preisbildung*), recalled after the war, price controls already reached their limits in the cases of razor blades and women's hats.[76] Similar problems are documented for the field of wage controls.[77] Public procurement agencies like the *Wehrmacht* or the Office of the Four Year Plan often tolerated violations of the rules to secure sufficient numbers of workers for their projects. Rationing was another measure. By controlling the allocation of raw materials or building materials, the state was able to influence the industry's production

71 Scherner, 'Keine primäre moralische Verantwortung?', 176–77.

72 On price regulation in the Third Reich, see Wolfgang Bopp, 'The Evolution of the Pricing Policy for Public Orders During the Third Reich', in *After the Slump. Industry and Politics in 1930s Britain and Germany*, ed. Christoph Buchheim and Redvers Garside (Frankfurt am Main: Peter Lang, 2000), 149–60; André Steiner, 'Der Reichskommissar für die Preisbildung – "eine Art wirtschaftlicher Reichskanzler"?', in *Hitlers Kommissare: Sondergewalten in der nationalsozialistischen Diktatur*, ed. Rüdiger Hachtmann and Winfried Süß (Göttingen: Wallstein Verlag, 2006), 93–114; André Steiner, 'Industry and Administrative Price Regulation 1933–1938/39', in *German Industry in the Nazi Period*, ed. Christoph Buchheim (Stuttgart: Franz Steiner Verlag, 2008).

73 Ziegler, 'Regulated Market Economy', 145.

74 Streb, 'Preisregulierung', 30. But even at the Flick-owned armament and steel producer Maxhütte, the state failed in its attempts to control prices: Frei et al., *Flick*, 199.

75 Markus Albert Diehl, *Von der Marktwirtschaft zur nationalsozialistischen Kriegswirtschaft. Die Transformation der deutschen Wirtschaftsordnung 1933–1945* (Stuttgart: Franz Steiner Verlag, 2005), 72.

76 *Bundesarchiv Berlin-Lichterfelde* (hereafter: BArch), R 26-II ANH./1, Hans Dichgans, 'Zur Geschichte des Reichskommissars für die Preisbildung' (Unpublished manuscript, 1976), fol. 26.

77 Rüdiger Hachtmann, *Industriearbeit im 'Dritten Reich'. Untersuchungen zu den Lohn- und Arbeitsbedingungen in Deutschland 1933–1945* (Göttingen: Vandenhoeck & Ruprecht, 1989), 121–24.

programmes. However, despite the fact that sometimes the companies themselves wanted rationing measures to prevent overcapacity,[78] corporations could still decide what they wanted to use the allocated materials for to a certain degree.[79] On several occasions, state regulations came into conflict with each other, leading to unintended effects. The textile industry's development towards autarky and bilateral trade agreements, for example, was quite successful, insofar as it unburdened the German trade balance, but at the same time it destroyed any attempts to keep the prices of textile products at a stable level.[80]

Perhaps most importantly, several studies have shown that a chronic lack of personnel within the monitoring state authorities rendered the enforcement of all kinds of regulations problematic. The authorities monitored firms only seldom and at random, and depended heavily on information provided by the industry itself.[81] State attempts to limit industrial profits or to collect them retroactively were rather unsuccessful, too, at least compared to the huge administrative effort made.[82]

Likewise, many studies emphasise the companies' ample scope of action during direct negotiations with the state. Christoph Buchheim and Jonas Scherner have argued that companies were still autonomous in their decisions on production and investments in most cases.[83] The regime clearly preferred to use incentives, like favourable contract types, rather than force to secure the German industry's cooperation. In particular, Buchheim and Scherner convincingly demonstrate that the National Socialists normally left basic features of modern market economies like the freedom of contract[84] and private property[85] untouched. Freedom of contract

78 Hensler, *Stahlkontingentierung*, 23–24.

79 Höschle, *Textilindustrie*, 69–71, 167; Hensler, *Stahlkontingentierung*, 129,141,157.

80 Höschle, *Textilindustrie*, 318.

81 Höschle, 57, 173; Streb, 'Negotiating Contract Types', 368; Hachtmann, *Industriearbeit*, 118–19.

82 Streb, 'Preisregulierung'; Streb, 'Negotiating Contract Types'; Rauh-Kühne, 'Hitlers Hehler', 25.

83 Christoph Buchheim and Jonas Scherner, 'Anmerkungen zum Wirtschaftssystem des "Dritten Reichs"', in *Wirtschaftsordnung, Staat und Unternehmen: Neue Forschungen zur Wirtschaftsgeschichte des Nationalsozialismus*, ed. Werner Abelshauser, Jan-Otmar Hesse, and Werner Plumpe (Essen: Klartext Verlag, 2003); Christoph Buchheim, 'Unternehmen in Deutschland und NS-Regime: Versuch einer Synthese', *Historische Zeitschrift* 282, no. 2 (2006): 351–90; Jonas Scherner, 'Das Verhältnis zwischen NS-Regime und Industrieunternehmen – Zwang oder Kooperation?', *Zeitschrift für Unternehmensgeschichte / Journal of Business History* 51, no. 2 (2006): 166–90; Christoph Buchheim and Jonas Scherner, 'The Role of Private Property in the Nazi Economy: The Case of Industry', *The Journal of Economic History* 66, no. 2 (2006): 390–416; Scherner, *Logik der Industriepolitik*; Jonas Scherner, 'Anreiz statt Zwang. Wirtschaftsordnung und Kriegswirtschaft im "Dritten Reich"', in *Unternehmen im Nationalsozialismus: Zur Historisierung einer Forschungskonjunktur*, ed. Norbert Frei and Tim Schanetzky (Göttingen: Wallstein Verlag, 2010), 140–55.

84 Buchheim and Scherner, 'Anmerkungen zum Wirtschaftssystem des "Dritten Reichs"', 86–87.

85 Buchheim and Scherner, 'The Role of Private Property'. Dietmar Petzina stressed already in 1968 that private property was still respected under the National Socialists: Dietmar Petzina, *Autarkiepolitik im Dritten Reich. Der nationalsozialistische Vierjahresplan* (Stuttgart: Deutsche

especially included a "negative freedom of contract", that is, the option to say no to a contract offered by the state.

It is one aim of this thesis to investigate to what degree freedom of contract was preserved also under the OT in occupied Europe. However, I argue that – maybe more than the freedom to eventually turn down a contract – the crucial entrepreneurial scope of action for OT contractors opened up already at an earlier stage. As Fabian Lemmes[86] and Marc Buggeln[87] have noted, OT contracts were no fixed and unmodifiable papers. Rather, the firms were able to take part in their formulation and to negotiate on modifications. Lemmes therefore assumes that the OT could hardly decide anything against the explicit will of the major German construction firms. Especially in chapter V, I will draw upon this perspective and provide an in-depth analysis of how the construction industry had the opportunity to protect their interests during the formulation of EW's contracts.

Buchheim and Scherner's position in particular has been criticised by Peter Hayes.[88] Over the years, however, various studies have provided examples confirming this position. Only some of them are mentioned here. On several occasions in the 1930s, the steel giant and arms manufacturer Friedrich Krupp AG passed on investment costs for plants and machinery to the state when the investments ran counter to the company's long-term business strategy.[89] In May 1942, Opel successfully refused to erect a truck plant in Riga, even though it was the wish of Speer and Hitler.[90] Analysing the outsourcing decisions of German companies, Elena Dickert has documented state attempts to press companies to relocate their production to occupied territories. The most promising strategy was to single out companies and contact them directly. But this was a time-consuming procedure, and thus Dickert remains sceptical whether a significant number of companies was affected by these threats.[91] Neil Gregor has emphasised that throughout the 1930s, Daimler-

Verlags-Anstalt, 1968), 196–98. The banking sector even saw the partial re-privatisation of Deutsche Bank in 1933, and the complete re-privatisation of Commerzbank and Dresdner Bank in 1936–37: Ziegler, 'Regulated Market Economy', 141–43.

86 Lemmes, 'Arbeiten für das Reich', 382, 404.

87 Buggeln, 'Menschenhandel', 2010.

88 Peter Hayes, 'Corporate Freedom of Action in Nazi Germany', *Bulletin of the German Historical Institute* 45 (2009): 29–42.

89 Harold James, *Krupp. A History of the Legendary German Firm* (Princeton and Oxford: Princeton University Press, 2012), 184.

90 Anita Kugler, 'Airplanes for the Führer. Adam Opel AG as Enemy Property, Model War Operation, and General Motors Subsidiary, 1939–1945', in *Working for the Enemy: Ford, General Motors, and Forced Labor in Germany during the Second World War*, ed. Reinhold Billstein et al. (New York, Oxford: Berghahn Books, 2000), 63.

91 Elena Dickert, *Die 'Nutzbarmachung' des Produktionspotentials besetzter Gebiete durch Auftragsverlagerungen im Zweiten Weltkrieg: Staatliche Regulierung und Verlagerungsverhalten von Maschinenbau- und Automobilunternehmen* (Trondheim: NTNU, 2014), 232–34.

Benz " – like most other companies – continued to invest primarily in the flexible multi-purpose machine tools traditionally preferred by German industry." Also during the war, "the company's investment strategy was generally characterised [. . .] by a clear desire not to disadvantage itself in the long term for short-term gains." Daimler-Benz "[. . .] was happy to participate in and profited greatly from war production [. . .]." Nevertheless, it "[. . .] did not convert capacity to armaments production in such a way as to undermine its ability to reconvert to peacetime production after the war."[92] Similar results are reported from companies of the electrical industry.[93] Examples can even be found for non-German businesses. In occupied Norway, to give just two examples, German authorities guaranteed the mining company Sulitjelma Gruber A/S that they would buy its complete output at a profitable price. When the company was reluctant to expand their crude ore production in 1943 because it doubted the long-term value of this expansion, the Germans agreed to bear all expenses for the investment.[94] In the case of the Norsk Aluminium Company (NACO), the state agreed in 1942 to subsidise the company's bauxite imports when NACO was struggling to meet the fixed output price for aluminium and thus had been unable to make a profit.[95] In addition, Mark Spoerer has pointed out the high profits of German companies far into the war that compensated the industry for the risks of war and political interference and are incompatible with the oversimplifying interpretation of a "command economy".[96]

It is reasonable to assume that the size of a company and the structure of its sector were important variables impacting a firm's room for manoeuvre. Economic historians, however, have come to ambivalent conclusions in this respect. While large companies, or even monopolists, can certainly be expected to have had a stronger bargaining position, smaller companies were subject to controls to a lesser extent.[97] A high degree of organisation in a sector could be an advantage, too. The iron and steel sector, for example, was dominated by only a few, large concerns with homogenous interests. As a result, "the authorities in charge of steel rationing were frequently unable to oppose the industry [. . .]."[98] Similarly, Germany's three powerful brown coal cartels successfully opposed two attempts by the Economics

92 Neil Gregor, *Daimler-Benz in the Third Reich* (New Haven, London: Yale University Press, 1998), 247–48.

93 Wilfried Feldenkirchen, *Siemens 1918–1945* (München, Zürich: Piper, 1995), 283–84.

94 Rinde, *Det moderne fylket*, 185–86.

95 Hans Otto Frøland, 'Facing Disincentives? Norwegian Aluminium Companies Working for the German Aircraft Industry', in *Industrial Collaboration in Nazi-Occupied Europe: Norway in Context*, ed. Hans Otto Frøland, Mats Ingulstad, and Jonas Scherner (London: Palgrave Macmillan, 2016), 355–56.

96 Spoerer, *Eigenkapitalrentabilität*, 169–70.

97 Scherner, 'Keine primäre moralische Verantwortung?', 166.

98 Ziegler, 'Regulated Market Economy', 147; Matt Bera, *Lobbying Hitler. Industrial Associations between Democracy and Dictatorship* (New York: Berghahn Books, 2016).

Ministry to re-structure the market in favour of the hard coal industry in 1933 and 1936.[99] On the other hand, it would not be implausible to assume that a highly fragmented sector was more difficult to regulate, which could open up additional scope of action for firms.

c Principal-Agent Theory and Contract Types

Einsatzgruppe Wiking's cooperation with the German construction industry manifested in the contracts that the organisation entered into – individually – with every company for every construction project. While various laws, ordinances and other regulations provided the general framework for state-industry relations in the construction sector, it was during the contract negotiations, that state and industry faced each other directly. In this thesis, the contracts as well as the negotiations leading up to their signing are thus considered crucial for an understanding of the relation between the OT and its contractors.

This perspective rests on two basic assumptions: Firstly, German construction firms were neither conscripted into, nor absorbed by the OT, before they almost "dissolved" in the Organisation's apparatus. Rather, they remained independent economic entities with separately identifiable motives and strategies, also under the para-military OT.[100] Secondly, I assume that the OT contractors' decisions were guided by economic rationality, meaning that cost-benefit considerations and motives such as profit seeking and the wish to secure the firm's long-term survival remained central to entrepreneurial decision-making processes. These positions will be substantiated especially in chapters IV.1.b and V. This does not rule out that the firms' employees or managers were influenced by National Socialist ideology, became part of criminal occupation regimes and committed war crimes.

This thesis can draw upon a strand of literature that has highlighted the importance of contracts in the German war economy. British economic historian Richard Overy was the first to stress that contracts are a key to understand state-business relations in the Third Reich. He identified two main types of contracts, which were used in armaments production and construction alike: cost-plus contracts and fixed-price contracts.[101] Cost-plus contracts (*Selbstkostenerstattungsverträge*) were directly awarded by the state to single firms, without inviting offers. Awarding authorities like the OT reimbursed all of the company's expenditures on wages, equipment, overheads, and so on, and then granted a mark-up on the sum. The latter

99 Ziegler, 'Regulated Market Economy', 150–51.
100 This does not exclude the fact that they took over important tasks at various levels of the OT hierarchy.
101 Richard Overy, *War and Economy in the Third Reich* (Oxford: Clarendon Press, 1994), 348–49, 357–58.

was the company's profit. Cost-plus contracts made it possible to start production immediately, without lengthy calls for tenders, pre-calculations and price negotiations. However, the costs of a construction project were not certain until it had been completed and there was no incentive at all for a company to reduce costs. Quite the contrary: lavish use of resources and higher building costs actually increased the companies' profits.

Less problematic, in this respect, was the fixed-price contract (*Festpreisvertrag*) where the contractual partners agreed on a fixed price before the project started. Under this contract regime, the company was interested in working more efficiently, as it could keep the saved costs as an additional profit.[102] The Award Regulations for Construction Works (*Verdingungsordnung für Bauleistungen*, VOB) of May 1926 had set out that fixed-price contracts should be awarded ideally after a process of open competitive bidding.[103] Thus the awarding authority could compare prices and select a firm – not necessarily the cheapest, but the cheapest among the ones promising the contract would be executed professionally and reliably. The precondition for this method was of course that companies did not engage in illegal price-rigging.[104]

It is no surprise that German authorities were aware of the negative effects of cost-plus contracts and in principle agreed that cost-reducing fixed-price contracts were preferable.[105] As we will see below, no one in the German administration willing to see could close their eyes to the devastating budgetary effects of cost-plus contracts in the overheated construction boom of the late 1930s. Representatives of the Reich Ministry of Finance (RFM), and the German Court of Audit (*Rechnungshof des Deutschen Reiches*) in particular, seldom missed an opportunity to warn against the use of this type of contract in public procurement.[106]

In fact, a decree abolishing cost-plus contracts was issued by Hermann Göring in October 1941 and ultimately implemented in early 1942.[107] When Richard Overy ried to explain the alleged German "production miracle" under the newly appointed armaments minister Albert Speer, he drew upon this document and argued that the general transition from cost-plus to fixed-price contracts in the armaments industry played a crucial role in increasing the industry's efficiency from 1942 onwards. According to Overy, fixed-price contracts "under Speer [. . .] became one of the main

102 Streb, 'Preisregulierung', 37.
103 Part A, paragraph 3, in: Reichs-Verdingungs-Ausschuß, *Verdingungsordnung für Bauleistungen (VOB)* (Berlin: Bauwelt-Verlag, 1926).
104 Streb, 'Preisregulierung', 37–38.
105 Streb, 'Negotiating Contract Types', 366.
106 See below, chapter III.2.b.
107 Overy, *War and Economy*, 357; Jochen Streb and Sabine Streb, 'Optimale Beschaffungsverträge bei asymmetrischer Informationsverteilung: Zur Erklärung des nationalsozialistischen "Rüstungswunders" während des Zweiten Weltkriegs', *Zeitschrift für Wirtschafts- und Sozialwissenschaften* 118 (1998): 288.

instruments in encouraging greater efficiency."[108] While other economic historians have indeed acknowledged the crucial role that a profound understanding of procurement contracts plays for the analysis of the Nazi economy, they provided evidence of the use of fixed-price contracts before 1942. In doing so, they helped to undermine the idea that 1942, or the change from Todt to Speer, marked a turning point in the German war economy.[109] Obviously, the Göring decree did not have the impact that Overy ascribed to it and his claim does not hold for the construction industry either.[110] In occupied Norway, the EW still used cost-plus contracts to a considerable extent even in the last year of the war. In addition, the OT never introduced a true fixed-price contract, but designed a so-called efficiency-output contract (*OT-Leistungsvertrag*), which combined features of both fixed-price and cost-plus contracts.[111] But why?

There are several possible explanations for this. Based on the positions that have been discussed in chapter I.1.b, stressing German entrepreneurs' autonomy and ample scope of action, one could assume that the National Socialist state was not powerful enough to force private construction companies into signing an unwanted type of contract. According to this assumption the OT was "weak" and so dependent on the expertise of the private industry that the latter was able to dictate the terms of cooperation. An alternative explanation is that Overy's understanding of the advantages and disadvantages of the two contract types is too short-sighted and there were rational arguments for German procurement agencies like the OT to continue awarding cost-plus contracts after 1942. Throughout this thesis, I will use PA theory as a tool to analyse the contractual relations between the OT and the private companies and to gain a better understanding of the motives and conflicts between the contractual partners.

A Principal-Agent relation can take many different forms: a boss and her workers, a physician and his patient, or a state agency awarding a contract to a private company. The case of governments and state agencies that delegate the implementation of construction projects to private construction companies is a classic

---

108 Overy, *War and Economy*, 348–49, 357–58. For a similar interpretation, see Avraham Barkai, *Nazi Economics. Ideology, Theory, and Policy* (New Haven, London: Yale University Press, 1990), 237; Streb and Streb, 'Optimale Beschaffungsverträge', 277.

109 Lutz Budraß, Jonas Scherner, and Jochen Streb, 'Fixed-Price Contracts, Learning and Outsourcing: Explaining the Continuous Growth of Output and Labour Productivity in the German Aircraft Industry during the Second World War', *Economic History Review* 63, no. 1 (2010): 107–36; Rauh-Kühne, 'Hitlers Hehler', 21–22; Jonas Scherner and Jochen Streb, 'Das Ende eines Mythos? Albert Speer und das sogenannte Rüstungswunder', *Vierteljahresschrift für Sozial- und Wirtschaftsgeschichte* 93, no. 2 (2006): 186–90; Adam Tooze, *The Wages of Destruction: The Making and Breaking of the Nazi Economy* (London: Penguin Books, 2007), chap. 17, especially pp. 564–66; Scherner, *Logik der Industriepolitik*, 41, 301–5.

110 Cf. Streb, 'Negotiating Contract Types'.

111 The *OT-Leistungsvertrag* was a *Selbstkostenfestpreisvertrag*.

example of PA theory.[112] The assumption that principal and agent have differing interests forms the core of this theory. Providing the most thorough analysis of contract regimes in construction under the Third Reich, Jochen Streb has modelled the National Socialist construction industry as a three-staged PA problem.[113] The companies as agents are interested in maximising their profits with as little effort as possible, whereas the principal (in our case the OT) tries to find the optimum balance between reducing costs and carrying out a project as fast as possible. However, procurement agencies like the OT can be agents, too. While the OT's focus was on the realisation of the construction projects, the agencies within the German Reich that were concerned with public spending acted as principals when they controlled and attempted to limit the OT's expenses. A third PA relation exists between companies and their workforce, this time with the companies as principals who want to reduce production costs and maximise their worker's efforts, and the workforce as agents who try to earn as much money as possible with as little effort as possible. Throughout this thesis, when speaking about firms, about their motives, strategies and scope of action, I always refer to the firms' managements. Their motives, strategies and scope of action, especially under the para-military OT, do not necessarily coincide with those of their workforce. For now, I will focus on the relation between procurement agencies (principal) and construction companies (agents).

A PA relation always represents a conflict of aims in which the agent's room for manoeuvre originates from *information asymmetries*. That means that the contracting partners do not share the same relevant information about a project's costs. Building companies usually have a better idea of how long it will take to complete a particular construction project, how experienced they are in the specific area in question, and hence what the real costs will most likely be. When negotiating with the principal, the entrepreneur can therefore either understate the company's efficiency or overstate the project's complexity. Principals run the risk of selecting the agent who was most successful in deluding them – a problem known as *adverse selection*. After the conclusion of the contract, too, a company will not necessarily act in the interest of the principal (*moral hazard*). As examples for this phenomenon, Jean-Jacques Laffont and Jean Tirole mention "[. . .] purchase of materials and

112 For the following brief overview over PA theory's basic assumptions, cf. Jean-Jacques Laffont and Jean Tirole, *A Theory of Incentives in Procurement and Regulation* (Cambridge/Mass.: The MIT Press, 2002), 1 ff; David E. M. Sappington, 'Incentives in Principal-Agent Relationships', *The Journal of Economic Perspectives* 5, no. 2 (1991): 45–66; Mark Spoerer and Jochen Streb, *Neue deutsche Wirtschaftsgeschichte des 20. Jahrhunderts* (München: Oldenbourg Verlag, 2013), 161–71; Streb, 'Preisregulierung'; Rudolf Richter and Eirik G. Furubotn, *Institutions and Economic Theory. The Contribution of the New Institutional Economics*, 2nd ed. (Ann Arbor: The University of Michigan Press, 2010), 199–290.
113 Streb, 'Preisregulierung', 36.

equipment at high prices, and hoarding of engineers or machines not required under current contracts but useful for commercial profits or for winning future contracts [. . .]".[114]

The principal has options and tools at its disposal to reduce the negative effects of information asymmetries. Their implementation, however, comes at a price: *transaction costs*.[115] In the case of construction projects, the principal could first try to gather detailed information on the real construction costs, ideally of every single work step, and the real productivity of the agent, before concluding a contract. Second, the customer could *ex post* try to implement tight price controls, monitor the construction sites and scrutinise the company's book-keeping, or – third – enter renegotiations on certain contract clauses. All these measures create considerable transaction costs, as they require a staff of experts and controllers and – not least – time. Depending on the type of transaction, the costs are often labelled as information costs, control costs and bargaining costs, respectively.[116] The principal always faces a trade-off between lower costs of the overall project and lower transaction costs.

When dealing with this dilemma, the different types of contracts used by procurement agencies play a crucial role. As mentioned above, German state authorities in the 1930s and 1940s often used cost-plus or fixed-price contracts. Both contract types come with advantages and drawbacks and offer the agent different incentives.[117] Cost-plus contracts are prone to moral hazard because the firms do not bear the risk of rising costs during the construction phase and have no incentive to reduce the building costs. Since the profit is calculated as a percentage of the total building costs, there is even an incentive to inflate the costs. A fixed-price contract, on the other hand, reduces the costs of moral hazard. If the building costs turn out to be lower than the fixed price agreed upon beforehand, the agent can keep the margin. Therefore, the agent should be interested in finishing the project faster and at a lower price to maximise his or her earnings. The clear disadvantage of fixed-price contracts, however, are high transaction costs. Before being able to agree on a fixed price, both principal and agent have to define work steps and gather detailed information on the construction process. Moreover, an open and competitive bidding process has to be initiated to reduce the problem of adverse selection. Under certain conditions, the VOB also permitted single tender procurement, that

114 Laffont and Tirole, *Theory of Incentives*, 1.

115 PA theory is rooted in New Institutional Economics one cornerstone of which is the perception that every single transaction in an economy inevitably incurs costs. See Oliver E. Williamson, *The Economic Institutions of Capitalism. Firms, Markets, Relational Contracting* (New York: The Free Press, 1985), 15 ff; Richter and Furubotn, *Institutions and Economic Theory*, 47–77.

116 Richter and Furubotn, *Institutions and Economic Theory*, 51 ff.

117 Streb, 'Negotiating Contract Types', 365 ff; Patrick Bajari and Steven Tadelis, 'Incentives versus Transaction Costs: A Theory of Procurement Contracts', *RAND Journal of Economics* 32, no. 3 (2001): 387–407.

is, awarding contracts to a firm directly, without inviting tenders.[118] This speeds up matters, but the agent, who bears the risk of increasing costs under a fixed-price contract, could try to capitalise on information asymmetry and push through a higher fixed price. The more complex a project is, the harder it becomes for the principal to assess the real costs of a construction project, thereby reduce information asymmetry and negotiate a lower fixed price.[119] After the start of construction, changes in the project's design and unforeseen events may lead to expensive and time-consuming renegotiations.

The set-up of cost-plus contracts, in contrast, is simpler and there is no need for complicated renegotiations. Therefore, Patrick Bajari and Steven Tadelis conclude that fixed-price contracts are more suitable for simple, standardised projects where there is some time for planning and later design changes are unlikely, while cost-plus contracts can be the better choice for getting complex projects started as fast as possible.[120] In his analysis of contract types and negotiations in the German construction industry during the Third Reich, Jochen Streb points out accordingly that there is the paradox that a cost-plus contract is better "to get a construction project started quickly" whereas a fixed-price contract sets incentives "to get the construction project finished quickly".[121] Streb notes several cases in his sample where this paradox was solved by starting construction under a cost-plus contract and then switching to a fixed-price contract later on.[122] Thus, based on the assumptions of PA theory, there are rational arguments for the continuous existence of both types of procurement contracts.

Under a cost-plus contract the risk of delays is borne by the principal: under fixed-price contracts, however, the company bears the risk of financial losses when works stop. Therefore, PA theory assumes that companies try to conclude cost-plus contracts or – if that is not possible – demand that high-risk premiums are included in the contracts to cushion the risks of *ex ante* uncertainty.[123] The effects of this strategy should not be underestimated. When companies push through a high-risk premium and a high overall fixed priced for the construction project and also manage to include various exceptions in the contract clauses, the cost-reducing effects of fixed-price contracts may vanish completely. Accordingly, Bajari and Tadelis point out the importance of competitive bidding during the procurement process.[124] Only by comparing bids from

118 Part A, paragraph 3, in: Reichs-Verdingungs-Ausschuß, *VOB*.

119 Competitive bidding can help the principal to detect excessive prices. By comparing the offers of two or more companies, it becomes easier to get a picture of the "real" building costs: Sappington, 'Incentives', 56–59.

120 Bajari and Tadelis, 'Incentives versus Transaction Costs', 392.

121 Streb, 'Negotiating Contract Types', 373.

122 Streb, 373.

123 Streb, 365 and 374; Sappington, 'Incentives', 49.

124 Bajari and Tadelis, 'Incentives versus Transaction Costs', 390.

several companies can a procurement agency be sure of sorting out overpriced offers.[125] Competitive bidding was still the rule in the early days of the Third Reich, for example for the *Autobahn* projects. However, from the mid-1930s onwards, with construction companies overwhelmed with state orders and state agencies actually competing for contractors, the amendatory mechanism of competitive bidding was used less and less and was indeed often suspended completely. In other words, it is highly questionable whether the necessary preconditions for the effective functioning of fixed-price contracts were still given from the second half of the 1930s onwards. This holds true for the OT's projects in the occupied territories in particular.

In an attempt to find a middle ground between cost-plus and fixed-price contracts, National Socialist authorities introduced so-called *Selbstkostenfestpreisverträge*, literally translating as costs-fixed-price contracts, in 1940.[126] In the OT universe, this contract type became known as efficiency-output contract (*OT-Leistungsvertrag*) . Its main feature was that firms had to disclose their price calculations in a standardised schema and estimate the overall price of the project. This did indeed facilitate the subsequent control of prices, but increased the control costs, too.[127] The profit was now preferably to be calculated as a fixed sum in order to eliminate the incentive to increase costs deliberately. However, with the permission of the awarding authority, calculating the profit as a percentage of the construction costs was still possible.

Table 1 summarises the effects of different procurement procedures as described by Streb.[128] Given that the ideal – fixed-price contracts resulting from a

Table 1: Effects of different procurement procedures on information asymmetry and control costs.*

	Reduction of information asymmetry		**Reduction of control costs**
	Reduction of adverse selection	**Reduction of moral hazard**	
Competitive bidding resulting in a fixed-price contract	high	high	high
Single tender procurement resulting in a fixed-price contract	low	high	moderate
Cost-plus contract	high	low	low
Efficiency-output contract	moderate	high	low

*Streb, 'Preisregulierung', 43.

125 The precondition is of course that the firms do not engage in collusive pricing.
126 RGBl. 1940, I, 851–59.
127 Streb, 'Preisregulierung', 42.
128 A fifth option, group price menus (*Gruppenpreismenüs*), is omitted in this table and discussion, because it was hardly used in the sector beyond the construction of wooden barracks.

competitive bidding process – was often no longer an option, awarding authorities had to choose between several procedures and contract types that all had their drawbacks. For the firms, we can assume that they preferred either cost-plus contracts or fixed-price/efficiency-output contracts with high-risk premiums. Over the course of this thesis, I will take a closer look at contract negotiations and the firms' motives, particularly when working for the OT in occupied Europe.

2 Argument and Structure of the Thesis

The thesis conceptualises the OT as one of the institutional hybrids, located between traditional bureaucracy, the powers of plenipotentiaries and public-private partnerships, which have been identified as characteristic of the political system of the Third Reich.[129] During the Second World War, no other industrial sector relocated its activities to the occupied territories to such a degree as construction. I argue that in this setting, German construction firms were not only subject to, but an integral and formative part of the German construction administration. They shaped and influenced the OT's policy, rather than being subjugated by a military, dictatorial organisation. Thus, the National Socialist plans to literally *rebuild* Europe would have been unfeasible without relying on private business. On the construction sites all over occupied Europe, firms not only retained a great deal of entrepreneurial freedom, but took over administrative tasks that kept the OT machinery running. Together, the OT and its contractors laid the foundations of occupation.

To provide a background to my analysis, chapter II will briefly outline the structure and economic development of the German construction sector during the Third Reich. The sector experienced an unprecedented boom particularly in the second half of the 1930s, before domestic demand declined during the war. This put construction firms in a special position, as – more than in any other sector – German projects in the occupied territories became a chance to keep up their level of activity. Furthermore, chapter II will provide an overview over the history of the GdSt and the OT in pre-war Germany. The objective is to gain an understanding of the relation that Fritz Todt built with the German construction industry during the 1930s. When companies entered negotiations with the OT during the war, their expectations were of course influenced by the industry's earlier encounters with Todt's apparatus.

In order to operationalise and answer the research questions outlined above, in chapter III, I will look at crucial points of contact between state and construction

129 Rüdiger Hachtmann and Winfried Süß, eds., *Hitlers Kommissare: Sondergewalten in der nationalsozialistischen Diktatur* (Göttingen: Wallstein Verlag, 2006), 12. Similarly: Martin Broszat, *Der Staat Hitlers. Grundlegung und Entwicklung seiner inneren Verfassung* (München: Deutscher Taschenbuch Verlag, 2000), 328–32.

industry throughout the 1930s and 1940s. Such points of contact included, for example, the state's various attempts at regulating the construction sector and steering the companies in the direction that the regime desired. Regulations concerned the control of wage rates and entrepreneurial profits among other things. Construction materials were rationed, labour mobility restricted and construction sites closed down. While construction certainly was a highly regulated sector in National Socialist Germany, I argue that the detection and description of regulations alone tells us little about entrepreneurial room for manoeuvre in the sector and day-to-day work on the construction sites. Chapter III will therefore assess the regulations' implementation, limitations and sometimes unintended effects. An equally important point of contact were the contract negotiations between the OT and the firms. They, too, must be central to any analysis of the Organisation and will be covered in chapter III.

While certain state agencies relentlessly insisted on observing the National Socialist price and wage stop, the OT preferred to keep the industry on a long leash. More often than not, the OT and its contractors forged an alliance to ward off interference by the German Court of Audit and the Reich Price Commissioner, for example. In this context, it was of paramount importance that Fritz Todt was able to rely on the personal authority of Adolf Hitler, which also found its expression in the OT's status as a Supreme Reich Authority (*Oberste Reichsbehörde*). Chapter III shows how this contributed to rising profits and exploding costs. I argue, however, that this should not be misinterpreted as an organisational failure of the OT. It was its very nature – and it did in fact yield results, as attested by the countless massive constructions all over Europe.

Chapter IV then follows the OT to the occupied territories and Norway in particular. The chapter will analyse the OT's *modus operandi* in occupied Europe and particularly enquire what consequences the OT's increasing militarisation and status as *Wehrmacht* auxiliary force had for the OT's contractors and their entrepreneurial scope of action. The argument is that the OT contractors – although integrated into an increasingly militarised framework – did not lose their identity as independent economic entities. Moreover, they often played an active role within the criminal National Socialist occupation regime.

With regard to occupied Norway, chapter IV will show why EW was established in spring 1942 and how the German construction activities impacted the economy of the occupied country. To this end, EW's funding mechanisms will be investigated. The financial aspects of the OT's activities in occupied Europe and their impact on the occupied countries' economies are among the biggest desiderata in the research on the OT. In examining them, the chapter will add new, important findings to the history of occupied Norway. The chapter will also address the German attempts to regulate the construction sector in occupied Norway, again stressing the importance of control deficits and inefficiencies in particular.

Finally, chapter V will analyse in detail the motives of German construction firms for accepting an OT order in occupied Norway. It rejects the assumption that the OT conscripted German firms against their will. Thus, incentives provided by the Organisation and the companies own motives must become the centre of focus. The chapter will provide detailed analyses of the contract negotiations between the EW and its contractors and of economic aspects of EW's system of forced labour.

Moreover, two major findings of the chapter will contribute not only to the history of the OT, but also to the research on economic and business history as well as to the research of National Socialist forced labour: first, I argue that the OT's so-called efficiency-output contract was a pioneering attempt to raise companies' performance in a setting of high risk and time pressure in the second half of the war. Second, I regard the introduction of so-called performance factors for entire ethnic groups of forced labourers as an equally pioneering attempt to make the exploitation of forced labour manageable under the unpredictable conditions of war and occupation. In both cases, I will demonstrate that private German construction firms played a crucial and active role in their formulation and implementation.

As the Organisation relied heavily on the firms' expertise and support, the latter were able to profit financially from the information asymmetry between them and EW. German construction firms in occupied Norway were not subject to a command economy and they were not taken over by the OT. If anything, it was private business that more and more often assumed responsibilities previously held by the state. As stated before, this is not the story of a failing state organisation. A high degree of industrial self-responsibility was the precondition for the realisation of OT's projects all over Europe.

Combining the arguments outlined above, the thesis' main argument is that this industrial self-responsibility made it possible to lay the foundations for an entire continent under German domination in the first place. The OT did *not* carry out countless projects in Germany and occupied Europe for almost seven years by erecting a huge quasi-military apparatus with a rigorous contract regime, extensive controls and draconian penalties. Quite the contrary: the Organisation used incentives rather than force, let its contractors participate in the formulation and implementation of its policy, delegated tasks to private business and allowed its contractors to realise respectable profits. Moreover, steering a sector as large and fragmented as construction only by means of commands and controls would have been unfeasible. This opened up a scope of action for private business to pursue its own goals and safeguard its interests not only by eluding the Organisation's monitoring efforts, but also by exerting influence on the OT "from within".

3 Sources

As shown in the literature review, the relation between construction industry and state agencies in the Third Reich calls for an exhaustive analysis. There is no doubt that the unique and extensive records of *Einsatzgruppe Wiking* in the Norwegian

National Archives in Oslo are crucial for reconstructing the history of the EW. The approximately 450 linear metres of documents of the OT's Nordic branch are probably the most important and promising source for historians of the OT, as they have survived the war largely intact.[130] In contrast, the holdings of documents on the OT at the German Federal Archives in Berlin are fragmentary and contain little material on the mission in occupied Norway.[131]

Besides OT's *Einsatzgruppe Wiking*, the thesis puts the focus on the offices that were closely connected to the OT both institutionally and at the staff level: the GdSt and the GB Bau. While the GdSt has left behind self-contained holdings (R 4601), the history of the GB Bau had to be reconstructed from sources spread over various archives. Significant holdings furthermore exist at the *Hessisches Hauptstaatsarchiv Wiesbaden*,[132] as the *Reichsautobahnen* unit from Frankfurt am Main organised the mobilisation of firms and machinery transports in the beginning of EW's mission in 1942. The holdings at the National Archives in Oslo also contain material of GB Bau's representative in occupied Norway.

For the thesis, further material has been used more extensively and under different angles than existing research has done so far: several state agencies criticised the OT's financial conduct, the violent construction boom and the attempts to regulate the sector in the late 1930s. Therefore, I have supplemented existing literature and the sources left behind by the OT itself with seldom-used material from the German Court of Audit, the Reich Price Commissioner and the RFM.

Most importantly, however, records of private German construction firms have been used throughout this thesis. There are strong methodological reasons for this. As outlined above, one of the thesis' main arguments is that the relation between the OT and its contractors – and more generally between state and construction industry – was characterised by information asymmetries and control deficits. Trying to unveil the companies' motives and strategies during their negotiations with the OT and their reactions to state regulation exclusively on the basis of sources produced by state agencies would thus have inevitably missed important parts of the picture.

Two groups of internal wartime records are particularly relevant in this respect. First, the firms' local construction site managers in occupied Norway reported on a regularly basis to their head offices in Germany. Furthermore, contracts with EW usually had to be checked and approved by the firms' managers in Germany. At the National Archives in Oslo we find a great many documents that private German construction firms (for example Steffen Sohst, Kiel) had to

---

130 *Riksarkivet Oslo* (hereafter: RAO), RAFA-2188.
131 BArch, R 50-I.
132 Hereafter: HHStAW.

leave behind in the last days of the war. They can today be found among the records of the EW. Furthermore, substantial holdings of documents are available at the *Archiv des Instituts für Stadtgeschichte, Frankfurt am Main* (Philipp Holzmann AG), *Thüringisches Hauptstaatsarchiv, Weimar* (Straßenbau AG), and *Landesarchiv Berlin* (Allgemeine Baugesellschaft Lorenz).[133] Second, the companies discussed the terms of their mission in Norway with *Wirtschaftsgruppe Bauindustrie* (Wigru Bau), the construction industry's association, which also established a liaison office in Oslo. Material of Wigru Bau can be found both in the OT archive at the National Archives in Oslo and at the Federal Archives in Berlin.[134]

In addition, post-war statements in memoires and anniversary *festschriften* as well as the construction industry's publications between 1933 and 1945 have been evaluated. They have of course to be read with particular care. After the war, firms were presumably interested in downplaying their own participation in the exploitation of occupied Europe and presenting themselves as victims of a dictatorial organisation. Publications from the period before 1945, on the other hand, did certainly not display the construction industry's unvarnished opinions. Nevertheless, Wigru Bau's periodical *Die Bauindustrie* has proven to be a helpful and extremely rich source. In addition, business reports of 26 construction companies from the 1930s and 1940s have been evaluated.[135]

In general, however, only few construction firm records have survived the decades. Enquiries at almost 100 German state, regional and town archives regarding firms that were active in Norway yielded only few results. This can be partly explained by the fact that construction is a sector that experienced countless mergers, take-overs and bankruptcies in the decades following the Second World War. This holds true for small, often family-run businesses in particular, in addition to the fact that these firms have produced only few records in the first place. Therefore, large firms are clearly overrepresented also in the sources used in this thesis – a well-known problem for many business historians. The conclusions drawn in this thesis should thus be applied to small construction businesses only with reservations.

Finally, enquiries with a dozen of construction firms that are still operative today mostly remained unanswered, even though some of them verifiably possess records on their history during the Third Reich.[136] The question is whether the

133 Hereafter abbreviated IfS, ThHStAW and LAB, respectively.

134 BArch, R 13-VIII.

135 Unless otherwise stated, the business reports have been gathered from the digitalised press archives of the *Institut für Weltwirtschaft*, Kiel, and the *Hamburgisches Welt-Wirtschafts-Archiv*. See Sources for a reference.

136 The notable exception is the *Unternehmensarchiv der Bilfinger SE,* Mannheim (hereafter: Bilfinger Archive).

opening of corporate archives during the 1990s and early 2000s – a reaction to the public debate on National Socialist forced labour – really marked a change in corporate culture or in fact was merely a temporary concession to public pressure.[137]

137 On the motives behind German companies' contribution to the compensation fund for former forced labourers, see Mark Spoerer, 'Moralische Geste oder Angst vor Boykott? Welche Großunternehmen beteiligten sich aus welchen Gründen an der Entschädigung ehemaliger NS-Zwangsarbeiter?', *Perspektiven der Wirtschaftspolitik* 3, no. 1 (2002): 37–48.

II The German Construction Sector between Crisis and War

1 Construction in the Third Reich: An Overview

The construction sector has always been sensitive to cyclical ups and downs, but the twelve years of National Socialist rule were certainly an age of extremes for the German construction industry. Few sectors were hit as hard by the economic crisis of the early 1930s and few experienced such an overwhelming boom later on. Initially fuelled by public work creation measures, the construction sector increasingly profited from the regime's policy of rearmament that was to make the armed forces, the armaments industry and other infrastructure ready for war within the shortest possible time. According to Adam Tooze, between the National Socialist seizure of power and the autumn of 1938, the share of national output spent on the military rose from one to almost 20 percent: "Never before had national production been redistributed on this scale or with such speed by a capitalist state in peacetime."[1]

Between the disastrous year 1932 and 1938, construction output in the *Altreich*, that is Germany in its pre-1938 borders, rose by more than 400 percent. Even compared to 1928 – a strong pre-crisis year – output had almost doubled by 1938 (Figure 1). This almost matched the growth of the 10-year period between 1950 and 1960, when an entire country had to be rebuilt after a devastating war.[2] Especially from the mid-1930s onwards, the construction projects managed by Fritz Todt contributed greatly to this boom: first the *Reichsautobahnen*, which Todt planned in his position as GdSt, and from June 1938 onwards the *Westwall*, which would become the birthplace of the OT and probably the largest construction undertaking Germany had ever seen.

After the outbreak of the war, however, domestic demand for construction works was cut back considerably and under the direction of the OT, construction firms relocated many of their activities to the occupied territories. Within Greater Germany, construction output fell by more than 70 percent between 1939 and 1944.

In the following, part a will provide a brief overview of the structure and organisation of the German construction sector in National Socialist Germany. The aim is to identify peculiarities of the sector that impacted the state's ability to steer and regulate German construction firms. Furthermore, I will introduce key terms and definitions that were used in contemporary statistics and politics to describe the sector, many of which will also be used throughout this thesis.

---

1 Tooze, *Wages*, 659.

2 Erhard F. Knechtel, *Auf Dialog gebaut: 50 Jahre Hauptverband der Deutschen Bauindustrie; Eine Chronik* (Wiesbaden, Berlin: Bauverlag, 1998), 31.

https://doi.org/10.1515/9783110694291-002

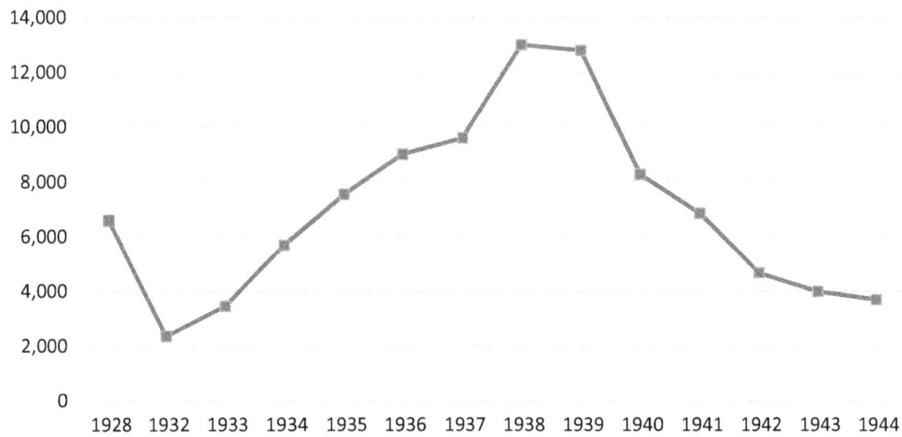

Figure 1: German construction output 1928/32–44 in million RM, current area of "Greater Germany", constant 1938 prices. Source: appendix, Table A1.2.

Part b will cover the period between the economic crisis of the early 1930s and the end of the decade, when the massive state demand overburdened the construction sector. In particular, I will investigate the effects of the work creation measures on construction and discuss the construction industry's leading role in lobbying for work creation. Although the first work creation measures neither originated from National Socialist policy nor were responsible for sparking the upswing in the construction sector, I argue that they decisively shaped the construction industry's attitude towards the new regime.

Finally, part c will ask about the consequences that the considerable drop in domestic demand had for construction firms after 1939. The situation in construction differed from that in other sectors of the German war economy and thus helps us to understand construction companies' particular motives and options during their negotiations with the OT.

a Structure of the Sector

Measured by workers employed, construction was Germany's largest industrial sector on the eve of the Second World War. Between June 1933 and June 1938, the number of workers in the construction sector had risen from 1,745,000 to 2,071,000.[3]

3 *Statistisches Jahrbuch für das Deutsche Reich 1939/40* (Berlin: Schmidt, 1940), 379. The figures include unemployed construction workers and refer to Germany's pre-1938 territory.

However, the occupation census of 1938 still underestimated the number of persons working in this fragmented sector, as the German Statistical Office (*Statistisches Reichsamt*, SRA) only counted employees that were required to have labour books, the obligatory employment record books that had been successively introduced from 1935 onwards.[4] In its industrial census of 17 May 1939, the SRA registered almost two and half million Germans that were working in construction. This remarkable growth within less than a year can be mostly explained by the annexation of Austria and the Sudetenland and the fact that this time, also the owners of construction businesses were counted.[5] Moreover, the labour offices sent many workers from other occupations to the *Westwall*, whose construction had started in June 1938. Construction was, however, not only one of the largest, but also one of the most fragmented sectors of the German economy: applying a very broad definition, the SRA listed 266,803 independent production units in May 1939.[6] This huge sector was usually divided along three axes, establishing several categories of construction firms and activities that often overlapped and were not always clearly defined.

First, the number of 266,803 production units included almost 88,000 firms of the construction sector in the narrow sense, also called the main trades contractors (*Bauhauptgewerbe*), and almost 145,000 special trades contractors (*Baunebengewerbe*) such as painters, roofers, chimney builders or insulation contractors. Second, around 28,000 firms were industry firms, defined as such by their membership in the association of the German construction industry, Wigru Bau. More than 238,000 production units, however, were handwork firms, organised in one of the various construction handwork organisations.[7] The handwork businesses were often smaller and at times only consisted of the owner, some family members and a handful of employees. More than 200,000 of these production units employed ten persons or less.[8] Third, construction activities in Germany were traditionally divided into *Hochbau* and *Tiefbau* works.[9] The terms were also commonly used in German statistics. *Hochbau* refers to structures with a mainly vertical extent, mostly works above ground level, such as residential

4 Mason, *Arbeiterklasse und Volksgemeinschaft*, 1247–48.

5 *Statistisches Jahrbuch für das Deutsche Reich 1941/42* (Berlin: Schmidt, 1942), 181.

6 *Statistisches Jahrbuch für das Deutsche Reich 1941/42*, 181; *The Effects of Strategic Bombing on the German War Economy (European Report #3)*, The United States Strategic Bombing Survey (1945; repr., New York: Garland Publishing, 1976), 235. The regional establishments of larger firms were counted as independent production units.

7 The term "handwork" is used in the 1945 American survey on the German war economy and is a direct translation of the German *Handwerk*, which covers work carried out by tradesmen and (artisan) craftsmen. See *The Effects*. *Handwerker* were usually members of "handwork" guilds. This thesis follows the abovementioned survey and uses "handwork" instead of the German term.

8 *Statistisches Handbuch von Deutschland 1928 – 1944*, 242–43.

9 Paul Gorgass, *Das Baugewerbe und seine Eingliederung in die Organisation der gewerblichen Wirtschaft: Eine Untersuchung der Gliederung und betrieblichen Struktur des Baugewerbes* (Berlin: Triltsch & Huther, 1937), 38–42.

construction, factories and public buildings. *Tiefbau* refers to construction with a mainly horizontal extent, sometimes including work below ground level, like tunnels, roads, railways, harbours, fortifications, and canals. Traditionally, *Hochbau* was dominated by handwork firms, while *Tiefbau* projects were planned by engineers. The existence of a separate employers' liability insurance association for *Tiefbau* (1887) and the foundation of a separate *Tiefbau* association (1900) bear witness of this division within the sector.[10] *Tiefbau* workers were covered by the nationwide employers' liability insurance association for *Tiefbau* (*Tiefbau-Berufsgenossenschaft*) . Workers of handwork firms, mostly engaged in *Hochbau*, were covered by twelve regional associations (*Baugewerks-Berufsgenossenschaften*). However, the division between *Tiefbau* and *Hochbau* activities had already lost much of its meaning in the 1920s and 30s. Due to advances in the field of structural steel engineering and the increasing use of reinforced concrete, for example, *Hochbau* was no longer the exclusive domain of tradesmen and craftsmen. Many companies were engaged in both fields of construction.

Two major shifts in the sector's structure occurred over the course of the 1930s. On the one hand, the National Socialist building boom was primarily driven by a rising state demand. While approximately half of the German construction volume had been publicly financed during the 1920s, the share rose to two thirds in the second half of the 1930s. Residential building suffered most from the shifting priorities, especially because publicly financed housing programmes were cut back (Figure 2). By absolute numbers, *private* residential building activity actually remained quite stable in the second half of the 1930s, and even above the level of the boom years of 1927–29. Timothy Mason has interpreted this fact as evidence for a run on tangible assets, motivated by a persistent fear of inflation among the German population.[11] However, private building activity alone could not countervail the increasing housing shortage, and by 1939, there was an uncovered demand of 2.5 million homes.[12] The situation worsened when residential construction came to an almost complete halt during the war.

Dependence on public orders was even higher among the larger building firms that were organised in Wigru Bau. In mid-1935, the association started collecting data on orders placed with their member firms and published these figures for the years 1936 to 1938. As it becomes clear from Table 2 below, orders by the *Reichsautobahnen* Corporation and the *Wehrmacht* (mostly hidden in "non-recurring public investments") were extremely important for the construction industry firms. But even for Wigru Bau's members, private industry as well remained an important customer

10 Cf. Erich Reuss, *Die Verbände der Bauindustrie* (Wiesbaden: Bauverlag, 1966), 23–26. The present thesis has decided to use the German terms as the English terms "civil engineering" and "structural engineering" do not cover precisely the same set of activities as *Hochbau* and *Tiefbau*.

11 Mason, *Arbeiterklasse und Volksgemeinschaft*, 205.

12 Tilman Harlander, *Zwischen Heimstätte und Wohnmaschine. Wohnungsbau und Wohnungspolitik in der Zeit des Nationalsozialismus* (Basel, Berlin, Boston: Birkhäuser, 1995), 180.

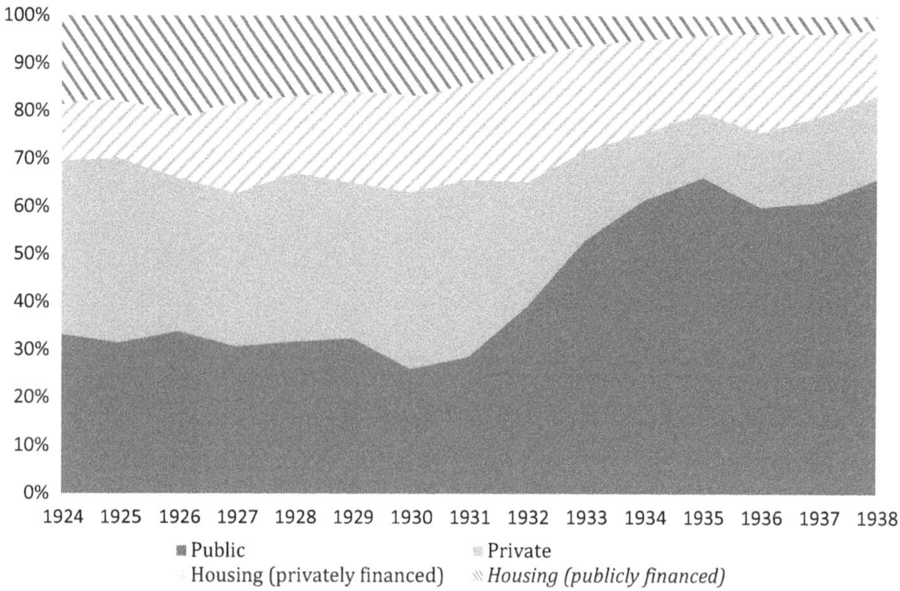

Figure 2: Proportion of the production volume in public and private construction and housing, 1924–38. Source: appendix, Table A3.1.

throughout the 1930s. The figures confirm Jonas Scherner's observation, according to which private sector investments even increased during the rearmament boom, instead of being crowded out by public investments.[13] In the case of the construction industry, however, the *Westwall* construction shifted the ratio between public and private construction in favour of the former from June 1938 on. The 1938 figures of Wigru Bau did not yet include the orders by the OT at the *Westwall.* The fortification line included, the association estimated that the share of public construction among the member firms exceeded 80 percent already in 1938.

The second major change in the sector's structure concerned the share of *Hochbau* and *Tiefbau* works. As will be shown below in section II.1.b, the government's work creation measures already targeted primarily at *Tiefbau.* The Institute for Business Cycle Research (*Institut für Konjunkturforschung*) even arrived at the estimate that 90 percent of the growth in the building production of 1933 could be attributed to *Tiefbau* projects for public customers.[14] In the following years, firms of this sub-sector continued to profit from the upswing to a larger degree than *Hochbau* firms. The latter

13 Jonas Scherner, '"Armament in Depth" or "Armament in Breadth"? German Investment Pattern and Rearmament during the Nazi Period', *Economic History Review* 66, no. 2 (2013): 511.
14 *Vierteljahreshefte zur Konjunkturforschung* 8, vol. 3, part B (1933), 171.

Table 2: Orders placed with Wigru Bau member firms, in %, 1936–38.*

	1936	1937	1938
Public construction	**74.4**	**72.8**	**68.8**
Reichsautobahnen	*17.7*	*16.2*	*16.0*
National Railways	*4.4*	*3.7*	*4.5*
Non-recurring public investments	*32.9*	*27.4*	*21.3*
Post Office	*0.2*	*0.3*	*0.2*
Reich building administration	*1.8*	*3.6*	*3.5*
Reich waterway administration	*4.2*	*2.9*	*4.9*
German states	*5.1*	*6.1*	*5.9*
Cities and municipalities	*7.0*	*7.6*	*6.3*
NSDAP and its divisions	*1.0*	*4.9*	*5.6*
Other organisations	*0.1*	*0.1*	*0.6*
Private construction	**25.6**	**27.2**	**31.2**
Industry	*19.4*	*22.3*	*24.3*
Trades and transport	*0.7*	*0.4*	*0.5*
Banks and insurance companies	*0.5*	*0.5*	*0.4*
Agriculture	*0.1*	*0.2*	*0.2*
Housing (privately financed)	*4.9*	*3.8*	*2.8*

*'1938–1939. Rückblick und Ausblick', *Die Bauindustrie* 7, no. 1 (1939): 6–10; Karl Ehrler, *Die statistische Erfassung der Bauwirtschaft. Methode und Auswertung* (Dresden: M. Dittert & Co, 1940), 117.

were increasingly hit by cutbacks in public spending, especially after the outbreak of the war. In contrast, major projects like the *Reichsautobahnen*, the *Westwall* and many of the OT's projects in the occupied territories were *Tiefbau* works. Therefore, the situations of, for example, a road construction company and a firm specialised in the building of dwellings could be very different.

While for large companies that were engaged both in *Hochbau* and *Tiefbau*, the latter had always been an important field of business,[15] the 1930s saw an increasing number of small firms and craftsmen focusing on *Tiefbau*. Indeed, among the "dubious elements" that sought their fortune in the booming *Tiefbau* and depressed the price level with frivolous tenders – a perpetual source of irritation for established *Tiefbau* companies during those years – were often craftsmen formerly engaged in *Hochbau*. As a *Tiefbau* entrepreneur complained, roadworks had been awarded to a mason, in another instance to a car rental company owner together with a carpenter. A stonemason carried out cable-laying works for the telegraph office while another mason was awarded land improvement works.[16]

15 See for example the enduring importance of *Tiefbau* works for Philipp Holzmann: Pohl, *Philipp Holzmann*, 120–21, 171, 223.
16 'Leidensweg eines Unternehmers!', *Die Bauindustrie* 3, no. 30 (1935): 445.

It is difficult, however, to quantify exactly to what extent *Tiefbau* superseded *Hochbau* during the Third Reich. The last figures on the sub-sectors' production volume were published by the non-governmental Institute for Business Cycle Research in 1935 and covered the year 1934.[17] After then, all information that has been found thus far is fragmentary, conflicting and uncertain to such a degree that it does not allow any qualified statements to be made on the development throughout the 1930s or even in wartime.[18] Apart from the abovementioned indistinct definition of *Hochbau* and *Tiefbau*, the construction volume of these two parts of the sector was only measured indirectly.[19] Estimates based on the monthly industrial reports (*Industrieberichterstattung*) to the SRA, published by Wigru Bau, show that the share of hours worked in *Tiefbau* rose from 50 percent in 1930–31 to almost 70 percent in 1938.[20] The monthly industrial reports, however, seem to have covered only a part of the sector.[21]

A third change in the sector's structure is not clearly identifiable. During the war, particularly from 1943 onwards, small construction firms were threatened by concentration and rationalisation measures that aimed to comb out workers and close down inefficient firms. The sector's "big players", which were represented in various industrial rings and committees, decided on closures and were thus often suspected of using these measures to swallow up smaller competitors.[22] However, although many small firms complained of losing workers to larger companies and the *Wehrmacht*, of limited access to building materials and of being ignored by procurement agencies, it is unclear to what extent wartime measures changed the

---

17 Ernst Wagemann, *Konjunkturstatistisches Handbuch 1936* (Hamburg: Hanseatische Verlagsanstalt, 1935), 50.
18 For a presentation and critical discussion of these estimates, see appendix A2.
19 The contemporary estimates on the *Hochbau* and *Tiefbau* construction volume are critically discussed in appendix A2.
20 Eugen Vögler, 'Die Lage der Bauwirtschaft. Rede auf der Tagung der Wirtschaftsgruppe Bauindustrie 1936 in Essen-Ruhr', *Die Bauindustrie* 4, no. 12 (1936): 205; Werner Glöckner, 'Die Bauindustrie im Spiegel der Beschäftigung des Jahres 1936 (Fortsetzung)', *Die Bauindustrie* 5, no. 7 (1937): 110; '1938–1939. Rückblick und Ausblick', 7.
21 The figures should be regarded as rough estimates, as the reliability of the industry reports is somewhat questionable. Wigru Bau noted in February 1936 that the reports covered only 17 percent of the workforce in the construction sector. Hence, the SRA demanded that more building firms should return the SRA's questionnaires: 'Klare Berichte über die Bauindustrie', *Die Bauindustrie* 4, no. 6 (1936): 97–98. The industry reports only included larger construction firms and excluded handwork firms that dominated residential construction, that is, the most important part of *Hochbau*. Therefore, the monthly industrial reports arguably overestimated the *Tiefbau* share and refer only to the construction industry in the narrow sense, rather than the whole construction sector. Cf. Adam Tooze, 'No Room for Miracles: German Industrial Output in World War II Reassessed', *Geschichte Und Gesellschaft* 31, no. 3 (2005): 454. This should not change the general trend suggested by the data, though.
22 On this topic, see chapter III.3.c.

sector's general structure. Reliable data are lacking, but in December 1943, Reinhard Jecht of Dyckerhoff & Widmann stated that the number of firms in the construction sector had remained more or less stable.[23]

b The National Socialist Building Boom: From Work Creation to Excess Demand

After the truly devastating economic crisis between 1929 and 1932, work creation measures stood central in the construction sector's recovery. This part of the chapter will investigate the effects of the work creation measures on construction and discuss the construction industry's leading role in lobbying for work creation. Although the first work creation measures were neither originating from National Socialist policy, nor responsible for sparking the upswing in the construction sector, I argue that they decisively shaped the construction industry's attitude towards the new regime.

It has become a truism that the Great Depression needs to be central to every account of the National Socialist society and economy. This economic downturn was truly devastating, and the share it had in dragging the Weimar Republic into the abyss can hardly be underestimated. Between 1929 and 1932, industrial output in the German Reich had plunged by 39 percent. Among the European economies, only Poland and Austria experienced similar contractions in industrial production.[24] In the winter of 1931–32, and again in the winter of 1932–33, the number of unemployed Germans exceeded six million. These figures, however, do not include those workers who did not register as unemployed when losing their job or were removed from the official register – many of them women and long-term unemployed. Therefore, the unemployment figures only insufficiently reflect how many Germans actually lost their jobs in the downturn. Between the middle of 1929 and January 1933, the number of German wage earners working full-time fell from 20 million to 11.4 million.[25] As Richard Overy points out: "The dimensions of the economic and political crisis in 1932 are easier to understand once it is recognised that the problem was not 6 but 9 million unemployed. Two out of every five Germans employed in 1929 were without work in the winter of 1932–3."[26]

23 Reinhard Jecht, 'Vereinfachung des Firmeneinsatzes. Zur Anordnung des Reichsministers Speer vom 28.10.1943', *Die Bauindustrie* 11, no. 19 (1943): 505.
24 Charles H. Feinstein, Peter Temin, and Gianni Toniolo, *The European Economy Between the Wars* (Oxford: Oxford University Press, 1997), 106.
25 Richard Overy, 'Unemployment in the Third Reich', in *War and Economy in the Third Reich*, ed. Richard Overy (Oxford: Clarendon Press, 1994), 38.
26 Overy, 38.

Unsurprisingly, the cyclically sensitive construction sector immediately felt the effects of the crisis. Companies and private housing builders postponed investments and repairs, while state and municipalities cut back on their construction projects in order to relieve their budgets. The public authorities' policy of belt-tightening was disastrous for an industry where public involvement was as great as in the case of construction.[27] The value of German construction output fell by more than 70 percent between 1929 and 1932, measured in current prices.[28] Unemployment figures, too, reflect how vulnerable this part of the economy was. In June 1933, the unemployment rate in the industry group of main and special trades contractors was at 44.8 percent: 897,170 construction workers and employees were without work. Among the larger industry groups, only the machine building and vehicle manufacturing sector showed a similarly high rate of unemployment.[29]

It is therefore hardly surprising that in business circles, it was industrialists from the construction sector and from parts of the economy closely connected to it, such as producers of construction materials and machinery, that became the prime advocates of work creation measures. Other parts of the German industry were sceptical about public work creation programmes and employment schemes. They often regarded them as unacceptable interference in the sphere of private business and saw credit expansion as a dangerous fiscal experiment.[30] Helmut Marcon has noted that the position of the Reich Association of the German Industry (*Reichsverband der deutschen Industrie*, RdI), for example, was "remarkably schizophrenic", as the RdI combined a general rejection of direct work creation with the repeated complaints about its insufficient extent.[31]

Only a handful of industrialists within the ranks of the RdI publicly advocated work creation. Among them was the Hessian cement producer Walter Dyckerhoff, whose plants were in serious economic trouble due to the crisis in the construction sector. Dyckerhoff published a memorandum on work creation in the summer of 1932, demanding "large public works based on state initiative, though executed by

---

27 Harold James, *The German Slump. Politics and Economics 1924–1936* (Oxford: Clarendon Press, 1986), 205.

28 See appendix, Table A1.1.

29 Statistisches Reichsamt, *Die berufliche und soziale Gliederung der Bevölkerung des Deutschen Reichs*, vol. 2, Statistik des Deutschen Reichs 453 (Berlin: Verlag für Sozialpolitik, Wirtschaft und Statistik, 1936), 11–12.

30 For examples on the opposition of big business, see Barkai, *Nazi Economics*, 46, 168; Dietmar Petzina, 'Hauptprobleme der deutschen Wirtschaftspolitik 1932/33', *Vierteljahreshefte für Zeitgeschichte* 15, no. 1 (1967): 27; Bera, *Lobbying Hitler*, 33–38, 79; Kurt Gossweiler, 'Der Übergang von der Weltwirtschaftskrise zur Rüstungskonjunktur in Deutschland 1933 bis 1934. Ein historischer Beitrag zur Problematik staatsmonopolistischer "Krisenüberwindung"', *Jahrbuch für Wirtschaftsgeschichte* 9, no. 2 (1968): 86–88.

31 Helmut Marcon, *Arbeitsbeschaffungspolitik der Regierungen Papen und Schleicher. Grundsteinlegung für die Beschäftigungspolitik im Dritten Reich* (Frankfurt am Main: Peter Lang, 1974), 157.

private business".[32] While he was an outsider within the RdI, Dyckerhoff's memorandum reflected the position of the construction sector in general, for which the situation in 1932 was far too dramatic to follow the reserved line of the RdI. As early as March, an alliance of about 50 industrial associations, handwork guilds and federations had addressed a memorandum to the Reich Chancellor. Large and small construction companies, craftsmen, architects and engineers, producers of construction machinery and building materials alike demanded public work creation schemes. The paper provided worrying figures: in January 1932, unemployment among union members in the construction sector was at 88.6 percent. It was estimated that construction output in 1932 would fall to 24 percent of the 1928–29 average in residential building, 28 percent in private construction and 26 percent in public construction. The alliance demanded a 1.5 billion RM investment programme in order to get the sector back on its feet. Because the share of wages and salaries in the total costs of construction projects was comparatively high, the memorandum prognosticated considerable effects on other sectors and overall consumption.[33] The calls for public work creation were repeated in the following months, for example by the Study Society for the Financing of German Road Construction (*Studiengesellschaft für die Finanzierung des deutschen Straßenbaus*) in June 1932,[34] or by the *Tiefbau* industry, which even predicted the complete collapse of the *Tiefbau* sector in 1933.[35]

At the height of crisis, the ranks of those rejecting the austerity policy of the Brüning government were swelling. Concepts were discussed that aimed at reducing unemployment with the help of large-scale work creation programmes that were to be financed by deficit spending.[36] Many of the propositions came from politicians and economic experts of the extreme right.[37] In May 1932, also the National

32 Quoted in: Wilhelm Grotkopp, *Die grosse Krise. Lehren aus der Überwindung der Wirtschaftskrise 1929/32* (Düsseldorf: Econ Verlag, 1954), 25.

33 BArch, R 13-VIII/27, Memorandum by the associations of the German construction industry to the Reich Chancellor, 17.3.32.

34 BArch, R 2/18648, Exposé by the Study Society for the Financing of German Road Construction regarding "The Execution of an Extraordinary Road Construction Programme as Part of a Work Creation Measure", 14.6.32.

35 BArch, R 43-I/900, fol. 186–95, Petition of the *Tiefbau* industry to the Reich Chancellery, 3.11.32. In his memoirs, the representative of the *Tiefbau* industry in the RdI, the entrepreneur Adolf Mast, recalls a meeting between the RdI and Brüning in 1932. There, Mast had presented road construction as the solution to unemployment in the Reich and, at least according to his own account, found acclaim among Brüning and the industry leaders present: Adolf Mast, *Unternehmer aus Berufung: Adolf Mast erzählt aus seinem Wirken als Bauingenieur und im öffentlichen Leben* (Wiesbaden, Berlin: Bauverlag, 1964), 90–91.

36 Marcon, *Arbeitsbeschaffungspolitik*, 37–86.

37 Silverman, *Work Creation Programs*, 50, 68; Petzina, 'Hauptprobleme', 25. One noteworthy exception was the *WTB Plan* of 1932, developed in Social Democratic circles and adopted by the Federation of German Trade Unions (*Allgemeiner Deutscher Gewerkschaftsbund*). Ironically, while

Socialists presented their answer on unemployment and economic crisis to the German electorate in the form of the Immediate Economic Programme (*Wirtschaftliches Sofortprogramm der N.S.D.A.P.*). The programme presented the party's ideas on job creation and more general economic measures and was to serve as a guideline for party speakers during the election campaign at the end of June. While previous economic policy would have concentrated on increasing German exports, the Nazis wanted "to seek increased sales where increased sales are to be found: in the domestic market."[38] Financed by credit expansion, housing construction, a 40 percent state subsidy to support "each worker willing and able to buy a single-family house"[39] as well as large infrastructure projects played a prominent role in the programme. Because the Immediate Economic Programme also contained some anti-capitalist statements, as well as plans on price controls and foreign trade restrictions, it was already withdrawn in September – after the intervention of the German business community, according to Avraham Barkai.[40] The programme that came to replace the Immediate Economic Programme was the Plan for Economic Reconstruction (*Wirtschaftliches Aufbauprogramm*) of October 1932. It toned down its predecessor's anti-capitalist and interventionist demands, but still promised work creation measures worth 5–6 billion RM, of which 3 billion would be deficit-financed.[41]

After the strictly deflationary policy pursued by Reich Chancellor Heinrich Brüning between 1930 and 1932, more and more voices in government circles, too, called for the state to take an active role in "kick-starting" the economy and creating jobs.[42] After Reich Chancellor Franz von Papen had drawn up a small plan of direct work creation measures in September 1932,[43] his successor Kurt von Schleicher appointed the president of the German Association of Rural Municipalities (*Deutscher Landgemeindetag*), Günther Gereke, as Reich Commissioner for Work Creation on 15 December 1932. Gereke was known as an advocate of direct work creation and the man behind the so-called *Sofortprogramm* of January 1933, which earmarked 500

---

the plan was controversial among the Social Democrats and trade unionists themselves, it was met with approval by the National Socialists and other right-wing groups. On the *WTB Plan*, see Silverman, *Work Creation Programs*, 50–53. Cf. also Michael Schneider, 'The Development of State Work-Creation Policy in Germany, 1930–1933', in *Unemployment and the Great Depression in Weimar Germany*, ed. Peter D. Stachura (New York: St. Martin's Press, 1986), 163–86.

38 Hauptabteilung IV (Wirtschaft) der Reichsorganisationsleitung der N.S.D.A.P., *Wirtschaftliches Sofortprogramm der N.S.D.A.P.* (München: Franz-Eher-Verlag, 1932), 9.

39 Hauptabteilung IV (Wirtschaft) der Reichsorganisationsleitung der N.S.D.A.P., 13.

40 Barkai, *Nazi Economics*, 43–46.

41 Gregor Strasser, *Das Wirtschaftliche Aufbauprogramm Der NSDAP* (Berlin: Hempel, 1932).

42 Birgit Wulff, *Arbeitslosigkeit und Arbeitsbeschaffungsmaßnahmen in Hamburg 1933–1939. Eine Untersuchung zur nationalsozialistischen Wirtschafts- und Sozialpolitik* (Frankfurt am Main: Peter Lang, 1987), 26–29.

43 Wulff, 29–30; Detlev Humann, *'Arbeitsschlacht'. Arbeitsbeschaffung und Propaganda in der NS-Zeit 1933–1939* (Göttingen: Wallstein Verlag, 2011), 43–47.

million for direct public investments. In a radio broadcast on 23 December 1932, Gereke explained that the austerity policy of the recent years had particularly harmed the industries that depended on public orders, leading to unemployment and declining tax revenues, which again limited the financial scope of action for vital public investments. To break the vicious circle, orders were to be placed by municipalities and districts all over the country, and not only with large, but also with mid-sized and small companies. To employ as many workers as possible, manual labour was to be preferred to machines and the work week limited to 40 hours. As far as possible, domestic building materials were to be used on the construction sites.[44] The *Sofortprogramm* was financed through credit creation: the companies were paid with work creation bills, and the Reich obligated itself to redeem them in the following years.

Like the Papen Plan, Gereke's programme burdened not the current, but future budgets. In doing so, the Papen and Schleicher governments were gambling that the economy would be back on track soon. However, they did not live to see the results of their policy. While the timing of the Papen Plan was unfortunate – it was introduced too soon before the winter when unemployment figures were naturally increasing –, Gereke's *Sofortprogramm* was passed only two days before Adolf Hitler attained power.

Although it has been argued that rearmament, not the battle against unemployment, became Hitler's top priority soon after he became chancellor on 30 January 1933,[45] fighting unemployment remained an important task for the new National Socialist government. In a sense, the two objectives were two sides of the same coin. On the one hand, work creation measures were often used to fund and camouflage rearmament projects, for example when airfields were built under the disguise of land ameliorations.[46] On the other hand, the "battle for work" was to consolidate the Nazi rule, which was the prerequisite for rearmament. Accordingly, in his radio broadcast on 1 February, Hitler committed Germany to an "all-embracing attack on unemployment" and proclaimed that "within four years unemployment must be finally overcome."[47] The circumstances were favourable, indeed. With two work creation plans in place that did not differ significantly from its own economic ideas and backed by a cyclical upturn, "Hitler's government was lucky in its timing."[48] To a certain

44 Document 36, Radio broadcast by the Reich Commissioner for Work Creation, Gereke, 23.12.32, in: *Das Kabinett von Schleicher. 3. Dezember 1932 Bis 30. Januar 1933*, Akten Der Reichskanzlei. Weimarer Republik (Boppard am Rhein: Boldt, 1986), 156–62.

45 Christoph Buchheim, 'Das NS-Regime und die Überwindung der Weltwirtschaftskrise in Deutschland', *Vierteljahreshefte für Zeitgeschichte* 56, no. 3 (2008): 387–91; Tooze, *Wages*, 38, 49–66; Silverman, *Work Creation Programs*, 61–65.

46 Humann, *Arbeitsschlacht*, 86–87, 718–19.

47 Max Domarus, ed., *The Years 1932 to 1934*, Hitler: Speeches and Proclamations 1932–1945. The Chronicle of a Dictatorship 1 (London: Tauris, 1990), 234.

48 Richard J. Evans, *The Third Reich in Power 1933–1939* (London: Penguin Books, 2006), 329.

degree, the new rulers could simply afford to wait and watch the plans of their predecessors take effect.[49] Planning and implementation of work creation schemes took time and as an examination of the work creation programmes showed, "the major part of the expenditure involved was made from nine to eighteen months after the drawing up of the programmes".[50] This considerable time lag should be kept in mind when assessing the causal nexus between work creation policies and their results. At least publicly, however, many in the German construction industry credited the National Socialists for kick-starting the upswing in sector. The most striking example in this respect is probably the speech of the head of Wigru Bau, Hochtief's Eugen Vögler, at the association's annual congress in Essen in 1936. There, Vögler directly linked the upswing in the sector to the National Socialist seizure of power.[51] The myth that the National Socialists had led the construction industry out of the crisis, was also perpetuated by some industrialists after 1945 in order to defend their early commitment to the Nazis. When Hans Meyer-Heinrich, board member of the Philipp Holzmann AG, was asked in 1946 why he had already become a member of the National Socialist German Worker's Party (NSDAP) on 1 May 1933, he claimed that he thought that the "new and radical methods" of the NSDAP were the construction industry's "only resort".[52] What he did not mention was that the upswing at Philipp Holzmann had already set in over the course of 1932.[53] Also other firms noted that the economy already picked up in autumn 1932.[54]

The National Socialist work creation plan that was finally implemented in June 1933 was the so-called Reinhardt Programme. It was named after Fritz Reinhardt, a

49 Wulff, *Arbeitsbeschaffung*, 41.

50 Leo Grebler, 'Work Creation Policy in Germany, 1932–1935', *International Labour Review* 35 (1937): 339.

51 Vögler, 'Lage der Bauwirtschaft'.

52 Dr.-Ing. Hans Meyer-Heinrich (1885–1976) had joined Philipp Holzmann in 1911 and became member of the board in April 1936. Meyer-Heinrich maintained close contact to the circle around Fritz Todt. He was responsible for the company's *Westwall* construction sites, led the negotiations with the OT, and visited Holzmann's sites in occupied Europe during the war. Politically, he was a committed National Socialist, a member of the NSDAP since May 1933 as well as of the mounted corps of the SA. After the war, Meyer-Heinrich was first suspended from his position and both the company's works council and Martin Arndt (the only board member who had not been member of the Nazi party) deemed Meyer-Heinrich politically implicated to such an extent that a return to his former position seemed impossible. However, after his denazification and Arndt's death in 1948, Meyer-Heinrich was able to return to his post as a board member. From 1956 to 1971, he was member of the supervisory board. See Pohl, *Philipp Holzmann*. Meyer-Heinrich is the author of Holzmann's 1949 anniversary *festschrift*, where he deals with the Nazi period only cursorily and apologetically: Meyer-Heinrich, *Philipp Holzmann Aktiengesellschaft im Wandel von hundert Jahren*.

53 Pohl, *Philipp Holzmann*, 207–8.

54 Habermann & Guckes AG, business report 1932; Julius Berger Tiefbau AG, business report 1932; Straßenbau AG, business report 1933.

state secretary in the Reich Ministry of Finance and member of the NSDAP since the mid-1920s. The programme was part of the Law on the Reduction of Unemployment and allocated the unprecedented sum 1 billion RM to direct work creation measures.[55] The Reinhardt Programme authorised the finance minister to issue work creation bills for the repair of private housing and public buildings, for suburban and agricultural settlements, river regulations, public utilities, earth moving works and benefits in kind for welfare recipients.[56] No funds were allocated for the construction of new dwellings and although the *Autobahn* had played a particularly prominent role in Hitler's remarks during a meeting with leading industrialists only few days before the Reinhardt Programme was announced, the programme also earmarked almost no funds for road construction (Table 3).[57] Thus, Silverman has argued that "during 1933, land reclamation projects connected with the resettlement of urban industrial workers to rural areas, not roads and housing, spearheaded the Nazi battle for jobs."[58] On the other hand, we should not forget that the effects of the previous governments' housing and road construction programmes were just starting to unfold.[59]

It was not until September 1933 that the Second Law on the Reduction of Unemployment moved the focus to housing repairs again. Its aim was to absorb some of the usual unemployment in construction during the winter, since housing repairs were mostly indoor works. This second Reinhardt Programme provided 500 million RM from the ordinary budget in the form of subsidies to house-owners. If house-owners financed further renovations out of their own pocket, the government supported them with interest subsidies worth another 332 million RM, which could be redeemed until 1939.[60] Detlev Humann has shown that in several regions, local craftsmen cooperated with building inspectors in order to urge reluctant house-owners to renovate their buildings: "Apparently, the main focus of the officials was work creation, rather than building safety."[61]

55 Reichsgesetzblatt (RGBl) 1933, I, 323–29.

56 RGBl. 1933, I, 323.

57 Document 147, Meeting with leading industrialists, 29.05.33, in: *Die Regierung Hitler. Band I: 1933/34*, vol. 1, Akten der Reichskanzlei. Regierung Hitler 1933–1945 (Boppard am Rhein: Boldt, 1983), 506–27.

58 Silverman, *Work Creation Programs*, 63.

59 Rainer Fremdling and Reiner Stäglin, 'Work Creation and Rearmament in Germany 1933–1938 – A Revisionist Assessment of NS-Economic Policy Based on Input-Output Analysis', *DIW Berlin Discussion Paper No. 1473*, 2015, 5; Grebler, 'Work Creation', 339.

60 RGBl. 1933, I, 651–53. René Erbe, *Die nationalsozialistische Wirtschaftspolitik 1933–1939 im Lichte der modernen Theorie* (Zürich: Polygraphischer Verlag, 1958), 29–30; Grebler, 'Work Creation', 337. Schiller gives a slightly higher figure, 360 million, for interest subsidies: Karl Schiller, *Arbeitsbeschaffung Und Finanzordnung in Deutschland* (Berlin: Junker und Dünnhaupt, 1936), 155.

61 Humann, *Arbeitsschlacht*, 111.

Table 3: Allocation of funds to direct work creation measures, divided into different purposes, up to 31 December 1934 (=98.9 % of the total appropriations), million RM, current prices.*

	Papen Plan	Sofort-programm	First Reinhardt Programme	Other pre-financed programmes	Ordinary budget	Total
I. Civil engineering, etc.:						
a) Construction of waterways	46.6	5.8	77.7	–	–	130.1
b) Roads	98.7	103.8	57.5	–	–	260.0
c) Bridges	9.0	17.3	20.1	–	–	46.4
d) Municipal public utilities	6.1	56.4	116.5	–	–	179.0
e) Maintenance and extensions of public buildings, bridges, etc.	–	–	169.2	–	–	169.2
f) Other civil engineering work (dams, harbours, river correction, etc.)	18.2	10.8	25.7	–	–	54.7
g) Miscellaneous measures	–	95.0	68.0	–	–	163.0
Total I.	178.6	289.1	534.7	–	–	1,002.4
II. Housing, etc.:						
a) Repairs	–	–	67.0	–	952.0	1,019.0
b) Suburban settlements	19.9	19.2	69.1	–	80.8	189.0
c) Small house building	–	–	19.1	–	26.5	45.6
d) Emergency, makeshift and refugee dwellings	–	–	12.0	–	–	12.0
e) Improving slum quarters	–	–	5.0	–	9.4	14.4
Total II.	19.9	19.2	172.2	–	1,068.7	1,280.0
III. Transport undertakings, etc.:						
a) National Railways and Post Office, navigation, narrow-gauge railways, etc.	24.1	110.6	97.2	937.0	165.0	1,333.9
b) Autobahn	–	–	–	300.0	50.0	350.0
Total III.	24.1	110.6	97.2	1,237.0	215.0	1,683.9

Table 3 (continued)

	Papen Plan	Sofort-programm	First Reinhardt Programme	Other pre-financed programmes	Ordinary budget	Total
IV. Agriculture and fishing:						
a) Land improvement, river correction, etc.	49.9	178.8	107.9	–	–	336.6
b) Agricultural settlement	10.0	–	34.1	–	–	44.1
c) Miscellaneous measures	5.0	–	3.5	–	–	8.5
Total IV.	64.9	178.8	145.5	–	–	389.2
V. Goods vouchers	–	–	–	–	70.0	70.0
VI. Initial contributions from the Reich Office	–	–	–	–	568.5	568.5
Grand total	**287.5**	**597.7**	**949.6**	**1,237.0**	**1,922.2**	**4,994.0**

*Schiller, *Arbeitsbeschaffung*, 158–59; Grebler, 'Work Creation', 340.

What all these plans and work creation programmes had in common was a clear focus on the depressed construction sector. After the Nazi's seizure of power, leading financial experts emphasised the construction sector's potential to pump-prime the economy. In the words of the Minister of Economic Affairs, Kurt Schmitt,

> the Reich Government has in all its economic political measures extended particular care towards agriculture and building. In reviving the construction industry it started out with the old experience that the revival of building activity is the decisive precondition for a general economic upswing and that the strongest impetus for the revival of general economic activity derives from the building market.[62]

Hjalmar Schacht, president of the Reichsbank from March 1933 onwards, and Schmitt's successor as Minister of Economic Affairs after July 1934, described the objective of the First Reinhardt Programme in his post-war memoires: "It was intended especially to benefit the construction industry – the hardest hit of all –

62 Speech by Minister of Economic Affairs, Dr. Schmitt, 20.9.33, as quoted in Sidney Pollard and Colin Holmes, eds., *The End of the Old Europe 1914–1939*, Documents of European Economic History 3 (London: Edward Arnold, 1973), 469–70.

which, as a key industry, is universally and at all times acknowledged as the employer of auxiliary industries."[63] In fact, the Reichsbank went along with the government's credit inflation policy under the aegis of Schacht. At the time, however, Schacht seems to have been not entirely convinced of the public work schemes' effect.[64] Like others, he doubted whether earth moving projects could stimulate the whole economy and do more than temporarily reduce the number of unemployed.[65] The construction industry itself shared these concerns, probably because it did not want to forego the opportunity to renew its machinery during the coming upswing. Projects exclusively based on manual labour could not stimulate machine builders, fuel producers or transport businesses: "[. . .] at most, the local hardware dealer sold some shovels and spades [. . .]."[66]

The recovery of the construction sector between 1932 and 1935, however, was definitely and inseparably linked to the work creation policy. Altogether, five billion reichsmark were appropriated for direct work creation measures, of which 4.8 billion had been spent by the end of 1935. Besides the Papen Plan, Gereke's *Sofortprogramm* and the two Reinhardt Programmes, considerable investments were made by the National Railways, the National Post Office, the *Reichsautobahnen* Corporation and the Reich Office for Labour Exchanges and Unemployment Insurances (*Reichsanstalt für Arbeitsvermittlung und Arbeitslosenversicherung*, RfAA). More than 60 percent of the funds were raised by credit creation, the rest came from the ordinary budget (mainly for the Second Reinhardt Programme and the expenditures of the Reich Office).[67] Almost all the money appropriated for direct work creation measures had been allocated to specific purposes by the end of 1934. The major part of the expenditures – 48 percent – was spent in 1934, 30 percent in 1933 and only 16 percent in 1935.[68] The approximately two and a half million RM spent in 1934 were a considerable sum, with the whole construction sector's volume being about five and a half million RM.

Table 3 illustrates that the predominant share of spending assigned to different direct work creation purposes in the early 1930s benefitted the German construction sector. It has been estimated that the construction sector met approximately 73 percent of the demand created by the programmes.[69] Especially the labour-intensive

63 Pollard and Holmes, 466–67.
64 Tooze, *Wages*, 41.
65 Wulff, *Arbeitsbeschaffung*, 66.
66 Theodor Krauth, 'Die Maschine im Arbeitsbeschaffungsprogramm', *Die Bauindustrie* 1, no. 21/22 (1933): 271–72.
67 Schiller, *Arbeitsbeschaffung*, 155. It should be noted in this context that the level of deficit spending was in no way extraordinary in the first years of the National Socialist rule, as Albrecht Ritschl has shown in a highly interesting comparison with other deficit ratios from the 20th century: Albrecht Ritschl, 'Hat das Dritte Reich wirklich eine ordentliche Beschäftigungspolitik betrieben?', *Jahrbuch für Wirtschaftsgeschichte* 44, no. 1 (2003): 132–33.
68 Grebler, 'Work Creation', 338.
69 Fremdling and Stäglin, 'Work Creation and Rearmament', 3–6.

Tiefbau with its ability to absorb unskilled labour was able to profit. Considering that the bulk of the sum in the category "civil engineering", as well as significant shares in the sums spent on transport undertakings and in the contributions from the Reich Office, can be attributed to *Tiefbau*, we can estimate that up to half of the direct work creation expenditures went to this part of the sector.[70] Investments in building new and – first and foremost – repairing dwellings and settlements, that is *Hochbau*, amounted to almost one quarter of the funds.

The effects on the *Hochbau* sector were greater still. When providing capital and interest subsidies for reconstruction and repair of dwellings, mostly under the Second Reinhardt Programme, the state demanded that the house-owners themselves invested a certain sum from their own savings or by borrowing. The ratio was fixed at 1:1 for reconstruction and 1:4 for repairs. Therefore, over the years, the 667 million RM allocated by the state for these purposes, resulted in an investment volume of nearly three billion RM for the *Hochbau* sector.[71]

Beyond the investments made as part of direct work creation and the investments induced by it, the state took further steps to help the ailing construction sector. Two laws of 15 July 1933 concerning tax reliefs for the repair of industrial buildings and for newly-built dwellings were to stimulate the construction industry further.[72] Moreover, under the Reinhardt Programme of 1 June 1933, machinery replacements were exempted from taxation on the condition that the machinery was a German product and did not replace human labour.[73] The tax reform of 1 January 1935 also exempted short-lived capital items with a life time of no more than 10 years from taxation. This last measure in particular was well received by the industry. It has been estimated that it made those short-lived capital items cheaper by 10 to 45 percent.[74] In late 1934, there were so many machine orders coming in that the delivery date that companies had to meet in order to gain the tax exemption had to be extended. Some machine producers were working double and night shifts to cope with the demand.[75] Domestic sales of newly produced construction equipment and machinery already

70 Grebler, 'Work Creation', 341.

71 Schiller, *Arbeitsbeschaffung*, 126–27; Grebler, 'Work Creation', 334–35. It has been suggested that the investment sums under the Second Reinhardt Programme should be regarded as an upper limit. The RFM informed local authorities about numerous instances where house-owners and entrepreneurs had obtained subsidies without investing them as intended by the programme. Comprehensive controls of the correct use of the subsidies, however, were hardly possible: Humann, *Arbeitsschlacht*, 112–13.

72 Law on tax reliefs, RGBl. 1933, I, 491–92. Law on tax exemption of newly-built dwellings, RGBl. 1933, I, 493.

73 RGBl. 1933, I, 324.

74 Erbe, *Wirtschaftspolitik*, 30.

75 'Verlängerung der Lieferschlußfrist für Inanspruchnahme der Steuerfreiheit für Ersatzbeschaffungen', *Die Bauindustrie* 3, no. 4 (1935): 57.

outstripped the pre-crisis levels in 1936.[76] The development was particularly strong in road construction, as this was a relatively new field of activity and hence the need for investments was higher. The number of domestic orders for road rollers already exceeded the 1929 level by 9.3 percent in 1934.[77]

While the work creation measures that had helped the construction sector to recover from the crisis had already contained hidden investments in military infrastructure, *Wehrmacht* orders gained even greater importance during the following years. They became the material expression of an increasingly militarised society. While the money from work creation schemes was still paid out to construction firms and craftsmen, the government was already irrevocably shifting its priorities. In fact, after 1933 no new funds were allocated to national work creation schemes.[78] Instead, enormous sums were granted to the *Wehrmacht* – beyond any budgetary oversight[79] – for the construction and expansion of armaments plants, airfields and barracks. Roads to military facilities had to be built, waterways enlarged. Under the Four Year Plan, the capacities of the armaments and autarky industries were expanded significantly, while in 1938 and 1939, the *Westwall* construction became a prominent factor. By March 1939, Fritz Todt could estimate that *Wehrmacht* projects amounted to 50 percent of the German construction production.[80] German construction output, that is, the gross production of construction industry and handwork firms, comprising new construction, repairs and maintenance, peaked in 1938. The same was true for the production of building materials.[81] The sector profited from work creation programmes and the build-up of the German war machinery under the National Socialist rule more than almost any other sector.

In 1941, the construction firm Dyckerhoff & Widmann drew its conclusion regarding the situation in the sector during the previous years: "For the company D. &W., the quantitative output of 1938 was, under the given circumstances which were not under its control, the economic maximum."[82] Comments stating that they were only able to meet the demand through maximum strain on machinery and

76 See below, Table 12.
77 Wagemann, *Konjunkturstatistisches Handbuch*, 254.
78 Tooze, *Wages*, 61.
79 Michael Geyer, 'Rüstungsbeschleunigung und Inflation. Zur Inflationsdenkschrift des Oberkommandos der Wehrmacht vom November 1938', *Militärgeschichtliche Mitteilungen* 30, no. 2 (1981): 122.
80 BArch, NS 26/1187, fol. 95–96, letter Fritz Todt to ambassador von Ritter, 3.3.39. According to Todt, the remaining shares were: 15 percent for housing, 15 percent for state construction, and 20 percent for the private sector. Todt's estimate is confirmed by an overview over the *Wehrmacht's* spending on construction in 1938, amounting to approximately 5.3 billion RM: Geyer, 179–80.
81 Rolf Wagenführ, *Die deutsche Industrie im Kriege 1939–1945* (Berlin: Duncker & Humblot, 1954), 161.
82 *Archiv des Deutschen Museums*, München (hereafter: DMM), FA 010/100, Capacity of the construction industry. Comment on the statistics 1935–1940, 4.7.41, fol. 4.

workforce became a commonplace in construction firms' business reports in the late 1930s.[83]

In the building boom of the pre-war years, labour quickly proved a bottleneck. Between January 1933 and March 1936, the number of unemployed had decreased by 74.3 percent in the construction sector (*Baugewerbe*) and 77.6 percent in the building materials industry (*Industrie der Steine und Erden*).[84] The number of employees in the construction industry quintupled between 1932 and 1936: "No other branch of industry even came close to achieving such positive results."[85] From the mid-1930s onwards, both companies and public customers began to complain about labour shortages. It is difficult to determine the extent of this development based on unemployment figures showing the annual average. Their information value is limited in the case of the building trades: because seasonal fluctuation was very high in this branch of the economy, monthly figures – or at least figures indicating the annual high and low point as in Figure 3 – give a better impression of the labour market situation.

Building activity was high in summer and autumn, while many firms laid off workers in wintertime. It was during the summer and autumn of 1936, however, that the Reich Labour Ministry (*Reichsarbeitsministerium*, RAM) first reported a "palpable labour shortage" in the building trades – a shortage that threatened the timely completion of important construction projects.[86] Alarming reports came in from Bitterfeld, where crucial chemical plants were being built under the Four Year Plan, stating that there was a serious shortage not only of skilled workers, but of unskilled labour, too.[87] In autumn 1936, fewer than 48,000 unskilled building workers were registered unemployed in the entire country.[88] The reintroduction of compulsory military service in 1935 had already exacerbated the situation and it just worsened in the following years. In 1938, the construction industry had already recruited every worker available on that season's labour

83 For example: Heilmann & Littmann AG, business report 1938; Julius Berger Tiefbau AG, business report 1938; Philipp Holzmann AG, business report 1937; Allgemeine Baugesellschaft Lenz & Co., business report 1937.

84 *Statistisches Jahrbuch für das Deutsche Reich 1936* (Berlin: Schmidt, 1936), 340.

85 Timothy W. Mason, *Social Policy in the Third Reich. The Working Class and the 'National Community'* (Providence and Oxford: Berg, 1993), 123–24.

86 Document 2, Letter by the Reich and Prussian Minister of Labour to the Head of the Reich Chancellery, 28.8.36, in: Mason, 194–96. Centres of military building activity already registered labour shortages earlier, like the small town of Neuruppin, northwest of Berlin, in 1935: Humann, *Arbeitsschlacht*, 721.

87 Document 37, Report by the Trustee of Labour for Central Germany to the Reich and Prussian Minister of Labour, 20.2.37, in: Mason, *Arbeiterklasse und Volksgemeinschaft*, 340–47.

88 *Statistisches Jahrbuch für das Deutsche Reich 1937* (Berlin: Schmidt, 1937), 350.

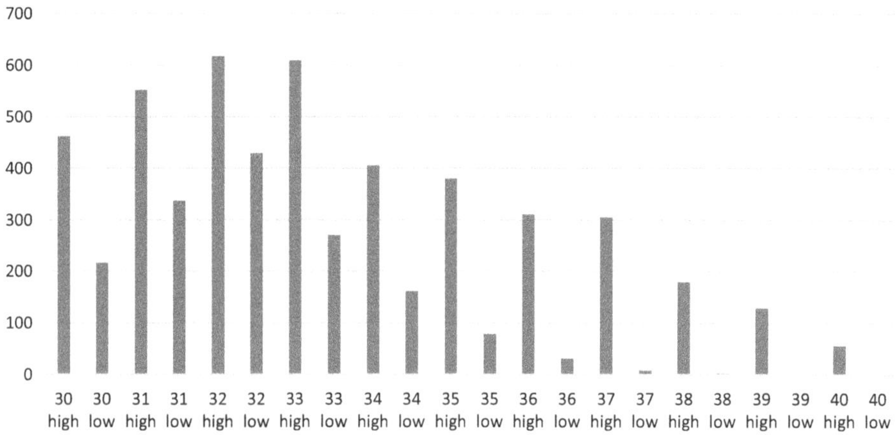

Figure 3: Annual high and low points of unemployment among skilled construction workers, in thousands, 1930–40. Source: 1933–39: Mason, *Arbeiterklasse und Volksgemeinschaft*, 1238. 1930–32 and 1940: Statistical yearbooks of the German Statistical Office, volumes of the particular years. Figures refer to Germany's pre-1938 territory.

market by April. This was two months before works on the *Westwall* started, the largest construction project the country had ever seen, which required half a million workers.[89]

c Drop in Domestic Demand: The Second World War

The years of 1938 and 1939 marked a watershed for the construction industry within the borders of the Reich. When the war broke out on 1 September 1939, the market situation of construction firms in Germany began to worsen, as figures on construction investments show: In Figure 4, I have attempted to estimate the composition of domestic German construction investments in the Third Reich.[90] They do not represent the overall construction volume, as investment figures do not include expenses for repairs and maintenance. From 1928 to 1940, the figures refer to Germany in its pre-1938 borders, from 1941 on, they cover Greater Germany in its shifting borders. Because of the fragmentary base data, Figure 4 has some shortcomings, particularly with regard to the area covered, and the numbers should be regarded as rough estimates only. As they refer to Greater Germany from 1941 on, the investment figures

89 Document 79, Decree by the President of the Reich Office for Labour Exchanges and Unemployment Insurances, 17.5.38, in: Mason, *Arbeiterklasse und Volksgemeinschaft*, 547–48.
90 For the estimation procedure and sources used, see appendix A4.

probably underestimate the extent of the contraction of building activity within the *Altreich* in the second half of the war. Since figures on *Autobahn* and railways construction already refer to Greater Germany from 1938 on, construction investments in these two categories might be slightly exaggerated for the years 1938 to 1941, compared to the other fields of investment. Nevertheless, Figure 4 displays the most important trends in the German construction sector and the additional information it provides outweighs its shortcomings.

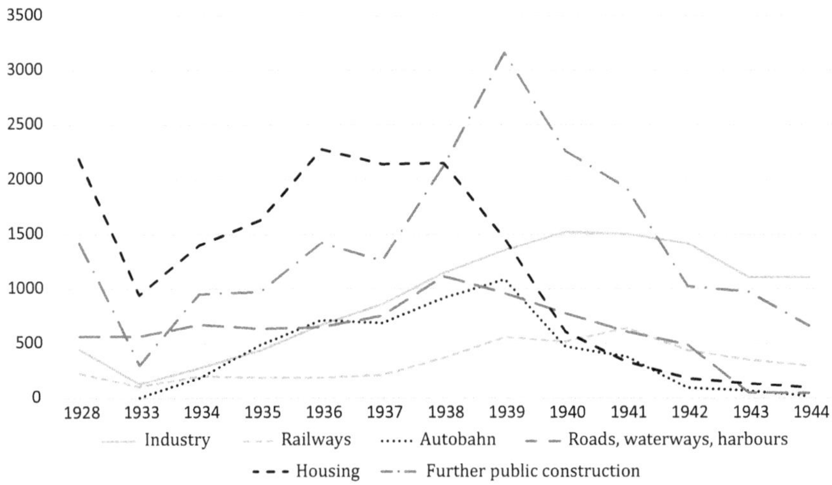

Figure 4: Estimate of construction investments in Germany, 1928/1933–44, million RM, constant prices. Source: appendix, Table A4.1, which has been deflated to 1938 prices by employing the construction price index of the GB Bau/Scherner. For the index, see Table 10. For better legibility, only the most important fields of construction investments are shown in this figure.

Public construction activity was cut back significantly, and most importantly, the OT finished works on the *Westwall* in 1940. This explains the massive decline in "further public construction" activity. Companies that were engaged in road construction struggled because investments in roads and the *Autobahn* network were reduced. After 1942, there was almost no money to be earned with the construction of roads in Nazi Germany. Residential building had already reached its high point in 1936 and declined quickly after 1938. Smaller handwork firms, for whom residential building had been their core business, were hit particularly hard. However, handwork firms were often engaged in repair and maintenance works, which are not included in investment figures and which were probably significant given the destructions due to Allied air raids.

Taken together, these figures show a peculiarity of the construction sector within the National Socialist economy. While many other industry sectors experienced excess demand after the outbreak of the war, domestic demand in construction fell. It is therefore reasonable to assume that from 1939 onwards, German construction firms lost some of their bargaining power and that the pressure on firms to seek orders in the occupied territories increased. At the same time, it must be said that the decline looked particularly dramatic in comparison to 1938 and 1939, two years when construction activity was at an exceptional level.[91]

Construction for private industry, in contrast, was a different case. It increased constantly during the 1930s, even after the war broke out. From 1942 on, industry actually became the most important field of construction investments within the borders of Greater Germany. Although much of it was state-induced, private industrial investment activity remained a cornerstone of the National Socialist economy.

The reason behind the cutbacks in construction was the state's wish to free up resources and manpower for the armaments industry and of course soldiers for the *Wehrmacht*. It is very difficult to draw a comparison between construction industry and handwork firms regarding the share of forced labourers and workers drafted into the *Wehrmacht*, because the available sources are fragmentary, partly based on estimates and often cover different areas and groups of workers. The major source on employment in the domestic German construction sector during the Second World War is the material collected for the *United States Strategic Bombing Survey*, a 1945 study that had the aim to assess the effects of Allied strategic bombing on the German war economy.[92] Its data on employment are mostly based on information provided by the SRA and *Reichsgruppe Industrie*, the peak organisation of the German industry, while figures for the handwork were prepared for the Allies by Karl Plümecke, the former head office manager of the Reich Association of the Building Handwork Guilds (*Reichsinnungsverband des Bauhandwerks*) in June 1945.

The figures in Table 4 cover industry firms and handwork firms within the area of pre-war Germany and do not include German construction workers that were working on OT construction sites outside the Reich. In the case of the industry firms, organised in Wigru Bau, it is possible to differentiate between German (male and female) workers, as well as foreigners, including Jews and prisoners of war (POWs). An according differentiation is not possible in the case of the handwork firms, although the figures include foreign workers.

In contrast, the figures in Table 5 provide a more detailed picture also for construction handwork firms. They refer to the territory of Greater Germany, however,

---

91 The lower construction output in 1940 may also be partly explained by the exceptionally long and hard winter of 1939–40, as several construction companies noted in their business reports: Allgmeine Baugesellschaft Lenz & Co, business report 1940; Allgemeine Hoch- und Ingenieurbau AG, business report 1940; Held & Francke Bau-Aktiengesellschaft, business report 1940.
92 *The Effects*.

Table 4: Construction workforce (Germans and foreigners) and workers of construction industry firms that have been drafted into the *Wehrmacht*, 31 May, area of pre-war Germany, in thousands, 1939–44.*

	1939	1940	1941	1942	1943	1944
Construction industry	890	684	711	507	438	448
. . . *of which: Germans*	*< 890*	*< 684*	*422*	*274*	*219*	*210*
. . . *of which: foreigners*	.	.	*288*	*233*	*219*	*238*
Drafted into the Wehrmacht	.	.	*162*	*245*	*276*	*289*
Construction handwork	1,624	1,008	1,036	721	701	682
Total	**2,514**	**1,692**	**1,747**	**1,228**	**1,139**	**1,130**

*Appendix tables 46 and 48, in: *The Effects*, 57, 236–37.

Table 5: Workforce in principal types of construction handwork firms and workers of these firms that have been drafted into the *Wehrmacht*, 31 May, area of Greater Germany, in thousands, 1939–43.*

	1939	1940	1941	1942	1943
Construction handwork	1,244	785	923	666	652
. . . *of which: Germans*	*< 1,244*	*< 785*	*729*	*517*	*456*
. . . *of which: foreigners*	.	.	*194*	*149*	*195*
Drafted into the Wehrmacht	.	*258*	*367*	*527*	.

*Appendix table 49, in: *The Effects*, 237. Figures for 1939 are of 30 June instead of 31 May.

they are still lower than the figures in Table 4 because they only cover five principal types of the construction handwork.[93]

Obviously, the material does not allow for a direct comparison, among other things, because the two tables cover different areas. It is possible, however, to identify some important trends with sufficient certainty. As Table 4 shows, the outbreak

93 These were the handwork firms of the five handwork guild associations that had been merged into the Reich Association of the Building Handwork Guilds (*Reichsinnungsverband des Bauhandwerks*) on 1 April 1942: the Reich Guild Association of the Building Trades (*Reichsinnungsverband des Baugewerks*), the Reich Guild Association of Carpentry Craft (*Reichsinnungsverband des Zimmerhandwerks*), the Reich Guild Association of Stucco Workers and Plasterers (*Reichsinnungsverband des Stukkateur- und Gipserhandwerks*), the Reich Guild Association of Stone Setters, Asphalt Contractors and Pavers (*Reichsinnungsverband des Pflasterer- und Straßenbauhandwerks*) and the Reich Guild Association of Roofers (*Reichsinnungsverband des Dachdeckerhandwerks*). Cf. Hermann Siedbürger, *Abgabe der Gewinnerklärung im Handwerk: Gewinnprüfung, Preissenkung und Gewinnabführung* (Wiesbaden: Betriebswirtschaftlicher Verlag Dr. Th. Gabler, 1942), 90.

of the war affected small handwork firms more than larger industry firms. In their business reports, the big players in the sector reported only a moderate reduction of activities in the first two years of the war. In contrast, it is much harder to tell how handwork firms reacted and adapted to the new situation. This is mostly due to the lack of archival sources from these small businesses.

When it comes to the German workforce, the two tables show that construction industry firms lost a higher share of their German workforce over the course of the war than handwork firms. This might be partly due to the fact that workers on OT and *Wehrmacht* construction sites outside Germany were not counted. The larger firms were probably able to relocate their activities to the occupied territories to a much higher degree than small, family-run handwork businesses.

The decline in the number of German workers, however, did not mean that all of these workers were called up by the *Wehrmacht*. A *Wehrmacht* survey on the German industry's manpower reserves from late 1944 shows that during the Second World War, the construction (and building materials) industry actually released more German workers to other parts of the industry, including the OT outside Germany, than it did to the *Wehrmacht*.[94] By 1942, the share of workers drafted into the *Wehrmacht* was significantly higher among handwork firms than among industry firms. But also handwork firms noted that many of their workers were in fact enticed away by industry firms, rather than called up by the military.[95]

At the same time, the industry firms employed a higher share of foreign forced labourers in Germany: on 31 May 1943, 30 percent of the construction workers in handwork firms were foreign civilian labourers, Jews and POWs, while they constituted 50 percent in industry firms.[96] The main reason for this is that the larger industry firms were more likely to possess the capacity to supervise forced labourers and that they more often worked on large-scale projects where large groups of unskilled labourers could be occupied with manual labour. Among the construction industry firms, the share of foreigners and POWs in the labour force had already reached almost 50 percent by 1942.[97] As the numbers were somewhat lower in handwork, the share of forced labourers in the entire construction sector was around one third in August 1944.[98] Overall, in few other sectors the share of forced

---

94 *Bundesarchiv-Militärarchiv*, Freiburg im Breisgau (hereafter: BA-MA), RW 6/416, *Wehrmacht-Ersatzplan 1945* by the Replacement Office (*Wehrmachtersatzamt*) in the OKW, fol. 41.

95 Adelheid von Saldern, *Mittelstand im 'Dritten Reich': Handwerker – Einzelhändler – Bauern* (Frankfurt am Main, New York: Campus Verlag, 1979), 52.

96 *The Effects*, 57, 236–37.

97 Herbert, *Fremdarbeiter*, 266.

98 Herbert, 314; Mark Spoerer, *Zwangsarbeit unter dem Hakenkreuz. Ausländische Zivilarbeiter, Kriegsgefangene und Häftlinge im Deutschen Reich und im besetzten Europa 1939–1945* (Stuttgart, München: Deutsche Verlags-Anstalt, 2001), 226.

labourers was as high as in construction. In addition, construction played a prominent role when it came to the deployment of concentration camp inmates.[99]

Furthermore, after an initial wave of draft calls immediately after the outbreak isof the war, there followed a period of stabilisation. Due to an agreement between the *Wehrmacht* and the Reich Labour Ministry of 19 April 1940, skilled construction workers were not drafted into the *Wehrmacht*. In fact, between 1 June 1940 and 31 May 1941, only 27.100 workers from companies of the construction and building materials industry were called up by the *Wehrmacht*.[100] Moreover, the decline in domestic construction activities could be partly compensated by projects in occupied Europe, while many of the drafted German workers were replaced with forced labourers. It is in this context, the words of one of Fritz Todt's close co-workers, Günther Schulze-Fielitz, have to be understood, when he reminded the readers of his paper on "the construction industry at war" of how fundamentally different this current war was from all earlier wars in 1941:

> For the first time in a war, the construction industry has become a factor of warfare. While in earlier wars, construction was regarded as the big reservoir from which army and industry could recruit their forces, and thus all construction activities were immediately shut down at the beginning of the war and the construction industry itself was restricted or even strangled, at the beginning of this war in 1939 it was necessary to hold a protecting hand over the construction industry, because under the requirements of the general reorientation of the economy towards the needs of the all-embracing war, the construction industry [. . .] had to take over tasks that paralleled its key position in economic life during peacetime.[101]

During the First World War, in comparison, construction was cut back massively, already leading to high unemployment in the sector during the first months of the war. When public demand increased from 1915 on, the sector had already lost such a high number of workers to the armed forces and producers of armament goods that it was no longer able to meet the demand.[102] Of course, what Schulze-Fielitz did not anticipate in 1941 was the *Wehrmacht's* enormous demand for soldiers in the years to come. Therefore, protecting the workforce from the grasp of the *Wehrmacht* remained a very important aspect of entrepreneurial decision-making throughout the war and a strong motive to take orders on construction sites that were regarded vital to the war effort.

Another interesting development can be seen from Table 4. Between 1940 and 1941, total employment in construction actually increased due to the large-

99 Herbert, *Fremdarbeiter*, 426.

100 BA-MA, RW 6/416, *Wehrmacht-Ersatzplan 1945* by the Replacement Office (*Wehrmachtersatzamt*) in the OKW, fol. 41. This stabilisation was of course also owed to the end of combat operations in Western and Northern Europe.

101 Günther Schulze-Fielitz, *Die Bauwirtschaft im Kriege* (Berlin: Junker und Dünnhaupt, 1941), 1.

102 Elisabeth Landmann, 'Der Einfluss des Krieges auf den Baumarkt' (Friedrich-Wilhelms-Universität Berlin, 1925), 116–20.

scale deployment of forced labourers. At the same time, the sector's output decreased, as can be seen from Figure 1 above. This indicates a decline in productivity. As young men were called up to the *Wehrmacht*, the average age of the German workforce increased.[103] Moreover, most of the forced labourers that came to replace Germans either lacked the skills or were treated too badly to reach the latter's productivity.

However, while construction output and employment numbers decreased, other characteristics of the pre-war building boom seem to have been not affected by the outbreak of the war to the same degree. As will be discussed in detail in chapter III of the thesis, wage levels, construction prices and corporate profits continued to rise. Despite the challenging market situation, German building companies obviously managed to prosper during the war, too, and this was a matter of concern to many National Socialists. The building boom of the 1930s burdened public budgets and threatened the price structure, while rising wages in construction bore the threat of both inflation and discontent among workers in other sectors of the economy. The profits of building companies after 1939, on the other hand, conflicted with the official principal that no one should profit from the war. Thus, from the mid-1930s onwards, constant efforts were made to regulate the construction sector and bring the situation back under control.

2 The Rise of Organisation Todt

The German construction sector not only experienced a massive boom in the course of the 1930s, it also witnessed the emergence of two powerful organisations: in 1933, the *Reichsautobahnen* Corporation was established under the leadership of the new General Inspector for German Roads, Fritz Todt, to manage the construction of a German network of motorways. In 1938, this organisation then became the blueprint for the *Organisation Todt*.

Besides giving an overview over the OT's history in pre-war Germany, its basic organisational structure and *modus operandi*, the objective of this chapter is to analyse the relation between the men around Fritz Todt and the German construction industry. A significant number of firms worked on the construction sites of the *Reichsautobahnen* and the *Westwall*. I argue that it was here the foundation for the subsequent cooperation in occupied Europe was laid. When entering negotiations with the EW in 1942, many firms were able to draw on the experiences they had gained with the OT during the previous years. Moreover, the OT's contractors already became partners of the regime within the framework of increasingly militarised work relations from 1933 onwards.

103 Cf. below, chapter III.3.c, fn. 207.

a Fritz Todt and the Reichsautobahnen

The history of the *Reichsautobahnen* is inseparably linked to the name of one man: Fritz Todt. In the crisis-ridden Germany of 1932, where dozens of work creation plans were discussed in newspapers, party conventions and government circles alike, the 41-year-old road engineer managed to attract Adolf Hitler's interest with his plan of a 5,000 to 6,000 km network of motorways stretching across the entire country, a plan that promised to merge the future needs of a motorised society, strategic considerations of the *Wehrmacht* and work creation.

Fritz Todt was born on 4 September 1891 in the town of Pforzheim. The First World War interrupted his education as a civil engineer at the *Technische Hochschule* in Munich, which is why he only received his diploma in 1920 after completing his studies at the *Technische Hochschule* in Karlsruhe. Todt worked for the construction company Grün & Bilfinger until August 1921, before joining Sager & Woerner in Munich in November the same year. He stayed there until July 1933, from 1927 onwards as head of the company's road construction department and from 1928 onwards as managing director. During these years, Todt developed his ideas not only on the practical aspects of road building, but also on how the entire system of planning, building and administrating roads in Germany should be reorganised and made more efficient.[104]

It would be misleading to characterise Todt, who loved nature and the music of Bach, Bruckner and Strauss, as an unpolitical man who just opportunistically followed the political movement that happened to allow him to realise his vision. Todt was deeply fascinated by the personality and ideas of Hitler, and together with his wife Elsbeth, he already applied for NSDAP membership in December 1922, at a time when this was a matter of conviction and belief, rather than opportunism. On 5 January 1923, he became member no. 2465 of the NSDAP and founded a local party group in the small village Eitting, northeast of Munich. Although his political commitment diminished in the following years, he became very active again from autumn 1931 onwards – as a member of the SA, head of the section of civil engineers in the Nazi Fighting Association of German Architects and Engineers (*Kampfbund Deutscher Architekten und Ingenieure*) and as an adviser in the NSDAP Office for Business Administration and Work Creation (*Amt für Wirtschaftstechnik und Arbeitsbeschaffung der NSDAP*).[105] Over the following ten years, Todt advanced to become one of the most powerful men in the National Socialist state, holding a total of eight offices. In December 1932, he finally brought forward the memorandum that was to secure Hitler's attention, "Road construction and road

104 Seidler, *Fritz Todt*, 17–35.
105 Seidler, 17–35.

administration" (*Strassenbau und Strassenverwaltung*), also called the "Brown Memorandum" (*Braune Denkschrift*).[106]

"A country's roads", Todt remarked in this memorandum, "are the lifeline of the nation."[107] Therefore, their construction and maintenance were crucial to the reconstruction and future of the Reich. Besides the modernisation of the existing road network, Todt envisaged a 5,000 to 6,000 km network of *Autobahnen*, intersection-free motorways that were to facilitate passenger and freight transportation.[108] He estimated that such a network could be realised within five years at the price of five billion RM. Since he identified the administration's fragmented organisational structure as an obstacle to modernising infrastructure – the responsibility for road construction in Germany lay with thousands of different administrative units on Reich, state and regional level – Todt recommended appointing a Commissioner for Road Construction to supervise and coordinate the project: "A capable expert with 2–3 staff members is absolutely sufficient."[109] It is interesting to see how Todt linked his programme to two key topics the political right was concerned with in the last years of the Weimar Republic.[110] Firstly, he emphasised the military benefits of the *Autobahn* network. On these motorways, an army of 300,000 men could be transported from the eastern to the western border of the Reich in only two nights. He was convinced that "such a road network can be more valuable at the negotiating table than the speech of a bowing ambassador."[111] Secondly, Todt linked the construction of roads to the question of work creation, estimating that every year, 600,000 workers could find work across the country. The use of machines should be reduced to what is absolutely necessary to maximise the projects' employment effects. By addressing the new regime's two central economic themes, rearmament and the fight against unemployment, Todt obviously struck a chord with Hitler who – in addition – was "an enthusiastic devotee of the automobile".[112] Hitler took up the plan and promoted it personally during the first half of 1933, promising that 5,000 km of *Autobahn* construction would create employment for 350,000 workers for the next four years in front of leading industrialists in the abovementioned meeting

---

106 Fritz Todt, *Strassenbau und Strassenverwaltung ("Braune Denkschrift")* (Munich, 1932). Printed in: Singer, *Organisation Todt*, 59–110.
107 Todt, *Strassenbau und Strassenverwaltung*, 81.
108 Today, it is well known that the *Autobahn* idea did not originate with Todt or the National Socialists. For a short overview over the genesis of the idea in the 1920s, see for example Thomas Kunze and Rainer Stommer, 'Geschichte der Reichsautobahn', in *Reichsautobahn: Pyramiden des Dritten Reiches. Analysen zur Ästhetik eines unbewältigten Mythos*, ed. Rainer Stommer (Marburg: Jonas Verlag, 1982), 22–48; Lärmer, *Autobahnbau*, 1–8. That Todt was not alone with his ideas of an *Autobahn* network is also shown by Seidler, *Fritz Todt*, 32–33.
109 Todt, *Strassenbau und Strassenverwaltung*, 95.
110 Ludwig, *Technik*, 307.
111 Todt, *Strassenbau und Strassenverwaltung*, 83.
112 Evans, *Third Reich in Power*, 322.

on 29 May 1933.[113] As early as September, he already turned the first sod near Frankfurt am Main, starting not only a construction project, but a huge propaganda campaign.[114]

On 27 June 1933, the government issued the Law on the Creation of the *Reichsautobahnen* Corporation, legally as a part of the National Railways.[115] The latter also provided the cadre for the new organisation. Fritz Todt was appointed General Inspector for German Roads on 30 June to oversee the construction of the *Autobahnen* and major provincial roads. Shortly after, Todt's position was further strengthened when he was made directly subordinate to Hitler and his agency became a Supreme Reich Authority (*Oberste Reichsbehörde*) – *de facto* a ministry for road construction.[116] This considerable room for manoeuvre beyond institutional boundaries constituted Todt's power in the following years.

The *Reichsautobahnen* were the most prestigious and glorified construction project of the early years of the National Socialist rule. During these years, however, their propaganda value probably outweighed the actual economic and employment effects. An extensive historiography has shed light on different aspects of the *Autobahn's* history and debunked more than one myth in the last decades. Contrary to the impression given by Nazi propaganda, the entire idea of a national motorway network did not originate from Todt or even Hitler, but had already been promoted for years by various societies, most prominently by the HAFRABA, the Association for the Preparation of the Motorway Hanseatic Cities – Frankfurt – Basel.[117] Moreover, the construction of the German motorway network did not contribute significantly to the reduction of unemployment in the Third Reich.[118] Nor did the works start early enough to be responsible for sparking the upswing in the construction sector. *Tiefbau* construction volume in Germany had already more than doubled compared to its 1932 low point when the *Autobahn* construction reached an appreciable volume in 1935.[119] A comparison of investments in the *Autobahn* and Reich and country roads during the first six years of the National Socialist regime reveals that the *Autobahn* alone hardly could have been responsible for the recovery of German road construction companies

113 Document 147, Meeting with leading industrialists, 29.5.33, in: *Die Regierung Hitler. Band I: 1933/34*, 1:506–27.

114 Schütz and Gruber, *Mythos*, 38–65. More generally on the Nazi propaganda with regard to work creation: Humann, *Arbeitsschlacht*, 593–702.

115 RGBl. 1933, II, 509–10. Only in June 1938, the *Reichsautobahnen* Corporation became independent from the National Railways: Seidler, *Fritz Todt*, 145.

116 RGBl. 1933, I, 1057; Ludwig, *Technik*, 330–32.

117 Kurt Kaftan, *Der Kampf um die Autobahnen: Geschichte und Entwicklung des Autobahngedankens in Deutschland von 1907–1935 unter Berücksichtigung ähnlicher Pläne und Bestrebungen im übrigen Europa* (Berlin: H. Wigankow Verlag, 1955).

118 Silverman, *Work Creation Programs*, 173–74, 261; Lärmer, *Autobahnbau*, 54–56; Ritschl, 'Beschäftigungspolitik', 128–29.

119 Cf. Table 6 and appendix, Table A2.2.

(Table 6). The judgement of Manfred Pohl, writing on Germany's leading road construction firm Straßenbau AG (Strabag), is representative of other firms in this part of the industry, too: "The overcoming of the crisis at Strabag in late 1932 and the upturn in the beginning of 1933 have absolutely nothing to do with the construction of the Autobahn."[120] The fact that the AG für Beton- und Monierbau and Julius Berger Tiefbau praised the *Autobahn* in their 1933 business reports, even though they had to admit that the project had not contributed anything to their company results, can be seen as a sign both of the high hopes that some construction firms pinned on the Nazis and of the power of the *Autobahn* propaganda.[121]

Table 6: Investments in *Autobahn* construction compared to construction of Reich, country and municipal roads, 1933–38, million RM, current prices.*

	1933	1934	1935	1936	1937	1938
Autobahn	3.1	178.3	467.4	692.2	679.2	916.4
Reich, country and municipal roads	360.1	431.3	437.3	471.5	559.1	867.6

**Statistisches Handbuch von Deutschland 1928 – 1944*, 606. The *Autobahn* figures do not include expenditure on land acquisition. Cf. Hartmannsgruber, 'Finanzierung', 675. The mere investment figures given in this table still overstate the importance of the *Autobahn*, because they do not include expenditures on the maintenance and repair of roads. While the *Reichsautobahnen* Corporation spent almost nothing on maintenance before 1936 and only 0.31 percent (of total expenditures) between 1936 and 1944, in every year between 1933 and 1937, more than 300 million RM were spent on the maintenance of Reich, country and municipal roads. In 1938, this sum rose to more than 420 million RM. See Robert Adamek and Friedrich Saake, eds., *Die Straßenkosten und ihre Finanzierung. Hauptband*, Forschungsarbeiten aus dem Strassenwesen 8 (Bielefeld: Kirschbaum-Verlag, 1952), 12, 149–54.

Moreover, the motorways were never able to make good on their promise and achieve the military importance suggested by Todt. The railway remained a more efficient means of transportation for *Wehrmacht* goods and troops. Moreover, the *Wehrmacht* regarded the ramified railway network to be less prone to disruption by air attacks.[122] Finally, the *Autobahn* proved an impressive example of how costs got

120 Pohl, *Strabag*, 143. Cf. also Pohl and Siekmann, *Hochtief*, 196–97.

121 AG für Beton- und Monierbau, business report 1933; Julius Berger Tiefbau AG, business report 1933.

122 Schütz and Gruber, *Mythos*, 88; Ludwig, *Technik*, 319–23; Seidler, *Fritz Todt*, 136–43. A notable exception was the road construction activity in preparation for the occupation of the Sudetenland, which was clearly motivated by military considerations. Karl Lärmer, in the typical manner of Marxist-Leninist historiography, has seen imperialistic and destructive motives as the main driving forces behind the whole *Autobahn* enterprise: Lärmer, *Autobahnbau*. Ludwig's and Seidler's argumentations, by contrast, are more convincing. However, Ludwig certainly underestimates the

out of hand over the course of the 1930s. 1933 estimates arrived at costs of 300,000 to 550,000 RM per kilometre, while Hitler even more optimistically claimed that 1.2 billion RM would finance the whole enterprise.[123] By 1942, when construction works on the *Autobahn* were stopped for the most part, more than 3,800 kilometres had been finished. The costs had piled up to more than five billion RM, far more than 1 million per kilometre (Table 7).[124]

Nevertheless, the *Reichsautobahnen* Corporation became one of the most important customers of the German construction industry in the second half of the 1930s. Most of the larger German construction companies carried out earth moving, road or bridge construction works in one of the many *Autobahn* construction sites around the country and gained their experience with the GdSt.

b The Autobahn Contractors

In 1933, however, the GdSt was a completely new institution. What were entrepreneurs to think of Fritz Todt, the man who suddenly appeared in such a powerful position in the summer of 1933? What could they expect – or maybe fear – from a man who demanded the "direction of the industry towards the principle: public interest before self-interest",[125] a principle so fundamental to Nazi economic thinking that it was impressed on the 1 reichsmark coin?

From the industry's point of view, the prime question in 1933 was whether the *Reichsautobahnen* Corporation would carry out the works with their own workers and equipment (*Regiearbeiten*) or award contracts to the private construction industry.[126] The construction industry had lobbied for public work creation programmes for years, and it is hardly surprising that many *Tiefbau* companies already showed a particular interest in the idea of building a national network of motorways during the 1920s. Many road construction firms, such as Fritz Todt's employer Sager & Woerner, had for example been members of the HAFRABA association.[127] Excluding the private construction industry from the *Autobahn* project would have represented a serious

ideological foundations of Todt's work, when he claims that only after May 1938, Todt "fully realised for the first time that engineering served political goals [. . .]": Ludwig, *Technik*, 321.

123 Hartmannsgruber, 'Finanzierung', 633–36.

124 Hartmannsgruber, 681. In his detailed analysis, Hartmannsgruber does not mention the initial estimate from Todt's 1932 "Brown Memorandum" (5,000 to 6,000 km at the price of 5 billion RM), but only the lower estimates from 1933. Todt's early estimate, however, proved to be the most realistic.

125 Todt, *Strassenbau und Strassenverwaltung*, 109.

126 The industry had taken a firm stand against *Regiearbeiten* in the previous years, as in the memorandum on work creation of March 1932: BArch, R 13-VIII/27, Memorandum by the associations of the German construction industry to the Reich Chancellor, 17.3.32.

127 Lärmer, *Autobahnbau*, 27; Kaftan, *Kampf um die Autobahnen*.

Table 7: Spending on *Autobahn* construction 1933–44, million RM, current prices.*

	1933	1934	1935	1936	1937	1938	1939	1940	1941	1942	1943/44	Total
Spending	3.1	178.3	467.4	692.2	679.2	916.4	1,116.5	505.2	505.6	146.3	134.6	**5,344.9**
. . . *of which on services by entrepreneurs*	*1.5*	*149.5*	*389.3*	*559.3*	*544.9*	*744.0*	*812.8*	*350.4*	*336.9*	*135.0*	*70.1*	***4,093.7***

*Hartmannsgruber, 'Finanzierung', 675. "Spending" does not include expenditure on the acquisition of land.

affront. However, the companies realised quickly that they had found an ally in Todt, who was himself an industry man. In May 1932, when he was still working for Sager & Woerner, Todt already provided an NSDAP member of the Bavarian *Landtag* with material, asking him to support the private industry in their fight against *Regiearbeiten*.[128] Todt had also criticised the practice in his "Brown Memorandum", because it "deprived the industry of the very last work opportunities", and called for its complete abolition.[129] In November 1933, Todt was able to reassure the member firms of the Reich Association of Civil Engineering (*Reichsverband des Ingenieurbaues*) that in road construction, "with the initiated measures, the Reich Government pursues two goals: work creation and stimulation of the economy. For work creation, it does not matter whether works are conducted *in Regie* or by entrepreneurs. For the stimulation of the economy, *Regiearbeiten*, apart from building material supplies, are pointless. [. . .] In the interest of stimulating the economy it is temporarily necessary to award to entrepreneurs works that in normal times without any problems could be conducted *in Regie*."[130] The Reich Association of Civil Engineering was also able to report on an agreement with the Reich Association of the German Industry (*Reichsstand der deutschen Industrie*, formerly the *Reichsverband der Deutschen Industry*) and the *Langnam-Verein*, a powerful association of Rhenish-Westphalian industrialists.[131] The latter requested that their member firms not use their own workforce for the construction and repair of industrial facilities, but commission proper building firms instead.[132]

In accordance with Todt's credo, the OBRs – administrative units established along the projected routes to supervise construction and manage logistics – relied on the private construction industry. The OBRs entered into contracts with larger construction companies that were able to again engage subcontractors in turn. All in all, around 1,000 companies are said to have participated in the construction of the motorways.[133] They brought along their regular workers and employees, as well as construction machinery, while the OBRs provided building materials and negotiated the allocation of workers with the labour offices. In the years to come, this system was to remain unchanged, first on the OT's *Westwall* construction sites and later all over occupied Europe.

---

128 Seidler, *Fritz Todt*, 29.

129 Singer, *Organisation Todt*, 103.

130 'Dr. Todt gegen die Regiearbeit bei Straßenbauten und Straßenunterhaltung', *Die Bauindustrie* 1, no. 18 (1933): 233–34.

131 The official name of the association, which was founded in 1871, was *Verein zur Wahrung der gemeinsamen wirtschaftlichen Interessen in Rheinland und Westfalen*.

132 'Keine Regie-Bauarbeiten der Industrie', *Die Bauindustrie* 1, no. 18 (1933): 234; 'Gegen artfremde Regiearbeiten industrieller Werke', *Die Bauindustrie* 1, no. 19 (1933): 247.

133 Seidler, *Fritz Todt*, 109.

After the National Socialists had seized power in the early 1930s, the time schedule for their prestige project, the *Autobahn* network, was tight. The earlier the works could contribute to relieving the labour market, the better. Under the present economic conditions, however, it was not necessary to force companies into contracts. The huge project sparked a "run on the customer",[134] while Todt assured those companies currently unable to participate in the construction works that their refusal would not lead to future discrimination.[135] Interested companies were asked to file tenders for fixed-price contracts – sometimes at very short notice. In the case of the surfaces for the new motorways, one OBR invited public tenders in late December 1934 and set 15 January 1935 as the deadline! Larger and more experienced road construction firms complained about this practice. Scores of smaller firms had filed tenders and depressed the price levels.[136] Besides, the short deadline made it very difficult to calculate the costs adequately, a crucial step when working under a fixed-price regime. The large companies made it clear that they would have preferred a restricted bidding process, which would have reduced the effort necessary to outdo their bothersome competitors.[137] However, given the complexity of the task, it is not surprising that it was the large road construction companies that ultimately prevailed.[138] Sometimes, as some reports suggest, companies tried to avoid the bidding process by using personal and political contacts or simply bribing OBR officials.[139] But the contracting practice of the OBRs also helped to dispel the larger companies' concerns about the new regime's economic thinking. Despite the Nazis' public attacks on "big business", Todt obviously preferred to see larger, experienced firms on his *Autobahn* construction sites.[140] The larger *Tiefbau* companies were reported to be operating at full capacity as early as late 1934.[141] Small companies had to content themselves

134 'Betrachtungen über die Ausschreibung und Vergebung der Reichsautobahndecken', *Die Bauindustrie* 3, no. 13 (1935): 174–76.

135 'Keine Benachteiligung wegen Nichtbeteiligung an Ausschreibungen', *Die Bauindustrie* 3, no. 1 (1935): 8. A few weeks later, the Reich Price Commissioner distributed Todt's decree to all other ministries in the Reich that were awarding construction contracts with the request to proceed likewise: 'Keine Benachteiligung wegen Nichtbeteiligung an Ausschreibungen', *Die Bauindustrie* 3, no. 13 (1935): 182–83.

136 Just two examples of companies complaining about the depressed price level are the Straßenbau AG and the Munich-based Heilmann & Littmann Bau-Aktiengesellschaft. On Strabag, see Pohl, *Strabag*, 191. On Heilmann & Littmann, see the annual reports from the first half of the 1930s.

137 'Betrachtungen über die Ausschreibung und Vergebung der Reichsautobahndecken'.

138 Lärmer, *Autobahnbau*, 82–83.

139 Lärmer, 83.

140 Silverman, *Work Creation Programs*, 161.

141 'Keine Benachteiligung wegen Nichtbeteiligung an Ausschreibungen', 1935. Cf. also Lärmer, *Autobahnbau*, 82–84. Due to the *Autobahn* project, the Leipzig branch of the German Federation of Technical and Scientific Organisations (*Deutscher Verband Technisch-Wissenschaftlicher Vereine*)

with the role of subcontractors, and only in 1935 were measures taken to give small handwork firms a bigger share in the *Autobahn* contracts.[142]

In order not to choke the frail economic upswing, investments by foreign-owned companies, or firms depending on foreign capital, were not be generally prohibited. For the same reason, some authorities initially continued to award contracts to "Jewish" firms.[143] According to the guidelines for the awarding of public contracts, presented to the Cabinet by the Reich Economic Minister on 14 July 1933, it was more important that a firm employed Germans, used German machinery and German raw materials. The proposal provoked one of Hitler's infamous laments about a global "Jewish boycott movement" against Germany. While foreign companies producing in Germany should be treated generously, Hitler stated that rigorous measures needed to be taken against "Jewish" businesses.[144] On the *Autobahn* construction sites, "Jewish" competition was eliminated in December 1934, when the head office announced that "the engagement of Jewish firms at the construction of the Reichsautobahnen should be avoided by all means".[145]

The Nazis' attacks on businessmen and employees, whether they were of the Jewish faith, merely defined as "Jewish", or persecuted for their political beliefs, were the most blatant and grave interferences with entrepreneurial autonomy in the early years of the regime. Many firms' behaviour was characterised not so much by open antisemitism as by opportunism. Sure enough, the consequences for their employees and board members were the same. The Hochtief AG dropped two board members in 1933, as the management feared to be discredited as a "Jewish firm".[146] Mostly due to political pressure, Julius Berger, founder and board member of the construction firm bearing his name, retired in 1933 and did not switch to the supervisory board, as could have expected. He and his wife were later deported and died in Theresienstadt.[147]

At Germany's largest construction firm, the Philipp Holzmann AG, the expulsion of Jewish members of the management and supervisory board was directly linked to the *Reichsautobahnen* project. In autumn 1933, Fritz Todt personally put pressure on the firm's management. Without staff changes in Holzmann's leadership team, the company would be excluded from future calls for tender. The firm felt compelled to conform to Todt's demands.[148] Fritz Todt and his men at the *Reichsautobahnen* Corporation were certainly not unpolitical technocrats. When it

---

already anticipated a shortage of road construction and civil engineers in November 1933 – before construction works had started – and initiated retraining courses for jobless mechanical engineers: 'Umschulungskursus für Straßenbau', *Die Bauindustrie* 1, no. 19 (1933): 249.

142 Silverman, *Work Creation Programs*, 161.

143 Silverman, 161.

144 Document 193, Cabinet meeting, 14.07.33, in: *Die Regierung Hitler. Band I: 1933/34*, 1:676–78.

145 Lärmer, *Autobahnbau*, 84.

146 Pohl and Siekmann, *Hochtief*, 133–47.

147 Stier and Krauß, *Bilfinger Berger AG*, 202–5.

148 Pohl, *Philipp Holzmann*, 193–207.

came to Jewish entrepreneurs and their businesses, very soon there was no more room for compromise.

Todt was more clement in the case of the entrepreneur Adolf Mast, who was not Jewish. After the war, Mast recalled that he had to step down from his position as board member of the employers' liability insurance association for *Tiefbau* (*Tiefbau-Berufsgenossenschaft*) because – at least according to his own account – he was a democrat and a Freemason. However, he continued his work in the association under the personal protection of Fritz Todt, who valued his expertise.[149] As far as the *Autobahn* construction project is concerned, there is no evidence that Todt's agency interfered directly with the entrepreneurial autonomy of the industry beyond the attack on construction firms labelled as "Jewish". Hence, accounts of threats and pressure on construction firms should be taken with a pinch of salt, especially post-war accounts like that of a Straßenbau AG employee, stating that "one day, the entire management appeared on the construction site with grave countenances, because they had received a letter from the OBR that threatened to send the entire management to the concentration camp if construction was not on schedule."[150] While there are some occasional reports of officials using drastic words to express their discontent with work progress, it is rather unlikely that the men at Strabag considered the OBR's letter a real menace. Detaining the entire management of Germany's leading road construction company because of one construction site behind schedule would certainly have done more harm than good to Todt's ambitious project.

Sometimes, entrepreneurial autonomy found its expression in episodes like one occurring in Austria in 1938. In order to make sure that construction works would start smoothly after the *Anschluss*, Todt's men had rashly bought construction machines and materials at overcharged prices. The Austrian companies, however, refused to take over the equipment at these prices. Thus, Todt had to redirect the machines to the *Westwall* and – much to the Court of Audit's chagrin – was stuck with the expense.[151] In general, however, privileged access to construction machinery became an important incentive for construction firms to work on the *Reichsautobahnen*, especially when steel and machinery produced from it became scarce in the late 1930s. In a special campaign in spring 1939, Todt ordered building machinery worth 20 million RM from several Czech producers such as the Škoda Works. Among other things, 20,000 roll cars and 430 locomotives, compressors and electric motors were to be sold to German construction companies.[152] Firms that had distinguished themselves as National Socialist model companies were contacted and granted privileged access to the machines.[153]

149 Mast, *Berufung*, 86–88.
150 Pohl, *Strabag*, 156.
151 Hartmannsgruber, 'Finanzierung', 645.
152 BArch, NS 26/1187, fol. 95–96, Letter Fritz Todt to Ambassador von Ritter, 3.3.39.
153 Ibid., fol. 424, Letter Fritz Todt to the construction companies Gebrüder Niemax (Neumünster), Stahlschmidt (Meppen) and Albert Dölling (Dresden-Bühlau), 25.5.39.

To maximise the employment effects of the *Autobahn* project, it was initially decreed that the share of regular workers should be limited to 10 percent for excavations and 20 percent for civil engineering works.[154] With the same intention, the General Inspector also restricted the use of construction machines, prices were to be calculated and contracts awarded on the basis of the cost of manual labour.[155] As the administrative board of the *Reichsautobahnen* Corporation noted in 1935, this approach led to a 35 percent price increase for excavations.[156] The restrictions on the use of machines could be interpreted as Todt's agency interfering with entrepreneurial freedom of action. However, Karl Lärmer has remarked that the propaganda at this point "made a virtue of necessity", since the German construction industry in 1934 had little experience in the construction of concrete roads and correspondingly only possessed few suitable machines. Building companies needed time to order and machine builders time to design and produce the new equipment anyway. When labour became a bottleneck for construction works in the following years, the use of machines on the *Autobahn* construction sites rose quickly,[157] and as has been mentioned above, the first construction machinery producers already struggled to meet orders in late 1934.[158]

Another aspect of the relation between the GdSt and its contractors regards the conditions in the *Autobahn* construction sites. The latter became the scene of increasingly harsh and militarised work conditions throughout the 1930s. The construction firms operated in a setting where the end justified the means and concerns for the individual worker's rights and safety eroded more and more. Also these developments were part of the experiences that contractors gained with the apparatus under Fritz Todt before the OT erected its empire of forced labour in occupied Europe after 1939.

Working conditions on the *Autobahn* construction sites were infamous.[159] The unemployed whom the labour offices allocated to the *Reichsautobahnen* Corporation either lived in simple camps where the food provision gave rise to constant complaints or had to put up with long daily journeys to and from the sites. For the former unemployed, earnings were hardly above the level of unemployment benefits, which led to unrest and protests resembling open strikes. Reports on the situation even reached the foreign press: in December 1934, the readers of the Manchester Guardian were able to read about open mutiny in the *Reichsautobahnen* camps around Breslau (Wrocław)

154 Hartmannsgruber, 'Finanzierung', 643.
155 Silverman, *Work Creation Programs*, 169.
156 Lärmer, *Autobahnbau*, 58.
157 Lärmer, 57–58. After the war, the Philipp Holzmann AG was able to reuse road construction machinery that it had acquired during the construction of the *Autobahn* when the company won the tender for a motorway stretch east of Aschaffenburg: Pohl, *Philipp Holzmann*, 324.
158 Cf. chapter II.1.b.
159 For an account of the conditions, see Lärmer, *Autobahnbau*, 60–70.

and workers singing the *Internationale*.[160] Many labourers unfit for the hard, physical work became ill due to insufficient food, clothing and sanitary facilities.[161]

We can assume that much of the pressure resulted from the fact that the firms could increase their profits under a fixed-price regime by finishing the works as fast as possible. This is exactly the intended incentive of a fixed-price contract. However, it could have fatal consequences in a system where employee rights were eliminated or systematically violated, especially for those workers allocated to the construction sites by the labour offices. After years of unemployment, they were in a particularly bad bargaining position. In 1934, the State Labour Office (*Landesarbeitsamt*, LAA) in Hesse was alarmed at a new premium system introduced by the construction firm Polensky & Zöllner on its construction site south of Frankfurt am Main. Their price calculation had – after consulting the OBR Frankfurt am Main – been based on the assumption that a worker could move 9–11 m^3 of earth per day. As experience showed, the workers actually managed 13–14 m^3 per day on average, which was already generating additional profits for the company. The new premium system now offered every man a 30 *Reichspfennig* (Rpf.) premium on top of the hourly wage for every cubic metre moved after the 11th cubic metre. For workers that normally earned between 50 and 65 Rpf. per hour, depending on location, this was of course tempting. Hoping to raise their income, the underpaid workforce self-destructively increased their performance up to 20 m^3 per day. The system was also introduced on the adjacent construction sites by Grün & Bilfinger and two other companies. The Hessian LAA complained about rising numbers of sick workers, many of whom were hospitalised due to overstrain. While accident and health insurance were burdened with high costs, as the Hessian LAA explained, the firms were able to realise "disproportionate" profits. The OBR Frankfurt am Main initially ignored the LAA's complaint, only taking action against this practice in mid-May after the LAA president had threatened to stop the payment of subsidies to the *Reichsautobahnen* Corporation.[162] Similar episodes were reported from other *Autobahn* construction sites all over Germany.[163] Members of the German Labour Front (*Deutsche*

160 Lärmer, 66–67.
161 For continued examples of bad working conditions and low pay for workers sent by the labour offices, cf. the *Sopade*-reports.
162 BArch, R 3903/132, fol. 1, Note of 22.5.34; ibid., fol. 3–8, Report from the president of the Hessian LAA to the president of the RfAA, 24.5.34. Cf. Silverman, *Work Creation Programs*, 172.
163 Silverman, 172; Sozialdemokratische Partei Deutschlands, *Deutschland-Berichte der Sozialdemokratischen Partei Deutschlands (Sopade). Dritter Jahrgang, 1936*, Deutschland-Berichte der Sozialdemokratischen Partei Deutschlands (Sopade) 1934–1940 (Salzhausen, Frankfurt am Main: Verlag Petra Nettelbeck, Zweitausendeins, 1980), 94; Sozialdemokratische Partei Deutschlands, *Deutschland-Berichte der Sozialdemokratischen Partei Deutschlands (Sopade). Vierter Jahrgang, 1937*, Deutschland-Berichte der Sozialdemokratischen Partei Deutschlands (Sopade) 1934–1940 (Salzhausen, Frankfurt am Main: Verlag Petra Nettelbeck, Zweitausendeins, 1980), 1289.

Arbeitsfront, DAF), who visited construction sites in 1934, were appalled by the "inhumane treatment" of the workforce, but after complaints of firms and OBRs about the "disruptive DAF activities", "Todt responded by banning DAF visits to construction sites", unless the respective OBR had granted them access.[164] Over and over again, various authorities reported almost unbelievably high amounts of overtime from the *Autobahn* sites, like those of an engine driver from Saxony who had worked 31 hours in a row or of eight workers under OBR Halle who had worked 557 long hours in 14 days.[165] This clearly was an illegal practice.[166] When confronted with the conditions on the construction sites, the General Inspector's answer was laconic: "It should not upset us in the central administration when hefty overtime is done in construction [. . .]."[167]

Under this high work pace, the number of severe accidents increased continuously. In June 1935, the alarming number of 15 workers who died on *Autobahn* construction sites that month was ascribed to the former unemployed's lack of experience and skills. But the firms were admonished to exercise more care, too. On one construction site, short breaks and insufficient provision of drinks had led to heat strokes among the workers three days in a row.[168] The lack of safety measures on many construction sites also claimed victims.[169] In the face of increasing numbers of sick workers, the director of the building company Funke & Co, Freital, which later also worked in occupied Norway, was said to have threatened his workforce. Should they continue to call in sick on his *Autobahn* construction site west of Dresden, he would report them for sabotage.[170] Altogether, the first 1,000 kilometres of the motorway network alone claimed 226 lives, and by the end of 1938, the number had risen to almost 500.[171] The number of fatal accidents in relation to the number of workers by far exceeded the rates in the rest of the construction sector and the German economy as a whole (Figure 5).[172]

---

164 Silverman, *Work Creation Programs*, 172–73.

165 Lärmer, *Autobahnbau*, 65.

166 According to German legislation, regular working hours were restricted to eight hours per day. Under certain circumstances, 10 hours per day were permitted temporarily: RGBl. 1934, I, 803–13.

167 Quoted from Lärmer, *Autobahnbau*, 65.

168 'In einem Monat 15 tödliche Unfälle auf Baustellen der Reichsautobahnen!', *Die Bauindustrie* 3, no. 31 (1935): 463.

169 *Sopade 1936*, 610.

170 1188.

171 Lärmer, *Autobahnbau*, 70.

172 Although it is plausible and likely that the increase in fatal work accidents in the construction sector after the National Socialist seizure of power is related to a more and more ruthless exploitation of the workforce, Figure 5 alone is unable to prove this causal nexus. Some of the increase stems from an increase in hours worked and from the fact that the share of unskilled workers in the sector rose significantly during the 1930s. This holds true especially for *Tiefbau*, where the share of unskilled labour was high. (As a matter of fact, during the crisis of the early 1930s, when companies laid off unskilled labour while keeping their skilled workers and white-collar employees, *Tiefbau's*

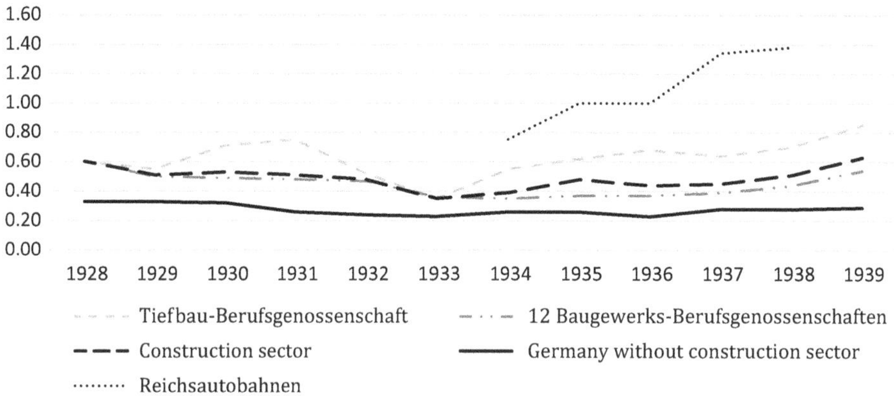

Figure 5: Fatal work accidents per 1,000 insurants, 1928–39.
Tiefbau workers were covered by the nationwide employers' liability insurance association for *Tiefbau* (*Tiefbau-Berufsgenossenschaft*). Workers of handwork firms, mostly engaged in *Hochbau*, were covered by twelve regional associations (*Baugewerks-Berufsgenossenschaften*). Their figures, as well as figures on the construction sector as a whole and the German economy excluding the construction sector are calculated based on the annual reports in the statistical yearbooks of the German Statistical Office. For 1937, see Reichsversicherungsamt, *Die deutsche Sozialversicherung 1937 – mit einem Blick auf das Jahr 1938*, Sonderveröffentlichung des Reichsarbeitsblatts und der Amtlichen Nachrichten für Reichsversicherung (Berlin: Reichsversicherungsamt, 1938). Note that we have to assume that from 1933 onwards, some fatal accidents that actually happened on *Tiefbau* construction sites were still reported to one of the twelve *Baugewerks-Berufsgenossenschaften*. Cf. appendix A2. For the *Reichsautobahnen* figures, see Lärmer, *Autobahnbau*, 70.

Hand in hand with entrepreneurs, who after years of economic depression grabbed any chance to earn money, Fritz Todt pursued his project regardless of the consequences for "his" workers. The reports from the *Autobahn* construction sites reveal the absurdity of one of the strongest myths established by Nazi propaganda around Todt – that of his constant concern for the welfare of the workforce – a myth even reproduced in post-war studies on the General Inspector and the OT.[173]

---

rate of work accidents went down to the level of the handwork firms that were mostly engaged in *Hochbau*). The accident rates of the *Reichsautobahnen* construction sites, however, are so high, even compared to those of the *Tiefbau* sector, that they cannot be explained by an increase in hours worked or unskilled labour alone.

173 Walter F. Renn, for example, in his otherwise thorough and critical study, claims that "Todt was especially skilled in handling labor problems, and he was genuinely concerned with the welfare of his workers as many of them who came from the Reichsautobahnen [to the *Westwall*, S.G.] knew. There Todt had made a name for himself in model social and material working conditions for his men." See Renn, 'Hitler's West Wall', 65. Speer biographer Dan van der Vat also echoes the propaganda of the 1930s too uncritically when he states that "countless German workers owed the

c The Westwall: Birthplace of the OT

By 1938, the General Inspector for German Roads had become the single-biggest customer of the German construction industry.[174] That year, the three-thousandth kilometre of the *Autobahn* was opened to traffic. On the back of the workforce and at the expense of exploding costs, Todt had managed to deliver what he had been asked for. It was no surprise then that the choice fell on him, when in May 1938, Hitler was looking for an energetic manager for the probably biggest construction project in the history of Germany, the *Westwall.*

The *Westwall* was a more than 600 km line of fortifications along the German-French border, consisting of bunkers, antitank obstacles and gun emplacements. The Fortress Engineers of the *Wehrmacht* had already been working on it since 1936. However, the project was not prioritised and the military struggled with gathering enough resources for the construction. After several postponements, the projected completion date was set to 1952, and still in early 1938, the amount of steel allocated to the Fortress Engineers was slashed.[175] But when Hitler decided in May 1938 to eliminate Czechoslovakia, he ordered a massive step-up in the construction effort. Until the planned date of attack on the eastern neighbour, 1 October, Germany's western border should be secured in a way that would make it impossible for French forces to come to its ally's aid and penetrate German territory. Within four months, 5,000 bunkers and armoured emplacements were to be built.[176] An extensive road construction programme should enhance infrastructure in the wall's hinterland.[177]

Obviously, Hitler had no trust that the Fortress Engineers were able to fulfil a task so extensive in such a short space of time. As Major Gerhard Engel later recalled, Hitler continuously attacked the construction units of the *Wehrmacht* for their alleged incompetence and dilatoriness: "If", Engel quotes Hitler, "he had given the task to the army alone, the *Westwall* would still not be finished by ten years' time."[178] Hitler's view was confirmed by a relentless report delivered by Göring who had inspected the construction sites in the West between 9 and 12 June, criticising the unsatisfactory progress of the project.[179] On his inspection, Göring was already accompanied by Todt and two of his closest staff members from the office of the General Inspector, Günther Schulze-

heroic engineer their liberation from unemployment and their families the very food on their tables.": Dan van der Vat, *The Good Nazi. The Life & Lies of Albert Speer* (Boston, New York: Houghton Mifflin Company, 1997), 109.

174 Bettinger and Büren, *Westwall*, 1:162.

175 Gross, *Westwall*, 27.

176 Renn, 'Hitler's West Wall', 24–34, 41; Bettinger and Büren, *Westwall*, 1:162–66; Seidler, *Fritz Todt*, 163–64.

177 Bettinger and Büren, *Westwall*, 1:165–71.

178 Hildegard von Kotze, ed., *Heeresadjutant bei Hitler 1938–1943. Aufzeichnungen des Majors Engel* (Stuttgart: Deutsche Verlags-Anstalt, 1974), 27.

179 Renn, 'Hitler's West Wall', 35.

Fielitz and Willi Henne.[180] Schulze-Fielitz established a head office in Wiesbaden to manage the entire *Westwall* construction, which was later led by Henne. Henne was one of only three men who came from Todt's office in Berlin to Wiesbaden. The remaining staff was recruited from civil engineering offices all over Germany. Although no written charter or order by Hitler existed, a new powerful organisation had come into being: *Organisation Todt.*[181]

In a meeting on 5 June 1938, the Inspector of the Western Fortifications (*Inspekteur der Westbefestigungen,* InWest) and the OT reached an agreement that the *Wehrmacht* should be responsible for planning the fortifications, determining their exact location, purchasing land for building, as well as for the final acceptance of the constructions. Todt, on the other hand, was responsible for the construction itself. The InWest assigned the construction responsibility to Todt on 8 June. To guarantee a close cooperation between the Fortress Engineers and the OT, they both had their headquarters in Wiesbaden's *Hotel Kaiserhof,* while local *Oberbauleitungen* (OBLs) were established alongside the existing Army Engineering Area Staffs.[182] The organisational structure of these OBLs, that supervised the single construction sectors between the German-Swiss border in the south and Kleve on the Lower Rhine in the north, was clearly modelled on the OBRs of the *Reichsautobahnen* Corporation – lean, decentralised and as autonomous as possible – and staffed with former personnel of those units.[183] Not surprisingly, the relation between the military and Todt was tense. Central responsibilities had been taken from the Fortress Engineers by a group of civil engineers that had never before been involved with military matters. Criticism grew among the men of the army. In their opinion, they just lacked the resources, manpower and building machines to execute the plans that they had drawn up thoroughly during the past years. Instead, Hitler was giving a blank cheque to the OT, which now sacrificed quality and the soldiers' security for speed. The former private Hitler himself came forward with countless ideas for "improved" bunkers. The Commander of the Army Group 2, General Wilhelm Adam, doubted the strategic value of the

180 Seidler, *Fritz Todt*, 164. Günther Schulze-Fielitz (1899–1972), was one of the leading officials working under Todt and later Speer, first as *de facto* leader of the office of the GB Bau in Berlin, later in the Ministry for Armaments and Munitions, and finally, under Speer, in the office of the General Inspector for Water and Energy. After the war, Schulze-Fielitz became a member of the management board and later supervisory board of the construction company Hochtief, Essen. See Christiane Botzet, 'Ministeramt, Sondergewalten und Privatwirtschaft. Der Generalbevollmächtigte für die Regelung der Bauwirtschaft', in *Hitlers Kommissare: Sondergewalten in der nationalsozialistischen Diktatur*, ed. Rüdiger Hachtmann and Winfried Süß (Göttingen: Wallstein Verlag, 2006), 119. On Willi Henne, see chapter IV.2.a, fn. 88.

181 Already in early June, the term *Organisation Todt* was used by the Commander of the Army Group 2: Bettinger and Büren, *Westwall*, 1:162. Hitler spoke of the *Organisation Todt* internally in July and publicly on the 1938 Nuremberg Rally in September: Seidler, *Fritz Todt*, 227.

182 Bettinger and Büren, *Westwall*, 1:163.

183 Dittrich, *Vom Werden, Wesen und Wirken der Organisation Todt*, 177.

Westwall in principle. In his eyes, the poorly manned army group would never be able to defend the fortifications and be overrun in no time in the case of a French attack. Many others in the ranks of the *Wehrmacht* shared Adam's doubts prior to the annexation of the Sudetenland.[184]

When confronted with criticism, Hitler usually answered by praising the ability of the OT to find a solution to any task: "Nothing is impossible for Todt."[185] Their shared aversion to bureaucracy is a factor that should not be underestimated when we ask why it was the OT that Hitler commissioned with the building of the *Westwall* and other projects later on.[186] Hitler's trust in Todt's ability to improvise, organise ruthlessly, and overcome any obstacles when fulfilling the orders of his "*Führer*", is one of the main reasons for the rise of the OT in the following years.

A second reason was the OT's and especially Todt's personal bonds with the German construction industry. The realisation of a project as huge as the western fortifications was unimaginable without taking private business on board. According to Xaver Dorsch, the army lacked the experience in large-scale construction and was "quite understandably and excusably at first unable to cope with the business site of the job", which was "foreign to its nature". Todt, on the other hand, "possessed the necessary reputation in the German construction industry, both professionally and personally".[187] And indeed did it take Todt only few days to gather a group of 16 construction business leaders in Wiesbaden to discuss their engagement at the *Westwall*. Hitler had granted Todt considerable decree powers over the construction industry, but the latter wanted to use persuasion and incentives, not force, to secure the industry's cooperation.[188] Also British intelligence came to the conclusion in March 1945, that "he had an extensive acquaintance among its [the construction industry's, S.G.] leading executives and was personally well-liked. When, therefore, he proposed a programme which, in the space of a little over two months, would provide a 24 hour working schedule for over a half-million men and one third of the entire German construction industry, the reaction

184 For these and similar examples of *Wehrmacht's* critique, see Telford Taylor, *Sword and Swastika: The Wehrmacht in the Third Reich* (London: Victor Gollancz Ltd., 1953), 194–205; William L. Shirer, *The Rise and Fall of the Third Reich: A History of Nazi Germany* (London: Secker and Warburg, 1961), 378; Renn, 'Hitler's West Wall', 90–120; Seidler, *Fritz Todt*, 178–84.

185 Taylor, *Sword and Swastika: The Wehrmacht in the Third Reich*, 203.

186 On Todt's aversion to bureaucracy, see Seidler, *Fritz Todt*, 307–8.

187 Dorsch, *Die Organisation Todt*. Quoted from the official translation of the report that can be found in the National Archives in Washington, DC: NARA, M-1035, fiche 1496A, Study MS-P037 by the Foreign Military Studies Branch of the Historical Division European Command, 3–7. However, whether the army was actually "relieved to see the work [. . .] assumed by *die Organisation Todt*", as Renn states based on the English translation of the report, is unclear. In the German original, it sounds like an assumption by Dorsch, rather than a fact. Cf. Renn, 'Hitler's West Wall', 49; Dorsch, *Die Organisation Todt*, 454.

188 Renn, 'Hitler's West Wall', 44.

of the latter was extremely favorable."[189] The 16, later 18, *Tiefbau* companies became general contractors, each responsible for one construction sector and attached to one OBL. They could hire smaller subcontracting firms, mostly from the western part of the Reich. The biggest incentive Todt could offer them was the use of cost-plus contracts. Although Willi Henne and his staff were said to have drafted and checked more than 60,000 construction drawings within two months, it would have been impossible to elaborate fixed-price contracts containing accurately defined work steps under the given time pressure.[190] Instead, the companies should start working immediately and the OT would reimburse all their costs. With this measure, Todt abandoned the procurement rules of the *Wehrmacht*, which had used fixed-price contracts. The risk of delays and exploding costs thus lay with the OT, not the construction firms. As an additional measure, Todt granted his general contractors an interested-free loan for the purchase of construction machinery, which they could also distribute among their subcontractors.[191] The loan amount depended on the construction volume: Todt granted the firms 2 RM per cubic metre of concrete. For a firm like Philipp Holzmann and its 85 subcontractors, pouring 440,000 m^3 of concrete at the *Westwall*,[192] this meant potentially an interest-free loan of 880,000 RM for machinery investments.

To minimise the risk of delays and to meet the ambitious deadline, the OT put significant effort in standardising construction. The idea was to agree with the Fortress Engineers on a list of bunkers and other constructions and then determine exactly how much concrete, wood, steel and labour input was necessary to build a certain object. The *Wehrmacht* could then order the desired type of fortification and the OT would immediately know how much building materials had to be provided. This would facilitate planning, supply and accounting significantly. To the British observers in 1945, "standardisation – always dear to the Germans and carried almost to the level of a religion by the Nazis – [was] the keynote of all OT construction work."[193] The effects of those standardisation measures should not be overstated, though. In the case of the *Westwall*, the existing catalogue of standard constructions (*Regelbauten*) of the Fortress Engineers had numbered 53 different constructions. But when the OT took over the organisation and supervision of works, it introduced 40 new standard constructions in 1938. Since these did not meet the requirements of the *Wehrmacht*, and some of them were regarded to be badly designed, another 90 types of bunkers and gun emplacements were introduced in

189 *Handbook of the Organisation Todt*, 6.
190 Seidler, *Fritz Todt*, 166; Renn, 'Hitler's West Wall', 94–95.
191 BArch, R 4601/3057, Letter by Schulze-Fielitz, GdSt, with attached contract, to all OBLs, 6.7.38.
192 Pohl, *Philipp Holzmann*, 242.
193 *Handbook of the Organisation Todt*, 51.

early 1939. Altogether, the number of constructions was rising significantly, which contradicted the very purpose of a *Regelbau*.[194]

The scope of the construction works and their impact on the western regions of Germany were unprecedented. In late September 1938, the number of workers under the OT, that is, the firms' regular workforce and conscripted workers, reached 241,000 and peaked at 342,000 shortly after.[195] In addition, 90,000 men under the Fortress Engineers and 300 units of the Reich Labour Service (*Reichsarbeitsdienst*, RAD) were working on the construction sites. Altogether, more than half a million people are said to have contributed to the construction of the *Westwall*.[196] The workday was characterised by military discipline: In the camps, workers had to muster every morning, hoist flags and march collectively to the construction sites.[197] Among the huge workforce, the OT tried to suppress any behaviour that could threaten the progress of the construction work, like political agitation, (alleged) idleness, or other lack of discipline. To detain, discipline and punish conscripted workers, the Organisation used first provisional jails, and from October 1939, a former DAF camp in the town of Hinzert. Hinzert became the centre of a network of several so-called *Westlager* along the fortification line. Guarded by SS units, the internees had to work on OT construction sites, in the armaments industry, for the *Wehrmacht* or formed work units that were used on *Autobahn* construction sites. Many firms regularly requested internees from the camps.[198] The construction company Krutwig from Cologne was working on an *Autobahn* stretch in the Eifel region and was very pleased with the workers that the nearby camp in Wittlich ad allocated to them. The performance of the workers was very good due to "intensive, non-stop work under strict discipline". "I am convinced,", the entrepreneur stated in a letter to the camp commander of Hinzert, "that in the course of time, [. . .] these elements will change their mind and realise that sluggards have no right to exist in our National Socialist state."[199]

---

194 Gross, *Westwall*, 117–235.

195 BArch, R 43-II/1293, fol. 13. Bettinger and Büren report a maximum of 56,000 workers for the Fortress Engineers, 203,000 for the OT and 100,000 for the RAD: Bettinger and Büren, *Westwall*, 1:187. Elsewhere in their study, however, also Bettinger and Büren quote the number of 342,000 OT workers: Bettinger and Büren, 1:178.

196 Seidler, *Die Organisation Todt*, 15; Seidler, *Fritz Todt*, 168; Lemmes, 'Arbeiten für das Reich', 56. The total number of more than half a million workers is also given in: *Handbook of the Organisation Todt*, 6.

197 Lemmes, 'Arbeiten für das Reich', 59.

198 Gabriele Lofti, 'SS-Sonderlager im nationalsozialistischen Terrorsystem', in *Ausbeutung, Vernichtung, Öffentlichkeit: neue Studien zur nationalsozialistischen Lagerpolitik*, ed. Norbert Frei, Sybille Steinbacher, and Bernd C. Wagner, Darstellungen und Quellen zur Geschichte von Auschwitz 4 (München: K. G. Saur, 2000), 209–29; Uwe Bader and Beate Welter, 'Das SS-Sonderlager/KZ Hinzert', in *Hinzert, Auschwitz, Neuengamme*, Der Ort des Terrors. Geschichte der nationalsozialistischen Konzentrationslager 5 (München: C.H. Beck, 2007).

199 Bader and Welter, 'Hinzert', 72–73.

Contrary to the *Reichsautobahnen*, no figures on work accidents and deaths were published for the *Westwall*. However, contemporary reports mention a rising number of accidents of overtired workers.[200] When workers died or were injured, the DAF organised their transportation home and took over provision for dependants.[201] The general contractor Grün & Bilfinger reported 14 casualties among its 4,600 workers (nine due to disease, four traffic accidents and one actual work accident), and 24 casualties (four actual work accidents) among its subcontractors' 6,700 workers.[202] The section south of Grün & Bilfinger's sites was led by the company Philipp Holzmann. Board member and head of Holzmann's *Westwall* construction sites, Hans Meyer-Heinrich, claimed in the early post-war years that there had been no casualties among the 14,000 workers of Philipp Holzmann and its 83 subcontractors. It is arguable how credible this statement is in the light of Grün & Bilfinger's figures.[203]

Most of the larger German construction firms could be found among the *Westwall* contractors. When works on the fortification line were finished in 1940, the number of individual firms had risen to 840, in addition to 110 working combines (*Arbeitsgemeinschaften*, Arge). Even if every Arge would have consisted of only two partners, the total number of firms on the *Westwall* would have been at least 1,060.[204] The *Westwall* became the single most important construction project in the Reich, but it did not completely supersede the countless undertakings of the *Wehrmacht* and private industry or the expansion of infrastructure all over the country. Until the war broke out in September 1939, the construction of the fortification line had on average consumed 5 percent, 8 percent and 20 percent of Germany's annual production of iron, timber and cement, respectively.[205] The by far greatest part of the transport capacities, however, was necessary to supply the sites with aggregate, that is, sand, crushed stones and gravel that was mixed into the concrete.[206] At the height of the construction activities, more than 15,000 trucks, up to 8,000 railway cars and one third of the Rhine shipping were deployed to supply the construction sites. Traffic in the West was close to collapse. 5,000 *Reichspost* buses were used to bring workers to their work sites every day.[207] Even

---

200 Gross, *Westwall*, 245.

201 Document 114, Speech of Göring on the 6th annual convention of the DAF, 10.9.38, in: Mason, *Arbeiterklasse und Volksgemeinschaft*, 677–86.

202 Bilfinger Archive, A 12, Report of the Kehl construction site, 1938–39.

203 IfS, W1-2/ 298, Draft of Hans Meyer-Heinrich's 1949 anniversary *festschrift*, 128.

204 BA-MA, RH 53-12/43, List over construction firms at the *Westwall* for the time between 1 July 1938 and 1 April 1941.

205 BA-MA, RH 11-III/170 K-4.

206 Until September 1939, 21.5 million tonnes of construction materials had been transported to the *Westwall*, over 17 million tonnes of which were aggregate: ibid.

207 Seidler, *Fritz Todt*, 168–69; Renn, 'Hitler's West Wall', 57–59; Bettinger and Büren, *Westwall*, 1: 178–79.

though the OT and the DAF erected large barracks, many *Westwall* workers had to live on farms, in pensions or in improvised camps in the ballrooms of restaurants. The fortification line cut through the landscape of Western Germany like a huge concrete scar, claiming the destruction of 5,600 farms and 120,000 hectares of land.[208]

The combination of the vast scope of the project, a tight time schedule and a contract regime that did not financially reward fast and proper work bred conflicts between the entrepreneurs and the OT. Todt was obviously dissatisfied with the quality delivered by some companies. Three months into the construction works, he already resorted to threats:

> Not only the expended wages are lost through improper work – the built-in materials (gravel, cement, iron) are devaluated, too. If my supervisory bodies or the supervisory bodies of the Pioneers detect that an entrepreneur delivers inferior results through improper work or even through gross neglect of the standards of concrete construction, I will ruthlessly secure the warranty by [. . .] stopping the payment for materials and wages for the respective building. Furthermore, I will file charges with the Secret State Police against the responsible bodies of the respective entrepreneur who damage the Reich through irresponsible neglect and gross fault during the production of the concrete.[209]

One of the firms confronted with the allegation of having done a botch job at the *Westwall* was the Cologne-based building firm Peter Bauwens. Todt complained to Bauwens that "I have been using your firm because I expected something better than what has been achieved." He criticised especially Bauwens's direction of his subcontractors and workforce, as well as questions of accounting. He deplored that "a company that had enjoyed a good reputation for decades" obviously had lost its position among the excellent building businesses of the Reich.[210] Another general contractor, the Julius Berger Tiefbau AG, was blamed for the tardy progress of construction in the vicinity of Karlsruhe. Todt warned the company that he was keeping an eye with the performance of the *Westwall* firms in order to assess their suitability for future tasks. "Unfortunately,", Todt concluded, "I cannot count Julius Berger among the successful companies so far."[211] Nevertheless, there is no indication that the General Inspector's discontent excluded the two companies from future OT orders, or even filed charges against firms with the Secret State Police. Their know-how seems to have been indispensable to the OT as numerous construction projects in Germany, Norway or other occupied countries bore witness to in the following years.

---

208 Hans-Erich Volkmann, 'The National Socialist Economy in Preparation for War', in *The Build-up of German Aggression*, ed. Militärgeschichtliches Forschungsamt, Germany and the Second World War, I (Oxford: Clarendon Press, 2015), 260.
209 BArch, NS 26/1189, fol. 99–100, Circular by Fritz Todt to all OBLs, entrepreneurs and subcontractors, 20.8.38.
210 Ibid., fol. 142–43, Letter Fritz Todt to Peter Bauwens, 30.9.38.
211 Ibid., fol. 120–22, letter Fritz Todt to *Baudirektor* Martens, Julius Berger Tiefbau AG, 14.9.38.

During the last days of September 1938, while Hitler, Chamberlain, Daladier and Mussolini were negotiating the fate of Czechoslovakia, Todt's men worked incessantly to meet the deadline and finish the 5,000 bunkers demanded by the "*Führer*" earlier that year. Works were still going on when the first soldiers moved into the bunkers and emplacements, where they had to find out that the construction shacks of the OT's building firms had not been removed yet and were blocking the line of fire.[212] Mostly due to transport problems and insufficient supply with building materials, even the acclaimed OT had not managed to build 5,000 bunkers. As Todt later recalled, "the first emplacement was poured concrete already in the beginning of June, the tenth on 25 July, the one hundredth on 12 August, the thousandth on 1 September 1938 and in the crucial days of late September, there were so many emplacements finished that on the opposite side, the number of guns necessary to their bombardment was not sufficient to defeat more than one third of them."[213] It is no coincidence that Todt avoided to state the number of bunkers finished by 1 October. Only 3,210 bunkers had been built, and of these were 75 percent reported to be only "provisionally defensible", meaning that they lacked fitments like camouflage or even armoured doors.[214]

The annexation of the Sudetenland in October 1938 was only one step in Hitler's expansionist plans. Bohemia and Moravia were already destined to become the next victims of German aggression. In the West, this meant the continuation of the construction works, especially the incorporation of Aachen and Saarbrücken into the defensive line. According to Walter F. Renn, Todt was genuinely surprised by Hitler's order to continue construction, which increased the number of projected bunkers and emplacements to 22,000. Todt had apparently hoped to return to his duties on the *Autobahn* project, instead, he saw the OT becoming a permanent organisation. On 9 October, in front of 10,000 *Westwall* workers in Saarbrücken, Hitler even promised the expansion of the fortification line "with increased energy".[215]

The pace of construction, however, and accordingly the number of workers, declined significantly during the winter of 1938–39. Besides the normal drop in construction activity during the winter months, the main problem was that the National Railways could not provide enough rolling stock to supply the construction sites with building materials. Railroad cars were urgently needed to bring in the autumn's harvest and to supply the cities with coal, crucial tasks, that already had been neglected

212 Renn, 'Hitler's West Wall', 180.
213 Fritz Todt, 'Die Neuordnung des deutschen Bauwesens', *Der deutsche Baumeister* 1, no. 1 (1939): 7.
214 Renn, 'Hitler's West Wall', 185.
215 Renn, 198–207.

during the past weeks.[216] In December, Berliners had to queue in front of the coal merchants' businesses because of the transportation problems.[217]

From January 1939 on, the number of workers increased again until the fortifications had reached a number and strength that justified the gradual reduction of the construction effort from summer 1939 on.[218] By the day of the German attack on Poland, the OT had managed to erect more than 16,000 of the projected 22,000 constructions in the West.[219] While the French did not launch a major attack on the fortification line, construction works were discontinued in the course of the first half of 1940. "The Westwall", as Todt wrote, "has fulfilled the task".[220] As the German Court of Audit assessed in November 1942, more than 2.9 billion RM had been transferred to the General Inspector's special account W (*Sonderkonto W*) to pay for the works in the West.[221] In 1944–45, the undermanned fortification line posed no major obstacle for the advancing Allied troops. It had been stripped of its equipment and the constructions were overtaken by technological advancements in weaponry.[222]

As this chapter has demonstrated, the state agencies under Fritz Todt were among the German construction industry's most important customers by the late 1930s. Most major companies had worked under the GdSt and the OT and their experiences were mostly positive. At no point had Fritz Todt questioned the autonomy of the German construction industry or denied firms the opportunity to earn good money on his construction sites. Thus, when the war broke out, the companies had no reason *not* to work for the OT in the occupied territories. When German construction firms entered negotiations with the OT on missions in occupied Europe from 1939 onwards, they were already able to look back on several years of fruitful and profitable cooperation with Fritz Todt and his staff.

216 Renn, 198–207; Alfred C. Mierzejewski, *The Most Valuable Asset of the Reich. A History of the German National Railway. Volume 2, 1933–1945* (Chapel Hill, London: The University of North Carolina Press, 2000), 49–50.

217 Document 151, Economic report for the fourth quarter of 1938 by the Mayor of Berlin, in: Mason, *Arbeiterklasse und Volksgemeinschaft*, 900.

218 Renn, 'Hitler's West Wall', 253–54; Bettinger and Büren, *Westwall*, 1:187.

219 Renn, 'Hitler's West Wall', 287.

220 Renn, 320.

221 BArch, R 2301/7172, fol. 286–92, German Court of Audit to the GdSt, Wiesbaden headquarters, 21.11.42.

222 Renn, 'Hitler's West Wall', 326.

III State Regulation of the German Construction Sector

1 Wages and Prices During the National Socialist Construction Boom

The construction boom of the 1930s was the result of the National Socialist state's policy of rearmament and highest possible economic self-sufficiency. This chapter will assess the consequences of this policy for the construction sector, which were rising costs and a considerable upward pressure on wages. Construction companies naturally tried to realise high prices and attract scarce workers with higher wages and various bonuses and allowances. These developments were problematic from the National Socialist government's point of view because they jeopardised the proclaimed stability of prices and wages.

Part a of this chapter will give a short overview over the institutions concerned with monitoring prices and wages in construction. Part b will then focus on one of the most important cost factors in construction, the workers' wages, before part c covers further factors that led to rising construction costs. I will focus on the situation in the construction sites of the GdSt and the OT in particular, in order to illuminate the experiences that construction firms gained with the OT as a customer. Finally, in part d, I argue that the official statistics on wages and prices in construction lost much of their information value due to a lack of control over the course of the 1930s.

a Institutions Monitoring Wages and Prices

The National Socialist state's aim of steering the economy towards rearmament and autarky required an expensive apparatus of monitoring agencies. Their work ranged from determining and controlling wages to carrying out price audits and revising contracts. In the PA model developed by Jochen Streb, which was presented in chapter I.1.c, these institutions rank at the top of the model and represent the National Socialist government.

After the unions had been eliminated in May 1933, wage regulation lay in the hands of the newly established regional Reich Trustees for Labour (*Reichstreuhänder der Arbeit*), who were directly subordinate to the RAM. Wage rates were no longer negotiated between unions and employer associations, but fixed and monitored by the state. In particular, every pay rise for a group of workers in a company depended on the trustees' approval. Moreover, they decided on collective dismissals, participated in the Social Honour Courts (*Soziale Ehrengerichte*) and reported regularly on

https://doi.org/10.1515/9783110694291-003

the developments in various industry sectors and the mood among the workforce.[1] Likewise in May 1933, the DAF was established as the central National Socialist labour organisation. It replaced the unions, but other than the Reich Trustees and in contrast to the traditional functions of a union, it never gained significant influence on wage policy.[2]

The monitoring and regulation of wages were complemented by price controls, one of the cornerstones of National Socialist economic policy. This became the task of the Reich Price Commissioner. The agency was initially established as Reich Commissioner of Price Controls (*Reichskommissar für Preisüberwachung*) under Reich Chancellor Brüning in 1931 and led by Karl Goerdeler. After Goerdeler's resignation in 1935, the office was re-established in 1936 as Reich Commissioner for Price Formation (*Reichskommissar für die Preisbildung*) under the Four Year Plan and headed by Josef Wagner. The new title underscored the state's aim not only to monitor prices, but to actively set their level.[3]

Finally, the German Court of Audit in Potsdam also increasingly tried to involve itself in the formation of prices and design of contracts, instead of just controlling projects *ex post*. Under its new president Heinrich Müller, the Court of Audit increasingly shifted its focus from monitoring to advising, a strategy for which Müller coined the term "*Beratungsrevision*". In this respect, the *Westwall* became a turning point in the Court of Audit's history: Potsdam sent officers to Wiesbaden to provide the OT with advice and prevent price increases before they occurred. This became common practice in the following years, when numerous new agencies were established in the occupied territories.[4] Major procurement agencies like the *Wehrmacht* had their own auditing units, too, which sought to control the calculations on which tenders by building firms were based. The general price stop of 26 November 1936, which forbade price increases for goods and services of all kind, formed the foundation of their work. Violations were punishable with imprisonment or an unlimited fine.[5]

The boom in the construction sector found its expression in rising entrepreneurial profits over the course of the 1930s. Excess profits thus increasingly concerned the abovementioned authorities, as they burdened the budget and ran counter to the National Socialist propaganda that sharply condemned all kinds of

1 AAndreas Kranig, *Lockung und Zwang. Zur Arbeitsverfassung im Dritten Reich* (Stuttgart: Deutsche Verlags-Anstalt, 1983), 168–89; Hachtmann, *Industriearbeit*, 32–36.

2 Tilla Siegel, *Leistung und Lohn in der nationalsozialistischen 'Ordnung der Arbeit'* (Opladen: Westdeutscher Verlag, 1989), 62–85.

3 Barkai, *Nazi Economics*, 188.

4 Franz-Otto Gilles and Gerhard Otto, 'Ordnungsgemäße Beuteverwaltung? Zur Finanz-, Verwaltungs- und Wirtschaftskontrolle in den während des Zweiten Weltkrieges von Deutschland besetzten Gebieten', in *Autonomie und Kontrolle: Beiträge zur Soziologie des Finanz- und Steuerstaates*, ed. Theo Pirker (Berlin: Schelzky u. Jeep, 1989), 43–45.

5 RGBl. 1936, I, 955–56.

war profiteering.[6] Levying the excess profits and fining the entrepreneurs that had realised them was an increasingly important task of the Reich Price Commissioner in those years. The Commissioner mostly relied on reports provided by various monitoring agencies like those of the three *Wehrmacht* branches and the German Court of Audit, which was responsible for controlling of projects paid for by the state. To this end, the agencies also requested special reports from tax authorities all over the country.[7] Firms often worked for various public customers and thus complained about frequent and seemingly uncoordinated audits by various agencies.[8] At least in construction, however, there was a visible effort to find a common line of action and to share the results of audits between the RFM, the Price Commissioner, the Court of Audit and the High Command of the *Wehrmacht* (*Oberkommando der Wehrmacht*, OKW).[9]

The monitoring of prices, wages and profits, however, was a particularly demanding task in the construction sector. Most structures were unique and calculations contained a multitude of uncertain cost factors, like the soil conditions, the remoteness of some building sites or the availability of building materials. This made it difficult to compare tenders, in contrast to mass products like ammunition. Moreover, as has been discussed in chapter II.1.a, construction was a large and fragmented sector. Many of the more than 200,000 production units were small, family-run handwork businesses, whose book-keeping hardly met today's standards.

Given the complexity of price monitoring in the sector and the sheer mass of construction projects, the lack of experienced auditors became a major problem for the German administration. Many officials of awarding authorities had been drafted into the *Wehrmacht*. As early as 8 September 1939, the GB Bau had to instruct public customers to pay construction firms based on "rough estimates" of the costs. Thorough controls of invoices would delay payments and jeopardise the firms' liquidity.[10] Although 1,600 auditors were checking books and monitoring prices around the country in 1940, the *Wehrmacht* was forced to admit that the number of auditors was insufficient to control all their suppliers and the construction prices of all state-owned armament plants.[11] The German Court of Audit reported that public

6 For a discussion of the term "excess profits", see chapter III.4.

7 A large number of these special reports (no. 500–1172) can be found in the files of the Court of Audit at the Federal Archives in Berlin: BArch, R 2301/7183–7186 and 9950–9951.

8 Rauh-Kühne, 'Hitlers Hehler', 24–25.

9 BArch, R 2301/7187, fol. 27, German Court of Audit to several of its departments, 19.3.40.

10 Notice of the GB Bau regarding payments in the building trades, 8.9.39, in: *Der deutsche Baumeister* 1, no. 10 (1939): 24.

11 Streb, 'Negotiating Contract Types', 368. In the *Inflationsdenkschrift* of November 1938, the contracts and price control department of the war economy staff unit of the High Command of the *Wehrmacht* (*Abteilung Vertrag- und Preisprüfwesen des Wehrwirtschaftsstabes des OKW*) had already conceded that they hardly had sufficient personnel to control the pricing of their armaments suppliers. Moreover, "it is absolutely impossible to scrutinise the pricing for sub-supplies, given the

demand for building works had risen so massively that the public administration had been unable to scrutinise the calculations as thoroughly as necessary. In the light of the vast numbers of projects, which were all regarded as crucial to the war effort, they simply lacked time.[12] However, the auditors' qualifications also gave cause for serious concern. Many auditors had little experience in calculating the price of a construction project because during the crisis years, companies underbidding each other had guaranteed low prices. Now, the auditors lacked the expertise to draw up their own calculations and compare them with the (potentially overpriced) offers.[13]

b The Upward Pressure on Wages in Construction

Among the factors that contributed to rising prices, wages were particularly important.[14] Before the war, the massive building activity triggered by German rearmament had far-reaching consequences, especially with regard to the labour market and wage levels. The unemployment figures in the sector dropped with stunning speed as the sector overcame the slump, and from 1936 onwards, construction companies began to absorb labour from other sectors, which led to a noticeable upward pressure on wages. Since construction was a labour-intensive business, the rising wage level was of great concern within National Socialist authorities. Depending of the field of construction, wages were either building prices' most important factor or the second most important after building materials (Figure 6).

Wages and prices were tightly connected, as every price calculation in the business was usually based on the wage bills of the respective construction site. From the perspective of the government, rising wages meant rising prices, which again could lead to inflation and destabilise the currency and the economy as a whole.

As a result of the labour shortage, companies competed fiercely for sought-after builders. Grün & Bilfinger reported in 1937 that the newspapers were full of job offers for construction workers, causing some disturbance in the sector.[15] The labour shortage increasingly came into conflict with the state's priority to keep wages down.

---

immense number of sub-suppliers and their volatile activity." The construction market was regarded as particularly overheated: Geyer, 'Rüstungsbeschleunigung', 157–58.

12 BArch, R 2301/3007, Memorandum of the President of the German Court of Audit on the Reich budget account for the budget year 1939, 19.4.43, fol. 260. The *Wehrmacht* came to the same conclusion: Geyer, 'Rüstungsbeschleunigung', 166.

13 BArch, R2301/7124, fol. 277. Cf. Streb, 'Preisregulierung', 45–46.

14 See below, Figure 6.

15 Grün & Bilfinger AG, business report 1937.

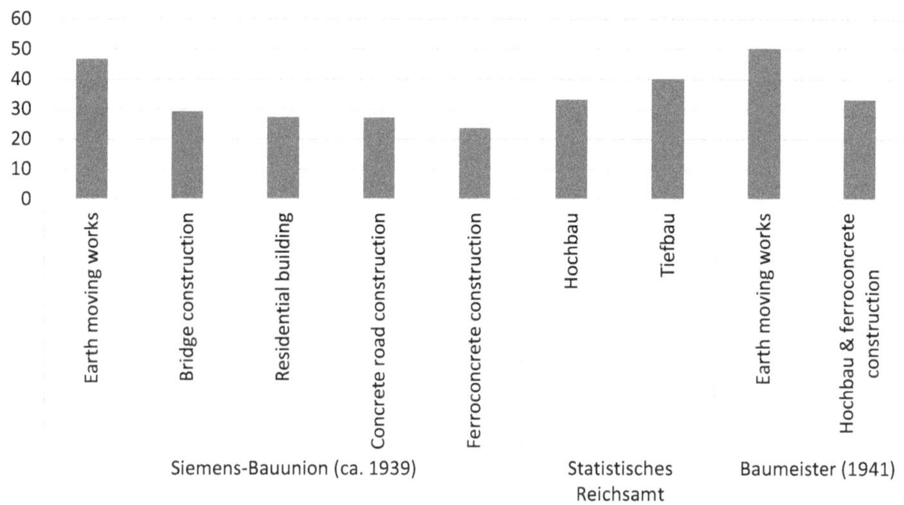

Figure 6: Various estimates on the percentage of wage costs in the total construction price. Source: Siemens-Bauunion: Georg Garbotz: 'Mittel und Wege zur rationellen Betriebsführung', in: *Der deutsche Baumeister* 1, no. 8 (1939): 22; German Statistical Office: BArch, R 3102/2701, confidential report on construction turnover 1929–1940, 10.9.41; Baumeister: Ludwig Baumeister, *Preisermittlung und Veranschlagen von Hoch-, Tief- und Eisenbetonbauten*, 8th ed. (Berlin, Heidelberg: Springer-Verlag, 1941), 28.

Higher building costs resulting from higher wages on the construction sites were one thing. But in an economy that was geared towards the production of armaments and autarky goods instead of consumer goods, every additional reichsmark in a worker's pocket also carried the danger of inflation. At the same time, the wages paid in booming sectors aroused discontent among workers in other industry branches. Construction began to absorb labour, especially from sectors like agriculture and building materials. In the case of the former, this not only threatened the domestic population's supply of food, but touched upon the core of National Socialist ideology, in which the farmer played a key role. By attracting workers from the building materials industry, on the other hand, the construction industry was effectively sawing off the branch it was sitting on.

In October 1936, the Reich Labour Ministry claimed that in some parts of the construction sector, entrepreneurs were paying wages three times higher than the standard wage.[16] Göring as Reich Minister for Economy and Labour Minister Franz Seldte both "observed with great concern" that in some parts of the Reich, building workers

16 Document 3, Explanatory statement on the draft of a Second Law for the Regulation of Labour Deployment, 6.10.36, in: Mason, *Arbeiterklasse und Volksgemeinschaft*, 198–216.

earned over 100 percent more than the standard wage in 1937.[17] In the Berlin area, standard wage rates had remained unchanged between 1935 and 1938. However, the *Wehrmacht* registered that on their building sites around the capital, real wages lay 30 percent above that level in 1935, and 75 percent higher in 1938. That means that actual wages in construction had risen by 35 percent in three years.[18] Even the prospering IG Farben complained that their rayon factory in Berlin (Aceta GmbH) struggled to compete with the capital's construction sector. Aceta was constantly losing workers to the large state construction projects, even though it paid its factory workers 90 Rpf. per hour. The management dismissed the idea of paying the workforce a bonus of 10 or even 20 Rpf. per hour because they thought that those bonuses would be outdone, too.[19] Around the same time, the regional Trustees for Labour were constantly complaining of the "lure wages" (*Locklöhne*) through which entrepreneurs were trying to woo workers away from other firms.[20] As a carpenter reported from a *Wehrmacht* building site in Silesia, "[. . .] skilled workers were handled like raw eggs."[21] In a speech on 26 January 1939 given in front of NSDAP functionaries at *Deutsches Museum* in Munich, Fritz Todt himself openly admitted and criticised the fact that in some parts of construction, "wage no longer matters".[22]

A post-war report of Bauunternehmung Gehlen KG, a mid-sized construction firm from Kaiserslautern, exemplifies how far standard wage rates and the wages actually paid diverged in the sector. The firm had carried out public projects in the Reich and the occupied territories, among others for EW in Norway. In 1947, in a report on the "social and economic efforts" of the firm, Gehlen stated that between 1939 and 1944, it had paid wages that were on average 15.5 percent and in the case of core workers (*Stammarbeiter*) even up to 38 percent above the standard wage rates. Salaries had been 32 percent higher than standard wage rates. Together with rewards, bonuses and pension provisions, during the war Gehlen had paid more than half a million RM per year in excess of the standard wage rates and compulsory social contributions.[23]

---

17 Document 72, Letter from the Reich Minister for Economy to various public authorities, 7.12.37, in: Mason, 522–25.

18 Geyer, 'Rüstungsbeschleunigung', 155.

19 *Archiv des Instituts für Zeitgeschichte*, München (hereafter: IfZArch), MA 1555/115, fol. 603–4, Letter Esselmann, IG Farben AG Wolfen, to director Gajewski, 21.7.38. At that time, the standard wage rate for Berlin was 108 Rpf. for bricklayers and 82.7 Rpf. for unskilled/*Tiefbau* workers. See *Statistisches Jahrbuch für das Deutsche Reich 1938* (Berlin: Schmidt, 1938), 341.

20 See the numerous reports by the Trustees for Labour in: Mason, *Arbeiterklasse und Volksgemeinschaft*.

21 *Sopade 1936*, 1203.

22 Todt, 'Neuordnung', 9.

23 *Stadtarchiv Kaiserslautern*, A04/310, Report on the social and economic efforts of the firms Bauunternehmung Gehlen KG and Eisenwerke Kaiserslautern, 1938–1945, 6.2.47. I would like to thank Björn Stemler for providing me with this report.

Christmas bonuses and a multitude of creative allowances became common measures for attracting or holding workers. A company working on the *Reichsautobahnen* ordered 100 *Kraft-durch-Freude* cars for their workforce, a quarry firm from Brandenburg bought motorcycles that the workers could pay off in small rates. Another company was reported to transport their personnel to the workplace by bus, while the skilled workers were driven in the entrepreneur's private car.[24] Workers received payments when they married, had children or fell ill. In addition, they were granted special leave when they married or there was a death in the family. Long-time employees could receive loyalty bonuses and company pensions. Contributions could also cover anything from work outings and newspaper subscriptions to sports equipment. The families of regular workers that were drafted into the military could receive additional benefits.[25] As a contemporary observer noted, payments to the companies' social funds increased by leaps and bounds in the construction industry during the late 1930s.[26] By doing this, firms could kill two birds with one stone. On the one hand, they were able to attract workers and partly circumvent the wage stop. On the other hand, allocations to social funds helped to keep profits below a critical level, above which the state tried to skim off any excess during the war years.[27] Smaller handwork firms in particular struggled to keep step with this development. In October 1937, handwork guilds in Northern Germany took matters in their own hands and dictated maximum wages to their member firms, even though they were clearly exceeding their competence as guilds in doing so.[28]

In contrast to earlier practice, more and more firms avoided laying off workers during the winter, because they feared that the men would take jobs on other construction sites and not return the following spring. Similarly, in anticipation of new contracts, the firms did not dismiss their workers after construction projects were completed.[29]

---

24 Mason, *Arbeiterklasse und Volksgemeinschaft*, 656, 790–91, 862.

25 Matis and Stiefel, *Porr*, 433; Sager & Woerner AG, *Sager & Woerner 1864–1964*, 2: Sammlung-Einsatz-Krise-Erstarkung, 1920–1945:195. Numerous business reports from the late 1930s also mention the allowances. See for example the business report 1938 of Heilmann & Littmann Bau-AG, Munich. The RAM confirmed that these kinds of allowances were permitted "as long as they remained within reasonable limits": Document 140, Decree of the RAM to the Trustees for Labour, 26.7.39, in: Mason, *Arbeiterklasse und Volksgemeinschaft*, 804–10, 807.

26 *Deutsche Zentralbibliothek für Wirtschaftswissenschaften, Leibniz Informationszentrum Wirtschaft* (hereafter: ZBW), Gustav Plum, 'Die Abschlüsse der Bauindustrie. 1933 bis 1939' (Unpublished manuscript, circa 1940), 84–85.

27 Matis and Stiefel, *Porr*, 432–33.

28 Document 47, Excerpt from the monthly reports of the Trustees for Labour, October 1937, in: Mason, *Arbeiterklasse und Volksgemeinschaft*, 401–8.

29 Document 39, Economic report from the *Oberpräsident* of Saxony for the months of February and March 1938, 25.4.38, in: Mason, 348–53.

Public customers abetted this practice as they insisted on building firms having enough building workers at their disposal before they would place an order.[30] This deprived the state of one of the few possibilities to steer and intervene in the construction sector before 1938. There was no longer a large reservoir of unemployed workers from which the labour offices could recruit men and allocate them to important construction projects, as they had done in the case of the *Autobahn*.

Understandably enough, the construction workers tried to capitalise on the situation. Companies complained about frequent job changes, and at times, workers tried to provoke the termination of their work contracts. They then took advantage of a peculiarity of the German wage system: standard wages could differ significantly between hundreds of local wage areas (in the construction sector, wages were – as a rule of thumb – higher in cities and the western parts of the Reich and lower in rural areas and the east). The result was a movement from rural areas and regions like Silesia and Saxony to areas with higher wages. In addition, this entitled many workers to allowances for working away from home.[31] This became a problem for the *Wehrmacht*, since their construction projects were often located in rural areas with lower wages. As the president of the Brandenburgian LAA reported in October 1937, it was almost impossible to find workers for military construction projects at the eastern border of the Reich, because the standard wage for *Tiefbau* workers in this area was 0.48 RM per hour, and 0.60 RM for skilled workers. At the same time, wages in Berlin were rising to 1.30 or 1.35 RM per hour.[32] When the navy invited tenders for a construction project at Flemhuder See in the summer of 1937, the Nordmark branch of Wigru Bau was afraid that the firms would not be able to attract any workers if they observed the low, local wage rates. Wigru Bau therefore asked the navy to only consider firms that had calculated on the basis of the higher wage rates paid in the neighbouring city of Kiel. When the navy refused to tolerate this interference, the association described its action as an "act of self-help". According to Wigru Bau, there was no longer a single construction project in the entire Nordmark that was executed on the basis of standard wage rates. The actual wage increase was said to be between 30 and 50 percent, in some extreme cases even several hundred percent. Eventually, the High Command of the Navy (*Oberkommando der Marine*, OKM) had to admit in mid-October that Wigru Bau had been right and accepted the higher wages.[33]

---

30 Document 69, Report from the president of the LAA Brandenburg to the president of the RfAA, 1.10.37, in: Mason, 507–13.

31 Document 33, Excerpt from the monthly reports of the Trustees for Labour, April 1937, in: Mason, 317–28. Document 96, Excerpt from the monthly reports of the Trustees for Labour, January and February 1938, in: Mason, 606–21.

32 Document 69, Report from the president of the LAA Brandenburg to the president of the RfAA, 1.10.37, in: Mason, *Arbeiterklasse und Volksgemeinschaft*, 507–13.

33 For the entire matter, see BArch, R 3101/10319, fol. 14–43.

In other industries, especially in the metalworking sector, piecework wages were said to cause rising wages. Piecework could provide workers with an incentive to workers to increase their performance; at the same time, however, unjustified pay raises were harder to detect. National Socialist propaganda was constantly alert to the danger of *Scheinakkorde*, that is, piecework wages without the output rising correspondingly, but it has been argued that these fears were unsubstantiated.[34] Due to the character of construction works, piecework pay remained the exception in the construction sector anyway. In March 1939, only 6.6 percent of the workers were paid by output, compared to 58.4 percent in metalworking and 45.4 percent in the textile industry.[35] Even in an industry like construction, where hourly pay dominated and wage rates were potentially easier to control, the state struggled to stop wages rising.

It is also worth mentioning that the wage policies at this time only targeted the workers, not the salaried employees. The building company Dyckerhoff & Widmann reported that between 1935 and 1940, the number of salaried employees had risen by 105 percent, their wage bill, however, by 160 percent.[36] The Trustee for Labour of the Saarpfalz had surveyed the salaries paid by one main contractor on the *Westwall* in 1938. Due to overtime and performance premiums, allowances and expenses, the salaried employees earned up to 200 percent more than their nominal salary.[37] For public procurement agencies like the *Reichsautobahnen* Corporation, the level of salaries became a problem – and not only because it contributed to rising construction prices. Private construction firms began to headhunt *Reichsautobahnen* officials by offering them higher salaries. At a meeting in Salzburg in 1939, OBR Frankfurt am Main reported that "because of the exceptional boom on the construction market, the construction industry is today paying 'fantasy salaries' to capable young employees." Private firms could easily outbid every pay rise or premium the *Reichsautobahnen* officials came up with.[38]

c Further Price-Increasing Factors

Besides wages, several other factors contributed to rising costs in construction. One parameter was the cost for leased building machines. Renting out temporarily idle

34 Hachtmann, *Industriearbeit*, 196–99.
35 Hachtmann, 173.
36 DMM, FA 010/100, Capacity of the construction industry. Comment on the statistics 1935–1940, 4.7.41, fol. 3. Similarly, the Court of Audit reported that most of the salaries of construction managers, book-keepers and civil engineers were greatly inflated ("*fast durchweg sehr überhöht*"): BArch, R2/19957, fol. 48.
37 Geyer, 'Rüstungsbeschleunigung', 157.
38 HHStAW, Abt. 485, 6, Conference with representatives of the *Reichsautobahnen-Direktion* and the OBRs on questions of personnel, finances, construction and operation in Salzburg, 1939.

machines was common practice in the construction sector. Firms that leased machines could then include the costs in their calculations. The machines from all over Germany that were allocated to the *Westwall* construction sites, for example, were normally leased and the OT refunded the costs to the contractors. Between 1936 and 1939, the shortage of building machines led to increasing lease prices. Rents for compressors were reported to be considerably inflated on the *Westwall*. As a reaction to this development, the Reich Price Commissioner enacted a decree on maximum rents for building machines on 16 June 1939. From this date onwards, companies and procurement agencies were no longer allowed to ask for, promise or pay rents that were higher than the official rates.[39] Since the lessors still realised excess profits of up to 50 percent of their turnover in the following years, the Price Commissioner reduced the rates further in 1941 and 1942.[40] It seems, however, that the administration was unable to enforce the regulations comprehensively. In April 1943, the Court of Audit still registered "unjustified increases of building costs due to the arbitrary interpretation and non-application of the regulations". Some companies were singled out and forced to reduce the leasing rates retroactively or pay a fine.[41]

The picture was more heterogeneous in the case of building materials prices. Both on the *Autobahn* and the *Westwall*, Todt's administration was able to negotiate cement prices that were lower than the wholesale price. The General Inspector paid 24 RM per tonne on the *Autobahn*[42] and 26 RM on the *Westwall*.[43] In August 1944, on their construction sites in Kaufering and Mühldorf, the OT paid 38.40 RM and 39 RM respectively to the *Zementgemeinschaft Südwest* in Heidelberg.[44] This was a considerable increase, but still corresponded to the wholesale prices for the Munich area that had been frozen in November 1936 on a level that was considerably lower than in the 1920s and early 1930s.[45] The price development for other materials was more alarming, though. Gravel cost 0.80 RM per tonne in late 1936, 1.02 RM in mid-1938 and 1.32 RM only two months later on the *Westwall* construction sites. Single firms charged the OT up to 1.55 RM for one tonne of gravel.[46] A *Wehrmacht* study comparing the prices for building materials between 1935 and 1938 in the Berlin

39 RGBl. 1939, I, 1043–44.
40 BArch, R 2301/3007, fol. 262.
41 Ibid., fol. 263.
42 BArch, R 50-I/334, fol. 108.
43 BArch, R 4601/3057, GdSt, Wiesbaden, to all OBLs, 2.9.38.
44 BArch, R 50-I/334. In Kaufering and Mühldorf, sub-camps of the Dachau concentration camp, the OT built massive underground bunkers which should protect the production of fighter jets. See Raim, *Kaufering und Mühldorf*; Schalm, *Überleben durch Arbeit?*
45 *Statistisches Handbuch von Deutschland 1928–1944*, 467; *Die Wirtschaftslage des In- und Auslands im Frühjahr 1938 – im Herbst 1938*, Halbjahresberichte zur Wirtschaftslage. 13. Jahrgang 1938/39 (Hamburg: Hanseatische Verlagsanstalt, 1938), 46.
46 Arthur Schweitzer, 'Plans and Markets: Nazi Style', *Kyklos* 30, no. 1 (1977): 94.

area arrived at similar results. While prices for some products like cement, roofing paper, roofing tiles and construction steel remained stable, other materials had become more expensive. For example, prices had risen for timber (+12 percent), bricks (+24 percent), gravel (+13 percent) and sand (+18 percent). The *Wehrmacht* estimated that those price increases were to a great extent responsible for the fact that *Wehrmacht* buildings around Berlin had become 20 percent more expensive within three years.[47]

The building materials prices were closely associated with the costs of transporting those materials to the construction sites. The GB Bau administration identified rising transport costs as one of the key factors when explaining the overall increase in construction prices. As many building materials producers in areas with a high building activity were at almost maximum capacity, procurement agencies often had to order materials from remote regions of the Reich, which naturally led to rising transport costs.[48] And yet again, the OT's billing practice on the *Westwall* intensified the problem. The building companies paid the haulage firms per kilometre, the OT then reimbursed the costs to the building companies, plus a 1.5 percent premium on the transport costs. In November 1938 the representatives of the Court of Audit criticised the fact that the haulage firms thus loaded their trucks more lightly in order to maximise their daily mileage, while the building companies covered up for the haulage firms to make money out of the inflated costs.[49]

For obvious reasons, it is hard to say to what extent corruption or collusive pricing, that is, companies capitalising on the excess demand and agreeing not to underbid each other, contributed to increasing building costs. Contemporary observers, however, were aware of cases of corruption. Fritz Todt and his officials in the *Reichsautobahnen* Corporation and the OT recognised the problem, too. In a letter to the RFM, Todt insinuated in January 1939 that it was impossible to limit Rhine shipping companies' excessive transport prices because "both the Reich Price Commissioner and the Reich Minister of Transport obviously do not have enough competent experts, except those who are again not independent."[50]

However, Todt was obviously unwilling to take a tougher line on corruption. He could not risk his amicable relations with the construction industry without jeopardising his ambitious construction projects. One of the main problems was the

---

47 Geyer, 'Rüstungsbeschleunigung', 153–55.
48 BArch, R 3/Film 77815, Joachim Steffens, 'Das Bauwesen – geschichtlich, technisch und wirtschaftlich gesehen' (Unpublished manuscript, circa 1944), fol. 53.
49 BArch, R 2/19957, fol. 47, Minutes on a meeting between representatives of the German Court of Audit, RFM and *Oberfinanzpräsident* Berlin in Wiesbaden, 17.11.38.
50 BArch, R2/19958, fol. 2, Letter Fritz Todt to *Ministerialrat* Gebhardt, RFM, 18.1.39.
Bundeskriminalamt Wiesbaden, ed., *Wirtschaftsdelikte (einschließlich der Korruption). Arbeitstagung im Bundeskriminalamt Wiesbaden vom 8. April bis 13. April 1957 über Bekämpfung der Wirtschaftsdelikte einschließlich der Korruption* (Wiesbaden: Bundeskriminalamt, 1957), 184.

auditors' relatively low salary and their prospect of returning to positions in private industry after the war. This made officials prone to corruption.[51] More and more cases were reported where construction firms simply handed over their unfinished calculations and invoices to the official supervising their construction project. This official then produced proper documents that complied with the rules of the awarding authority. As a member of an anti-corruption unit of the police noted after the war, the superiors of these officials at first did not see a need to intervene, since the officials offered those paid services in their free time. Of course, this left the door to corruption wide open.[52]

Another example was a ring of 32 construction firms that had rigged prices in the Hanover region and was discovered by the Gestapo in 1939. The case was closed quickly after the companies paid a collective fine of 500,000 RM to the *Winterhilfswerk*. Only one of the firms was prosecuted, because it had sent wine and other gifts to the official awarding the building contracts.[53]

Todt himself complained about collusive pricing in the case of steel work companies delivering bridges to his road construction sites. The state had tried to facilitate the procedure of awarding contracts in this field: the decision on which firm should be requested to make an offer lay with a committee consisting of representatives of the General Inspector, the Reich Ministry of Transport and the Steel Works Association (*Stahlbauverband*). However, representatives of the Steel Works Association constantly leaked committee decisions to the chosen companies, which then prepared overpriced offers. When the awarding authority tried to increase pressure on these companies by inviting offers from other firms, the Steel Works Association made sure that those firms did not underbid the initial offer. "In some cases," Todt reported, "the awarding authorities summoned some firms to their [the awarding authorities', S.G.] offices so as to draw up and submit offers. Here, they were compelled to prepare offers in conclave." However, this led the companies to later claim that they had not been able to study the bidding documents and inspect the building sites sufficiently. Therefore, Todt concluded that in his opinion, "this awarding process has to be rejected, because it is beneath the dignity of both the awarding agency and the entrepreneur."[54]

Problems continued on the *Westwall* construction sites, too. It is no coincidence that Germany's first national anti-corruption unit (*Reichszentrale zur Bekämpfung*

---

51 Streb, 'Preisregulierung', 46–47. See also below, chapter III.1.b, fn. 38.

52 Bundeskriminalamt Wiesbaden, *Wirtschaftsdelikte*, 184. The police officer did not mention names, but his remarks were obviously aimed at Todt and his staff. Cf. the half-hearted 1937 decree by the General Inspector, which describes this practice in detail, but still does not ban it categorically: HHStAW, Abt. 485, 251, Decree 14/37 of the GdSt, 7.4.37.

53 Schweitzer, 'Plans and Markets', 103–4.

54 BArch, R 2301/7144, fol. 134–37, Todt to *Oberregierungsrat* Burandt, RWM, and *Oberregierungsrat* Graf Yorck von Wartenburg, Reich Price Commissioner, 20.3.39.

von Korruption beim Reichskriminalpolizeiamt) was established in Wiesbaden in 1938. The unit not only denounced corruption, but uncovered hundreds of cases of deceptive business practices. However, not a single trial was opened. Instead, Todt held a protecting hand over "his" entrepreneurs and, as he had done in the case of the 32 construction firms from the Hanover region, he stopped all investigations upon a collective and voluntary donation of 500,000 RM to the *Winterhilfswerk*.[55]

d The Limitations of the Official Wage and Price Statistics

In the light of what has been said in the previous two parts of the chapter, it may surprise that the wage rises that the SRA registered in construction in the 1930s and 40s were lower than in other parts of the German economy. Similarly, the official construction price index only registered a moderate price increase before 1942. Were raising wages and prices not as big a problem after all?

In the case of wages, Rüdiger Hachtmann and Tilla Siegel maintain that the problem was not as great as often assumed. According to Hachtmann, the workers in the construction industry were an underprivileged group.[56] Bereft of any bargaining power by high unemployment rates and the dissolution of their unions, they saw their wages massively reduced in the early 1930s. Starting from this very low crisis level, earnings in the construction sector then rose by 14.9 percent between 1936 and 1944, which was below the average rise in the economy.[57] According to the Statistical Office, in 1944 brick masons received less than two thirds of the hourly wages they had earned in August 1929.[58] The wages of *Tiefbau* workers were even reported as declining again after 1940 (Table 8).[59]

Figure 7 shows the massive fall in wages in the construction sector during the economic crisis, both in terms of standard wage rates and actually paid wages. After the National Socialists came to power, the standard wage rates were frozen at crisis level and hence were no longer able to reflect the actual development in the

---

55 Bundeskriminalamt Wiesbaden, *Wirtschaftsdelikte*, 182–85. A contemporary British account mentions cases of fraud that the Gestapo uncovered: "As a result several army generals were withdrawn and over thirty contractors were arrested. This occurred at the end of July, 1939. The charge against these men was 'unintentional sabotage'. They came before the People's Court, which usually tries spies and traitors. All received heavy sentences.": James Eastwood, *The Maginot and Siegfried Lines: Walls of Death* (London: Pallas Publishing, 1939), 50. The author, however, does not provide any documentation and I was not able to find evidence in the literature or sources to confirm this account.

56 Hachtmann, *Industriearbeit*, 135. Similarly: Siegel, *Leistung und Lohn*, 127, fn. 5.

57 Gerhard Bry, *Wages in Germany 1871–1945* (Princeton: Princeton University Press, 1960), 250.

58 Hachtmann, *Industriearbeit*, 134.

59 Note that the wages specified here are in current prices. The picture would look even gloomier when considering the inflation.

Table 8: SRA statistics on average hourly earnings in construction in RPf., current prices, 1933, 1936, 1938–44. Figures for Germany with its shifting borders.*

1933	1936	1938	1939	1940	1941	1942	1943	1944
66.8	71.6	73.9	76.8	78.1	83.1	79.6	82.0	82.3

*Siegel, *Leistung und Lohn*, 290; Bry, *Wages*, 250.

Figure 7: Hourly standard wage rates and actual wages in construction according to the SRA, in Rpf., current prices, 1929–44. Figures for Germany with its shifting borders. Source: For actual hourly wages, see Hachtmann, *Industriearbeit*, 106–7. For standard wage rates, see the respective volumes of the statistical yearbooks of the German Statistical Office.

following years. Officially, they were minimum rates, but in 1932, authorities tolerated construction companies paying their workers below the standard wage rate. No data are available on the level of actual wages in construction for 1933 and 1934, but as evidence from other industry branches suggests, workers were forced to accept extremely low wages in those years, too. One notorious example were the starvation wages that the former unemployed received on the *Autobahn* construction sites. According to the SRA, the wages for skilled and unskilled workers were slightly above the agreed rates from 1935 on, and for *Tiefbau* workers from 1938 on.

At first sight, these figures seem to conflict with the countless contemporary complaints about rising wages in the construction sector. Primarily citing reports from the Trustees of Labour, Timothy Mason has argued that the National Socialist state lost control of the wage level in construction from 1936 onwards.[60] How can this be reconciled with the hourly real wages reported by the SRA (see above, Figure 7 and Table 8)? The SRA's own explanation was that its highly aggregated figures did not show the changes in the regional distribution of construction activities: the *Wehrmacht* and armaments industry built mainly in rural areas, where standard wages were lower. Building activity in regions with lower wages was thus growing faster than construction in the cities where wages were higher.[61] Moreover, the figures in Table 8 and Figure 7 refer to Germany in its shifting borders. With Austria, the Sudetenland and the former Polish territories that were annexed by the Reich, areas with lower wages had become part of what was then called "Greater Germany". This slowed the overall wage increase down. However, this did not mean that *individual* wages and the costs for single construction projects did not rise.

Apart from this explanation, however, there is some evidence suggesting that the SRA's data on the wage level lost some of their information value in the second half of the 1930s. Two major factors caused much of the discrepancy between these accounts and the wages that the SRA registered, the first being considerable control deficits. Information asymmetry was particularly high in the fragmented construction sector. German authorities had to rely on the companies reporting correct wage bills, but were only able to check the accuracy of this information to a limited degree.

The second and probably more important factor were the rewards and allowances. The SRA did not include the numerous allowances and premiums that were a common part of earnings in the construction sector in its statistics.[62] Separation allowances, for example, were paid out to 40 percent of construction workers in 1943–44. This was by far the highest share of any branch of the economy.[63] An overview of the wages on the *Westwall*, compiled by the *Wehrmacht* in late 1938, illustrates the gap between standard wage rates and actual wages (Table 9). The *Wehrmacht* pointed out that allowances, overtime pay and night shift premiums led to wages that were up to 100 percent higher than the base pay. The SRA, however,

---

60 Mason, *Arbeiterklasse und Volksgemeinschaft*, 500.
61 Document 15, Report of the Statistical Office on the development of actual earnings, 1936, in: Mason, 238–46, 242. Document 50, Report of the Statistical Office on the development of actual earnings, fourth quarter of 1937, in: Mason, 412–21, 416–17.
62 Document 132, Report of the Statistical Office on the development of actual earnings, second quarter of 1938, in: Mason, *Arbeiterklasse und Volksgemeinschaft*, 770–79, 775. Mason, 290, fn. 22.
63 Siegel, *Leistung und Lohn*, 291.

registered only the *Westwall* base pay, including overtime pay. It was lower than the average wage in the German construction sector in 1938.[64] Thus, the SRA statistics reflected neither the massive wage rise on the *Westwall* nor the general trend in the sector all over Germany.[65]

Table 9: Gross hourly wages in Rpf. without allowances, according to the *Westwall* pay scale, compared to gross hourly wages including allowances.*

	Gross hourly wage including allowances				Gross hourly wage without allowances
	8-hour shift	10-hour shift	12-hour shift	Nightshift	
Skilled worker	106	109	111	140	74
Concrete worker	98	101	103	128	67
Tiefbau worker	91	93	95	119	60
Unskilled worker	91	93	95	119	60

*Geyer, 'Rüstungsbeschleunigung', 156. The gross hourly wage without allowances of the *Westwall* pay scale matched roughly the wage of local construction workers that were working an 8-hour shift. Only allowances according to the *Westwall* pay scale are included. Since *Westwall* workers came from all over the country, especially separation allowances increased their wages. Voluntary gratifications by entrepreneurs are not included. For a list of the allowances and premiums according to the *Westwall* pay scale of 15 August 1938, see Gross, *Westwall*, 246–47.

Rüdiger Hachtmann has already pointed out that the industry used bonuses and allowances to circumvent the state's wage regulations.[66] However, he is right to emphasise that they can hardly be compared to regular wages as the workers were not legally entitled to receive them. The grants and bonuses could be cancelled any time, and thus they should not be interpreted as a strengthening of the working class.[67] Nonetheless, they had a *de facto* effect on construction prices, which is what we are interested in in this context.

It must also be stressed that the rising hourly wages in construction did not necessarily mean that the late 1930s were golden years for German construction workers. They paid for their higher earnings with overtime work, their health and the loss of their participation rights. Apart from some excesses, like the high sales figures for pralines and sparkling wine reported from the *Westwall* regions, the

64 Cf. Hachtmann, *Industriearbeit*, 106–7.
65 Mason, *Arbeiterklasse und Volksgemeinschaft*, 1256.
66 Hachtmann, *Industriearbeit*, 122–24. Cf. also: Scherner, 'Keine primäre moralische Verantwortung?', 171.
67 Hachtmann, *Industriearbeit*, 124.

"myth of the good life" (Christoph Buchheim) under the National Socialist regime has been thoroughly debunked.[68]

Similarly to the official wage statistics, also the information value of the SRA's index of construction prices is questionable (Table 10).[69] The index was constructed by combining prices for certain building materials and wages, a procedure with some serious shortcomings. First, it referred to prices in residential construction only, a sector that rapidly lost importance after the mid-1930s.[70] The development presumably depressed building prices in this field of construction. Second, the index did not reflect the increasing transport costs. The SRA calculated based upon local prices; however, local producers were often no longer able to supply all the construction sites in their area. Hence, materials had to be ordered wherever there were free capacities, even if this meant transporting them over long distances.[71] Third, and maybe most importantly, the index's wage factor was based on the standard rates, excluding the allowances, which did no longer reflect the actual wage costs in the construction sector.[72]

Table 10: Construction price index of the SRA and GB Bau/Scherner, 1928/33–44, 1938 = 100.*

	1928	1933	1934	1935	1936	1937	1938	1939	1940	1941	1942	1943	1944
SRA	*129*	*93*	*97*	*97*	*97*	*99*	*100*	*101*	*103*	*108*	*117*	*119*	*121*
GB Bau/ Scherner	*129*	*93*	*97*	*96*	*97*	*99*	*100*	*103*	*107*	*133*	*154*	*166*	*172*

*SRA construction price index: *Statistisches Handbuch von Deutschland 1928–1944*, 462. GB Bau/ Scherner construction price index, see appendix, Table A1.2.

An audit of the *Reichsautobahnen* carried out by the OBR Cologne, for example, examined 255 contracts and revealed considerable price increases between 1934 and 1940 (Table 11). Assuming that the numbers for the OBR Cologne were similar to those of other OBRs – and there is no reason why they should not have been – and given the fact that approximately 10 percent of German construction investments

68 Christoph Buchheim, 'Der Mythos vom "Wohlleben". Der Lebensstandard der deutschen Zivilbevölkerung im Zweiten Weltkrieg', *Vierteljahreshefte für Zeitgeschichte* 58, no. 3 (2010): 299–328.
69 Scherner, 'Preparation', 466.
70 Cf. above, Figure 4.
71 Steffens, 'Bauwesen', fol. 53.
72 In 1938, also the *Deutsche Bau- und Bodenbank* doubted the reliability of the official construction price index: Deutsche Bau- und Bodenbank, 'Die Entwicklung der deutschen Bauwirtschaft im Jahre 1937', 1938, 3.

between 1934 and 1940 were spent on the *Autobahnen*, this alone would have been sufficient to show that the official construction price index could hardly have been correct.

Table 11: Price increases 1934–40, OBR Cologne.*

Type of work	Price increase
Overpasses and underpasses	
Excavation for foundations	+ 156 %
Foundations of tamped concrete	+ 51 %
Rising abutments and wing walls of tamped concrete	+ 35 %
Earth moving works	
Topsoil	+ 114 %
Earth moving	+ 162 %
Compacting	+ 179 %

*BArch, R 2301/7174, fol. 271–72. Calculation of price increases for overpasses and underpasses based on 175 contracts, for earth moving works on 80 contracts.

One of the leading experts on the German construction sector, *Oberregierungsbaurat* Joachim Steffens of the office of the GB Bau, criticised these shortcomings and calculated construction prices based on own estimates for the years 1939 to 1943 (Table 10).[73] He also registered price increases due to decreasing work performance among all groups of workers. More men were necessary to achieve the same output: wage bills and construction prices were rising, even without individual wages necessarily rising.[74] Steffens' explanation, however, is only partly convincing, which may be explained by the fact that he at the time still planned to publish his manuscript. Steffens attributed the decrease in performance to the fact that Germans from different regions as well as foreigners were now working together on the construction sites. He claimed that different cultural backgrounds, habits and language problems led to conflict and discontent. These factors certainly played a role, but it is obvious that Steffens avoided calling a spade a spade. Much more importantly, workers began to work slower as a result of being constantly overworked and the workforce

73 Joachim Steffens, born in Berlin-Charlottenburg in 1897, was a close co-worker of Fritz Todt. Initially working in the field of hydraulic engineering, he worked for the *Reichsautobahnen* Corporation since 1934 and became later a leading official in the office of the GB Bau. He was a member of the NSDAP since April 1934. See BArch, R 4601/11395 and Botzet, 'Generalbevollmächtigter', 119.

74 Dyckerhoff & Widmann report this development. To generate 1 million RM in turnover, the company needed 132 workers in 1935, but 178 workers in 1940. This meant a price increase of 40 percent without the individual wages necessarily being higher: DMM, FA 010/100, Capacity of the construction industry. Comment on the statistics 1935–1940, 4.7.41, fol. 3–4.

in the sector became older, as the *Wehrmacht* drafted the young and able men. Instead, Steffens points out that far from home, workers were not always able to eat what they were used to. In the case of the German workers, references to diverging eating habits – "one region prefers a diet based on potatoes, while in other regions, meals are made of grain products like flour and semolina" – just appear odd. In the case of malnourished foreign workers or POWs, such comments were simply cynical.[75]

2 The Struggle for Wage and Price Controls

After having investigated the main reasons for why construction became more and more expensive over the course of the 1930s, the following chapter will focus on the National Socialist attempts to monitor, control and steer the German construction sector. The regulations defined the framework in which firms acted and in which the negotiations between procurement agencies, such as the OT, and their contractors took place. The aim is to assess German construction firms' entrepreneurial room for manoeuvre under the National Socialist rule. Moreover, the question will be where procurement agencies like the OT stood in this conflict: took they sides with their contractors or with the state agencies that tried to monitor the sector? I argue that it was not the amount of ordinances, laws and regulations alone that defined the construction industry's room for manoeuvre. Rather, the crucial question is to what degree the state was actually able to implement and enforce its policies.

Part a will map the most important attempts to contain the rising wages: the fixing of maximum wages, as well as the rules laid down in the Construction Price Ordinance (*Verordnung über die Baupreisbildung*) of 16 June 1939 and in the War Economy Ordinance (*Kriegswirtschaftsverordnung*, KWVO) of 4 September 1939. Part b will then focus on OT's *Westwall* construction sites. The Organisation was regarded as one of the main catalysts of rising construction costs, especially because of its use of cost-plus contracts at the *Westwall*. The change to fixed-price contracts was an attempt to stop the explosion of costs and entrepreneurial profits in the West. Both parts will highlight the limitations of the state's price controls and the companies' economic room for manoeuvre that resulted from these limitations. By providing an example from construction projects for the private industry, part c will show that control problems and rising costs were not just a problem in public construction projects.

75 Steffens, 'Bauwesen', fol. 50–51.

a Attempts to Limit Wage Rises

The National Socialist authorities were aware of the fact that rising wages contributed to rising costs in construction. A key role therefore fell to the Trustees for Labour. However, although they registered the rising wages, they initially lacked the means to stop the development. The problem was that the official tariff rates – as mentioned before – were minimum rates. In Lower Saxony, a quarry firm working on an *Autobahn* bridge paid their workers 20 RM per day. This upset the employees of neighbouring firms, however, "the Trustee for Labour cannot take action against these phenomena, because for the time being, excess wages are not prohibited. Neither expect the involved parties effective or fast success from the intervention of the pricing authorities, since re-audits of the calculation are complex and time-consuming."[76]

National Socialist authorities realised that any discussion of the trend on the basis of official tariff rates would have little to do with reality. The obvious alternative was the fixing of maximum wages. On 25 June 1938, Göring, in his function as head of the Four Year Plan, authorised the Trustees for Labour to fix maximum wages.[77] Whether the trustees introduced those maximum wages, however, depended on the region and could sometimes even differ between single construction sites. It turned out to be difficult to assess the effects of this measure. Some trustees reported that excess wages had been reduced in some firms and the constant migration of workers between construction sites had lessened. At the same time, the problems and limitations of the maximum wage rates shone through in the reports. Companies continued to circumvent the wage freeze by paying generous allowances.[78] There was also a fear that the workforce's performance would drop, and corresponding cases were in fact documented. Most importantly, the Trustees for Labour were forced to admit quite openly that they were unable to control whether their regulations were observed: "It is obvious that the implementation of the numerous measures can be monitored only to a limited extent by the Reich Trustee himself. The support of the DAF is thus necessary. Reports on unjustified and forbidden wage rises, however, are only seldom made by the DAF."[79] The DAF was indeed a rather unreliable partner in this respect. Instead of supporting the struggle for stable wages and prices, the DAF often sided with the German workforce by making "tactical concessions to the workforce", if only to strengthen its position

76 Document 96, Excerpt from the monthly reports of the Trustees for Labour, January and February 1938, in: Mason, *Arbeiterklasse und Volksgemeinschaft*, 606–21, 610.

77 RGBl. 1938, I, 691.

78 Document 136, Excerpt from the social reports of the Trustees for Labour, third quarter of 1938, in: Mason, *Arbeiterklasse und Volksgemeinschaft*, 787–91.

79 Document 137, Excerpt from the social reports of the Trustees for Labour, fourth quarter of 1938, in: Mason, 792–95, 793.

within the National Socialist state.[80] Interestingly enough, the Reich Economic Ministry (*Reichswirtschaftsministerium*, RWM) and the RAM had already identified the factors undermining the effectiveness of the measure in October 1936, when they had rejected the introduction of maximum wages. They argued that maximum wages would erode the workforce's motivation and that it would be almost impossible to prevent the regulations being evaded: both entrepreneurs and workers had an interest in higher wages, hence, no one would report them to the authorities.[81] Altogether, the Trustees for Labour never fully exploited the competencies they had been given in Göring's decree, and "there can be no talk of wage cuts or even an effective wage freeze" in pre-war Germany.[82]

Attempts were also made to lay down rules on how wages had to be disclosed in construction price calculations. The Construction Price Ordinance of 16 June 1939 prohibited wages that were higher than the official standard wage rates. The usual allowances for the journey to and from the construction site, weekend trips home, separation allowances and the like, had to be shown separately to make them easier to control.[83] However, these regulations had little effect as long as agencies like the *Wehrmacht* continued to nod through offers that included real wages in order to attract scarce labour to their construction sites.[84] As Tilla Siegel has put it, "[. . .] behind the veil of the relatively constant standard wages, a 'realignment of wages' took place until 1938–39, which was not state-driven, but resulted from market forces."[85]

The last major attempt to bring the price spiral back under control, particularly with regard to industries like construction or metal working, came with the War Economy Ordinance of 4 September 1939.[86] The Ordinance aimed to completely

80 Petzina, *Autarkiepolitik*, 166.

81 Document 3, Explanatory statement on the draft of a Second Law for the Regulation of Labour Deployment, 6.10.36, in: Mason, *Arbeiterklasse und Volksgemeinschaft*, 198–216, 203.

82 Mason, 798. An interesting comparison can be drawn with the textile industry, where wages were very low, especially due to the wage discrimination against female workers. Even in this sector, similar symptoms were registered: rising wages, the growing importance of allowances and bonus payments, and the decreasing validity of official wage rates. See Höschle, *Textilindustrie*, 262–66.

83 RGBl. 1939, I, 1041–43.

84 In particular, the Construction Price Ordinance did not apply to cost-plus contracts of public customers. Since the OT at the time used cost-plus contracts, the Construction Price Ordinance had no effect at all on the companies working under the organisation: Botho Bauch, 'Die Durchführung der Baupreisverordnung', *Die Bauindustrie* 8, no. 6 (1940): 121–23; Walter Daub, *OT-Verträge. Vertragsrecht der Organisation Todt, Stand 30. Juni 1942* (Wiesbaden: Schellenberg'sche Buchdruckerei, 1942), 11.

85 Siegel, *Leistung und Lohn*, 60. For a similar interpretation, see Tooze, *Wages*, 260–63.

86 RGBl. 1939, I, 1609–13. For a discussion of the KWVO from the construction industry's point of view, see 'Neue Regelung über Kriegslöhne', *Die Bauindustrie* 7, no. 43 (1939): 1309–11. The following section on the KWVO is largely based on the comprehensive discussion in: Marie-Luise Recker, *Nationalsozialistische Sozialpolitik im Zweiten Weltkrieg* (München: Oldenbourg Verlag, 1985),

subordinate price and wage policies to the war effort. In the field of wage policy, the KWVO ordered a general wage stop. Premiums for overtime and work on weekends and public holidays (usually 25 percent) were cancelled while the workday rose to 10 hours. The Trustees for Labour were authorised to fix maximum wages and reduce particularly high wages. Furthermore, employee salaries were to be cut back, even to the level of October 1936 according to internal plans of the RAM. The savings that the businesses would achieve through the lower wages and salaries had to be used to lower prices. Entrepreneurs could be punished with fines or even imprisonment if they violated the wage and price regulations of the KWVO.

The reports on the KWVO's effects, which came in from companies and party officials alike, were disillusioning. Workforce performance suffered from the increased work hours, reduced wages and the abolishing of overtime pay. Discontent rose among the workforce, as did the number of workers calling in sick. This situation provoked resistance within a broad alliance of industry, DAF, party officials and the War Economy and Armaments Department (*Wehrwirtschafts- und Rüstungsamt*) of the *Wehrmacht*. They feared that the mood of the German workforce would turn in view of the worsening work conditions, leading to a decline in output and delaying important armaments orders. In the weeks of autumn 1939, they argued for the regulations to be revoked. The GB Bau, in turn, tried to reassure construction companies that they had little reason to be concerned. As an official of the GB Bau assumed, the numerous regional regulations based on the decree of 25 June 1938 had already stabilised wages at an acceptable level. Hence, "construction may be one of the sectors of our economy where the wage structure least requires a fundamental modification."[87] The KWVO forbade the circumvention of the wage stop through gratifications and allowances, except when – and this was crucial for the construction sector – "they were customary in the company and the sector".[88]

Over the course of the following months, one part of the Ordinance after another collapsed under industry and party pressure. More radical ideas of rolling back wages to the standard wage rates had already been dismissed in the summer, on the one hand, because accompanying tax laws had not yet been finalised, and on the other hand, because the RAM feared the workforce's discontent. In some branches, real wages had doubled since 1933 while the official standard wage rates had remained more or less stable. In the RAM's eyes, cutting real wages back to the standard wage rates was far too risky in social terms.[89] By January 1940, premiums for work at weekends, on holidays and at night were being paid out again, as was

26–58. Cf. also Dietrich Eichholtz, *Geschichte der deutschen Kriegswirtschaft 1939–1945. Band I: 1939–1941* (1996; repr., München: K. G. Saur, 2003), 70–79.

87 'Was bringen die neuen kriegswirtschaftlichen Verordnungen für die Bauwirtschaft? Eine Übersicht, abgeschlossen am 10. September 1939', *Die Bauindustrie* 7, no. 37 (1939): 1198–1200.

88 RGBl. 1939, I, 1612.

89 Recker, *Sozialpolitik*, 28–29.

overtime pay. Firms were allowed to pay wages at the level of October 1939. In construction, this meant the level that had been reached after several years of a massive building boom. This also applied to salaries, and thus there was no longer any talk of reducing them to their 1936 levels. As a means of impacting and slowing down the price rises development by controlling wages, the War Economy Ordinance was little effective. An industry, party and military alliance had managed to avert what could have become the cornerstone of a state regulation of booming sectors like construction.

Of course, the Trustees for Labour still had the authority to fix maximum wages, and every wage rise had to be approved by them. The problem was not just the lack of regulations and sanctions, it was rather that the authorities did not have the capacity to record and monitor the actual wages in the first place. The Trustees for Labour admitted that they had great difficulties in registering the wages actually paid because companies used fake payrolls on a large scale. Firms recorded the official wage rates in their payrolls, while they actually paid out higher wages to their workers. The patchy book-keeping of smaller companies in particular made it difficult to uncover manipulations. Larger companies tended to use generous rewards and benefits to retain workers' loyalty.[90] The trustees complained that public customers tolerated and even encouraged companies to circumvent the wage stop in this way, because they did not want to jeopardise the timely completion of their projects.[91] Yet another problem was the changing classification of different types of skilled and unskilled workers, which became more fuzzier over the course of the 1930s. In order to circumvent the wage stop, companies reclassified unskilled workers as skilled workers. The Trustees for Labour normally did not have the capacity to control whether a worker's promotion was justified or just a hidden wage rise.[92]

In general, the available sources give the impression that the authorities only seldom prosecuted violations of the wage and price stop, resorting to single, exemplary punishments.[93] Companies were warned and asked to correct their prices instead of being placed under investigation, as in the case of four *Reichsautobahnen*

90 Hachtmann, *Industriearbeit*, 122–23.

91 Document 150, Excerpt from the social reports of the Trustees for Labour, fourth quarter of 1938, in: Mason, *Arbeiterklasse und Volksgemeinschaft*, 859–74, 862.

92 Hachtmann, *Industriearbeit*, 59.

93 Mason, *Arbeiterklasse und Volksgemeinschaft*, 799–800. I have reviewed several volumes of the bulletin of the Reich Price Commissioner. My findings support Mason's interpretation. In his *Mitteilungsblatt des Reichskommissars für die Preisbildung*, the Price Commissioner regularly published convictions for price stop violations. The judgments mostly dealt with cases of individuals selling goods on the black market. The situation resembles that of the Social Honour Courts (*Soziale Ehrengerichte*) that existed from 1934 onwards to provide a way for the Reich Trustees for Labour to take action against workers and entrepreneurs. The trials remained few in number and sentences were mostly mild: Kranig, *Lockung und Zwang*, 232–41; Diehl, *Wirtschaftsordnung*, 80.

contractors that had submitted offers with prices 30–50 percent above the average.[94] In the case of the building firm Liebergesell & Lehmann that had enticed away a worker from another firm on the *Autobahn* Vienna – St. Pölten by paying him 1.20 RM per hour instead of the standard wage of 0.86 RM, Fritz Todt ordered that the worker be fired. The company, however, was not prosecuted.[95]

b From Cost-Plus to Fixed-Price Contracts

As we have seen, the National Socialist state was able to limit construction prices to a minor degree by controlling the individual contributing factors, most importantly wages. Hence, a second option was to address the question of contracts. Maybe it was possible to use contract types that motivated building companies to reduce their costs. This brought the OT in the focus: the *Westwall* was Germany's most prominent construction project in the late 1930s and at the same time, its exploding costs were notorious. The cost-plus contracts, which offered the construction firms no incentive to reduce costs, were regarded as a prime example of bad practice. The OT used these contracts as they did not require complicated pre-calculations, so works could start immediately. However, as every cost increase resulted in rising profits for the companies, the latter had no incentive to work efficiently. The result was construction firms covering up excess transport costs (see above, chapter III.1.c) or manning the building sites with too many employees to increase the wage bill.[96]

A few weeks after the start of construction works, an alliance of the German Court of Audit, the Reich Price Commissioner and the Ministry of Finance already tried to intervene with the OT to stop the use of cost-plus contracts and limit the enormous profits of *Westwall* contractors. It was the *Wehrmacht* that triggered this intervention on 23 July 1938, when the pioneers in Freiburg asked the Price Commissioner for help, because the OT was using cost-plus contracts with excess prices to entice their suppliers away.[97] The information that prices on the would "increase costs by 300 % on average" alarmed even the Party Chancellery. A report compiled by party comrade Kehrwald, who had been visiting the *Westwall* headquarters in Wiesbaden, stated that entrepreneurs were playing the *Wehrmacht* off against the OT. Price controls were almost impossible due to lack of time and auditors. "Acquisitive contracts"

94 BArch, NS 26/1187, fol. 416–23.
95 Ibid., fol. 222–23.
96 Document 150, Excerpt from the social reports of the Trustees for Labour, fourth quarter of 1938, in: Mason, *Arbeiterklasse und Volksgemeinschaft*, 859–74, 868. In January 1939, Todt stated that while at the moment, there was work for 80,000 men on the *Westwall*, the companies were deploying more than 100,000: BArch, NS 26/1190, fol. 5–6, Todt to *Landesbaurat* Schmies, 19.1.39.
97 BArch, R 2/19957, fol. 1–8, Report by Kehrwald, NSDAP, 2.9.38.

were fostering the will to increase costs by deploying more workers than necessary. At the same time, the work performance would surely suffer because the entrepreneurs had no financial interest in urging their workers on.[98] The OT was granting the German construction industry huge profits at the expense of the Reich budget. The Court of Audit, the RFM and the Price Commissioner shared Kehrwald's concerns, and during the autumn of 1938, they tried to press the OT to abolish the cost-plus contracts.[99]

Todt and his staff had always tried to maintain close and amicable relations with the construction industry, but it was in the summer and autumn of 1938 that they definitely forged an alliance to ward off the attacks of officials concerned about prices, wages and budget balance. Willi Henne, head of the General Inspector's office in Wiesbaden, defended contracts, prices and profits on the *Westwall* against the critics in the financial administration.[100] Todt himself was furious about the Kehrwald report in particular. In a letter to the Staff of the Deputy Führer, he implied that Kehrwald had spied on Todt's officials in Wiesbaden and that he had – "instead of taking the comradely path of National Socialist cooperation" – passed on reports without having spoken with Todt. The General Inspector reserved his right to talk about Kehrwald with the head of the Party Chancellery, Martin Bormann.[101] Wiesbaden also tried to block inspections by the Price Commissioner. In a meeting on 24 October 1938, an official of the Reich Price Commissioner declared resignedly that Fritz Todt had been authorised to carry out the fortification works in the West by the "*Führer*" and Chancellor himself: "The General Inspector makes the contracts with the entrepreneurs. The agreed prices are beyond the control of the Price Commissioner. [. . .] The attempted intervention of the Price Commissioner in the Westwall project was rejected by Ministerialrat Schulze-Fielitz from the office of the General Inspector. Therefore, the Price Commissioner cannot comment on the costs of the Westwall. Particularly, it cannot be said whether the agreed prices are fair."[102] Todt stopped another initiative as well. At the end of October, Wigru Bau had formed a self-regulating committee to solve the most pressing problems on the *Westwall* construction sites.[103] There were plans that the committee's main task should be to collect unjustified profits from the entrepreneurs. However, Todt personally prohibited such a collection of excess profits.[104]

---

98 Ibid.

99 See the collection of documents in: BArch, R 2/19957.

100 BArch, R 2/19957, fol. 58–71, Memorandum on a meeting with an official of the GdSt in Wiesbaden on the organisation and pricing at the *Westwall*, 18.11.38.

101 Ibid., fol. 386–87, Letter Fritz Todt to *Ministerialdirektor* Sommer, Staff of the Deputy Führer, 2.12.38.

102 Ibid., fol. 9–10, Memorandum on a meeting at the office of the Reich Price Commissioner, 24.10.38.

103 Ibid., fol. 46–53, 51, Memorandum on a meeting with the representative of the German Court of Audit regarding the pricing at the western fortifications, 17.11.38.

104 Ibid., fol. 92–104, 94, Memorandum on a meeting between representatives of the Court of Audit and the GdSt, 21.11.38.

Nevertheless, Todt was also willing to make some concessions and called in officials of the German Court of Audit. They were supposed not only to control calculations and billing by the entrepreneurs, but also to advise the nascent OT and help to establish the administrative structures necessary for proper financial accounting.[105] Although there was now a satisfactory number of auditors at the Westwall (at least for a short period in the last weeks of 1938), the results of the audits were moderate, as the RFM had to admit.[106] It is unclear to what degree Todt used the authority he derived from Hitler's personal order to influence the auditors sent to Wiesbaden. But after months of confrontation, it is indeed conspicuous how cautious the reports suddenly sounded: "[. . .] The office of the General Inspector in Wiesbaden cannot be blamed at all."[107] The head of the Court of Audit's delegation to Wiesbaden, Friedrich Enderlin, conceded "that many drawbacks had emerged at the Westwall. The paid prices considerably exceed the normal rate." Yet his conclusion was remarkable: "In the light of the importance of the western fortifications, which have already been effective during the Czech crisis, one had to take the position that the Westwall expenses were in fact already saved costs of war."[108]

Around the same time, however, critics had managed to build enough pressure on the OT with regard to their most important demand, the abolishment of the cost-plus contracts. The OT was no longer able to defy attempts at reform, especially since the financial authorities seemed to have gained the support of Göring, the

105 The latter task became in fact a major field of activity for the Court of Audit during wartime, when various institutions were established in the annexed and occupied territories, like the Reich Commissariats or the OT's *Einsatzgruppen*. On this topic, see the results of a research project on the German Court of Audit in occupied Europe, conducted in the late 1980s at the Freie Universität Berlin: Gilles and Otto, 'Beuteverwaltung'; Franz-Otto Gilles and Gerhard Otto, eds., *Verwalteter Beutepartikularismus. Finanz-, Verwaltungs- und Wirtschaftskontrolle und nationalsozialistische Besatzungspolitik in den von Deutschland besetzten Gebieten: Ein Tagungsbericht* (Berlin: Freie Universität Berlin, 1991); Franz-Otto Gilles, *Effizienz ohne Ethik? Zur Praxis der Verwaltungs- und Finanzkontrolle in den während des Zweiten Weltkrieges von Deutschland annektierten bzw. besetzten Gebieten*, Berliner Arbeitshefte und Berichte zur sozialwissenschaftlichen Forschung 78 (Berlin, 1992).

106 BArch, R 2/19957, fol. 203–10, Summary of the official trip 'western fortifications' by *Ministerialrat* Gebhardt, RFM, 29.11.38. Facing critique from the NSDAP Party Chancellery, Todt tried to give the impression of a more successful auditing process: "[. . .] Overall, things have gone properly and the few rascals ["*die paar Spitzbuben*"] who had tried to make profits at the expense of the Reich were caught red-handed and led away by the secret police.": Ibid., fol. 386. As lists with auditing results show, far from all firms with excess profits seem to have been fined. At least one business owner, Carl Rose of Carl Rose Tief- und Betonbau GmbH, Berlin, was arrested in connection with the *Westwall* construction. Another business owner from Berlin, Ernst Rader, was arrested for faulty book-keeping, irregularities and false statements. His firm had worked at the Nuremberg party grounds and the new Reich Chancellery. For a reference to the lists, see appendix, fn. 34. The two arrests are mentioned in BArch, R 2301/7187, fol. 41 and 179.

107 BArch, R 2/19957, fol. 204.

108 Ibid., fol. 53.

head of the Four Year Plan. In June, Göring had still declared that "money *does not play such a decisive role* under the current situation [. . .]".[109] In his first speech in front of the Reich Defence Council (*Reichsverteidigungsrat*) on 18 November 1938, however, he made clear that the use of cost-plus contracts on the *Westwall* was an exception and that these contracts should be banned in future.[110] Too serious had the monetary situation of the Reich become in 1938.[111] Therefore, in late 1938, negotiations on the introduction of fixed-price contracts started between the OT and the *Westwall* entrepreneurs. The OT presented the entrepreneurs with two new contracts, one fixed-price contract and an "improved cost-plus contract" (*verbesserter Regievertrag*). The latter was only to be used wherever the introduction of a fixed-price contract was not possible. It spread the financial risk between the customer and the contractor and included incentives for working more cost-efficient. The new contracts were introduced on 1 January 1939.

According to Schulze-Fielitz, of the 13,000 bunkers that the OT had to build on the *Westwall*, 9,500 had already been finished by 1 January 1939. In the case of 500 of the remaining 3,500 bunkers, entrepreneurs refused to sign fixed-price contracts. Instead, the OT offered them improved cost-plus contracts.[112] Obviously, no company was forced into a contract against its will. As the president of the German Court of Audit recalled in 1944, some of the *Westwall* construction projects of 1939 were so urgent that the OT accepted companies starting to build without a contract after negotiations had been inconclusive. The OT reimbursed their costs later.[113] Moreover, the crucial transport contracts, identified as one of the main reasons for price increases in the sector, remained unchanged.[114]

The Court of Audit was thus sceptical about the new contracts' cost-reducing effect. The court's delegate in Wiesbaden, Enderlin, estimated that the price per cubic metre of concrete would also be 200 RM under fixed-price contracts, while the army's fortification pioneers in the east paid 105 RM. Some years ago, the price would have been 65 RM.[115] The problem was that the entrepreneurs calculated their

109 *Die Regierung Hitler. Band V: 1938*, Akten der Reichskanzlei. Regierung Hitler 1933–1945 (München: Oldenbourg Verlag, 2008), 463, fn. 2 (original emphasis).

110 Document 152, Speech of Göring in the first meeting of the Reich Defence Council, 18.11.38, in: Mason, *Arbeiterklasse und Volksgemeinschaft*, 908–33, 923. This shows that Göring had been aware of the problem of cost-plus contracts for almost three years, before he decreed their general abolishment in late 1941. Cf. chapter I.1.c.

111 Tooze, *Wages*, 285 ff.

112 BArch, R 2/19957, fol. 368–70, Memorandum on a meeting at the RFM on 6 January 1939, 7.1.39.

113 BArch, R 2301/3015, fol. 152, Memorandum of the President of the German Court of Audit, attached to the comments on the Reich budget account for the budget year 1940, 4.5.44.

114 BArch, R 2301/3007, Memorandum of the President of the German Court of Audit on the Reich budget account for the budget year 1939, 19.4.43, fol. 261.

115 BArch, R 2/19957, fol. 52.

new prices on the basis of the low work performance of the cost-plus period and added high-risk premiums for all imaginable difficulties. As long as the building firms were able to convince the procurement agencies to accept their calculations and bear the expenses, fixed-price contracts were unable to stop the upward spiral of prices. The whole industry was working at full capacity and the OT tried to hire any firms available, leading to a situation with virtually no competition. Some firms that were busy purposely submitted tenders with excessive prices, in order to be turned down by the OT. To their surprise, the OT accepted those tenders and awarded contracts to the firms.[116]

As early as in December, the RFM suggested adding a clause to the new contracts, according to which the construction firms could be made to repay excess earnings if an audit discovered that the prices had been too high. The firms flatly rejected the proposition and were supported in this by the OT.[117] However, they intimated that they might agree to a clause that would limit their profits on the *Westwall* construction sites to 10 percent of turnover.[118] Given the fact that the profit rate under cost-plus contracts had been 4 percent,[119] this meant that the new fixed prices could be more than halved compared to the old cost-plus prices without the firms making less profit. Such a clause would require laborious price audits, though, and was therefore unpopular both with the firms and the involved authorities. The companies were instead willing to reduce their prices to a certain extent. Hence, at the suggestion of the OT, the RFM, the Court of Audit and the Price Commissioner agreed in a meeting on 27 March 1939 to abandon the idea of a contract clause that would limit profits.[120] In exchange, the OT managed to reduce the fixed prices for the last 3,000 bunkers by one third. The time to concrete one bunker decreased from 101 to 72 hours on average.[121] This certainly showed the potential of fixed-price contracts. At the same time, however, the General Inspector agreed to cushion the difficulties of the transition from cost-plus to fixed-price contracts by only gradually reducing prices to this new level. The firms received a premium up to five, four and three RM per cubic metre concrete in January, February and March, respectively. The actual, negotiated fixed prices on the *Westwall* thus did not take effect before 1 April 1939.[122] This was two months before the OT began to reduce construction activity on the *Westwall* for good. The introduction of fixed-price contracts

116 BArch, R 2301/3007, fol. 261.
117 BArch, R 2/19957, fol. 368.
118 Ibid.
119 BArch, R 2301/3015, fol. 152.
120 BArch, R 2/19958, fol. 238–39, Memorandum on a meeting at the Reich Ministry of Finance on 27 March 1939, 28.3.39.
121 BArch, R 2/19957, fol. 368.
122 BArch, R 2/19958, fol. 238.

on the *Westwall* is therefore hardly a success story about rationalisation, price control and state action against the construction industry.

As in the case of building cost factors, the National Socialist authorities were forced to recognise that their attempts to limit prices in the construction sector by means of fixed-price contracts were failing. The precondition for reduced prices was either competition, which was eliminated under the excess state demand, or thorough price controls, which were impossible due to the lack of auditors. Hence the fixed-price contracts – although they were considered superior to cost-plus contracts – were also doomed to failure under the given circumstances.[123] Moreover, any regulations prohibiting price increases, like the Construction Price Ordinance of 16 June 1939, the War Economy Ordinance of 4 September 1939 or the LSBÖ (*Leitsätze für die Preisermittlung auf Grund der Selbstkosten bei Bauleistungen für öffentliche Auftraggeber*) of 25 May 1940,[124] remained ineffective as long as public customers were willing to pay any price to secure the realisation of their construction projects.[125]

The OT in particular constantly boycotted the financial authorities' attempts to bring the construction industry to heel. For the construction industry, a considerable economic room for manoeuvre thus resulted from the fact that procurement agencies like the OT served as a buffer between their contractors and state authorities such as the Price Commissioner and the Court of Audit. What prohibited a tight regulation of prices in construction was the conflict of interests between procurement agencies and their principals. This PA problem was more important than the one between procurement agencies and their contractors. For construction firms, working under the OT on the *Westwall* meant a protection rather than loss of their economic scope of action.

c Construction Procurement in the Third Reich: an Example from 1942

In the 1930s and 40s, the German Court of Audit in Potsdam was fighting a losing battle. An example from 1942 shows how wide the gap between ambitions and reality could be when it came to monitoring construction activities in National Socialist

123 R 2301/7124, fol. 275–83. In early 1938, the Reich Chancellery had already been sceptical about whether it was possible to control calculations before construction works started to such a degree that it would solve the problem of increasing building prices: Document 129, Note from the Reich Chancellery, 5.1.38, in: Mason, *Arbeiterklasse und Volksgemeinschaft*, 754–55. This step, however, would have been crucial to the functioning of fixed-price contracts.

124 RGBl. 1940, I, 851–59.

125 For the permanent complaints about procurement agencies rubber-stamping contracts that included excessive wages and prices, see Mason, *Arbeiterklasse und Volksgemeinschaft*, 352, 609 fn. 8, 611–12, 640, 657, 766.

Germany. One of the men in Potsdam who tried to document and oppose the development described in the previous parts of this chapter was *Oberregierungsrat* Friedrich Wilhelm Elfert, head of Court of Audit's department XI 1. The department was was responsible for auditing publicly financed construction projects. On 20 December 1942, Elfert received a confidential letter from a man called Erich Gebauer, who until June 1942 had worked as an internal controller for the Deutsche Bergwerks und Hüttenbau GmbH Berlin (D.B.H.G.), a subsidiary of the Reichswerke "Hermann Göring", and since 1 July 1942 had been employed at the Köthen factory of the Junkers Flugzeug- und Motorenwerke AG Dessau. Both companies awarded large contracts for the construction of new production facilities, paid for by the state. The letter of this "whistle-blower" is a truly remarkable document, illustrating the limits of the regulation of construction activities in the National Socialist economy.[126]

At the D.B.H.G., the controller reported, it was impossible to systematically register tenders, calculations and bills. According to his colleagues, "it was not worth worrying" about the proper documentation of the construction works since no one expected controls by the Court of Audit. In Gebauer's opinion, "the clerks had no more than nebulous ideas of the Award Regulations for Construction Works (VOB) , the Construction Price Ordinance, the Price Stop Decree, and last but not least the War Economy Ordinance (KWVO)." When confronted with these deficits and breaches of law, they simply used to refer to "P.P." or "Hermann" and made clear that they did not feel obligated to adhere to any regulations.[127]

The OT's method on the *Westwall* construction sites and later in occupied Europe of starting or even completing works without an existing contract was obviously being adopted by private industry, too. Both private firms and public awarding authorities increasingly lacked the time to invite tenders. More and more often, they replaced the competitive and thus potentially cost-reducing bidding process by inviting only a restricted number of firms or even awarded contracts with no bidding process at all. The Hamburg branch of Grün & Bilfinger reported in 1938 that they had not participated in a single public call for bids, but had still submitted 38 offers. The firm managed to reel in ten projects after restricted invitations and four by direct award.[128] At the D.B.H.G., invoices were retroactively declared tenders. Sometimes, when works were finished and construction firms submitted their invoices, the word "invoice" was simply crossed out and replaced by "tender".[129]

126 BArch, R 2301/7172, fol. 368–72, Letter Erich Gebauer to *Oberregierungsrat* Elfert, 20.12.42.
127 "P.P." referred to Paul Pleiger, head of the Reichswerke, "Hermann" to Hermann Göring, whose Four Year Plan office provided the funds for the construction works.
128 Bilfinger Archive, A 11, Annual report by Grün & Bilfinger's Hamburg branch, 1938.
129 The same practice was reported by Norwegian construction firms working for German authorities. Cf. chapter V.2.c.

After several confrontations with his superiors at the D.B.H.G., Erich Gebauer resigned and took up a job at the Junkers factory in Köthen. Here, Junkers was looking for an expert who would be able to get the company's accounts in order, because the company had been warned about the Court of Audit's plans to scrutinise the accounting of Junkers' large, state-financed new constructions.[130] However, "after a short time of familiarisation," Gebauer wrote, "I arrived at the belief that here, things are not substantially different than at the previous company." In general, the companies involved in the construction of air force armament plants did not adhere to most of the regulations. Gebauer attached several documents supporting his claims to the letter and asked Elfert to publish an article that could serve as a warning shot for companies like Junkers. Elfert's reaction, although he acknowledged the controller's good intentions, was rather reluctant. He did not pursue the matter further.

As this example shows, in the field of price and wage controls, the attempts of the National Socialist state to bring the development in the construction sector back under control were doomed to failure. There was no shortage of regulations, but the wage and price freeze was constantly ignored not only by the German construction firms, but also by their public and private customers. Moreover, without competition and with procurement agencies like the OT and the *Wehrmacht* branches, which valued speed more highly than cost efficiency, also the use of fixed-price contracts did not provide much relief for the Reich budget.

Everyone acknowledged the problem of rising building costs, but no one wanted to risk "his" construction project falling behind because of a lack of labour, materials or firms. It was not without reason a decree demanding "pricing discipline", that the Reich Price Commissioner issued in March 1943, targeted the procurement agencies rather than the contractors.[131] Fritz Todt, in particular, linked his own interests with those of the construction industry, concerning both the *Reichsautobahnen* and the *Westwall.* The consequences of Todt's policy were far-reaching, given that in 1938, approximately one fifth to one fourth of all construction investments in Germany were made on construction sites of the GdSt/OT. Time and again, Todt's staff had helped to undermine and fend off attempts at a state regulation of the construction sector by means of *ex ante* price monitoring, wage fixing and stricter contracts.

---

130 This is a remarkable case of indiscretion on the part of the Court of Audit, as Elfert, too, seems to have noticed, as he underlined this sentence in the letter.

131 The Reich Building Commissioner (*Reichswohnungskommissar*) referred to this decree when complaining in April that even in residential building, costs were rising because the officials of the awarding authorities were constantly ignoring the price regulations: BArch, R 3901/21510, fol. 21–22.

3 Attempts at Central Steering under the Four Year Plan

In October 1936, a new, powerful institution appeared at the core of the Third Reich's economy: the office of the Four Year Plan under Hermann Göring. In a secret memorandum, most likely from August 1936, Hitler had unequivocally expressed what he expected from his subordinates: "I. The German army must be operational within four years; II. The German economy must be fit for war within four years."[132] As the Four Year Plan included massive construction investments in industrial facilities of the armaments and autarky sector, many of Göring's subsequent interventions in the Reich's economy targeted the construction sector.

The initial approach chosen by the Four Year Plan office was to restrict labour mobility and to direct workers to those parts of the economy that were crucial to the plan's realisation. Part a of this chapter analyses the effects of the various ordinances and decrees that were issued between 1936 and 1938, culminating in the Ordinance for Securing Labour for Tasks of Special State Importance of 22 June 1938, often referred to as the *Dienstpflichtverordnung*.

However, part b will show that neither the *Dienstpflichtverordnung* nor attempts at steering construction activities by means of rationing construction iron and steel were able to solve the problems in the overburdened sector. Both state authorities and private businesses therefore called for a central authority to lead the German construction sector.

This central authority became the GB Bau, an office established in December 1938 and held by Fritz Todt. Todt's task as GB Bau was to curtail the number of construction projects to a level that corresponded to the given capacity of the construction sector by prioritising certain construction projects and steering the allocation of machines, workers and building materials towards these projects. In doing so, the GB Bau became the most distinct expression of a "steered economy" in construction. Part c will ask what competencies the GB Bau held, how effective its regulations were and to what degree the construction industry managed to protect its interests under this "minister for construction".

a The Restriction of Labour Mobility

As shown in the previous chapter, the Reich Trustees for Labour and the price auditors were trying to treat the symptoms, but were unable to fix fundamental problems in an increasingly overburdened sector. Thus, from 1936 on, Göring's Four Year Plan office attempted to steer the use of machinery and building materials (especially iron and steel) and to restrict labour mobility within the sector. Göring feared that labour

132 Tooze, *Wages*, 219–25.

shortages in particular would jeopardise the setting up and expansion of industrial capacities in his sphere of power. Therefore, on 18 October 1936, Göring secured extensive powers for himself with regard to the implementation of the Four Year Plan. The wording of the ordinance in question, signed by Hitler in Berchtesgaden, was vague enough to grant Göring a considerable room for manoeuvre.[133]

Concrete measures were made public in a set of six directives on 7 November 1936.[134] They targeted primarily the labour force in the metal working and construction sector, which shows clearly where the problem lay in Göring's eyes. Directive number three established that employers had to notify the labour offices within two weeks if skilled workers were executing the tasks of unskilled workers. The labour office could then demand that the worker be deployed in a position that matched his skill level either within the same firm or another business.[135] The directive aimed to prevent firms keeping skilled workers occupied with simpler tasks during periods with a lower workload instead of making them available to the labour market. It was well-nigh impossible to monitor whether the directive was adhered to, and examples reveal that the "hoarding" of skilled workers continued.[136] Directive number four determined that the labour offices had to authorise private construction projects with a wage bill of more than 5,000 RM and public projects with a wage bill of more than 25,000 RM.[137]

Another directive was issued on 22 December.[138] It allowed employers in the metalworking, construction and brick-making industry, as well as in agriculture, to retain the labour books of workers who terminated their work contract without authorisation. However, the Reich Trustees registered that especially in the agricultural sector, men and women left their workplaces for better positions in the industry: "The right to retain the labour book of the worker who breached the contract has remained mostly ineffective."[139] Understandably enough, the new employers of those

133 RGBl. 1936, I, 887, Ordinance for the Implementation of the Four Year Plan, 18.10.38.

134 Mason, *Arbeiterklasse und Volksgemeinschaft*, 223–31. Initially, the measures were to be laid down in a law, conjointly drafted by the RAM, the RWM and Göring's Four Year Plan: Document 3, Draft of a Second Law for the Regulation of Labour Deployment with explanatory statement and implementing rules, 6.10.36, in: Mason, 198–216. For reasons that are not entirely clear, Hitler refused to sign the law and it was reworked into directives: Mason, 217, 222.

135 Document 8, Third Directive for the Implementation of the Four Year Plan by the Return of Metal Workers and Skilled Construction Workers to their Occupations, 7.11.36, in: Mason, *Arbeiterklasse und Volksgemeinschaft*, 226–27.

136 Mason, 352, 511.

137 Document 9, Fourth Directive for the Implementation of the Four Year Plan on the Securing of Labour and Building Materials for Tasks of Special State Importance, 7.11.36, in: Mason, 227–28.

138 Document 12, Seventh Directive for the Implementation of the Four Year Plan by Prosecuting the Illegal Dissolution of Labour Contracts, 22.12.36, in: Mason, 230–31.

139 Document 36, Excerpt from the monthly reports of the Trustees for Labour, May 1937, in: Mason, 336. See also: Document 41, Excerpt from the monthly reports of the Trustees for Labour, June/July 1937, in: Mason, 368–69.

sought-after workers did not ask for the labour books either, even though this was unlawful.

The president of the RfAA, Friedrich Syrup, made a further attempt to remedy matters on 6 October 1937, ordering that masons and carpenters could only be hired with the written permission of a labour office.[140] In its wording, the directive was very similar to one of 27 April 1937 that had made the same demands concerning the hiring of workers in the chemical industry and construction around Bitterfeld, Halle and Wittenberg, a region massively affected by Four Year Plan projects.[141] Syrup's directive of October 1937, however, contained some exceptions. Permission was not necessary when hiring between 1 December and 28 February. Furthermore, permission should always be granted when the job change led to a family reunion. This applied to many workers from larger cities who had worked on public construction sites in rural areas in the previous years. Again, it seems that the directive's effect was limited – this time, because the labour offices themselves thought that it would be irresponsible to prohibit the employment of workers in the light of labour shortages.[142]

On 30 May 1938, Syrup felt impelled to extend the scope of his October directive. Labour offices now had to approve every job change of construction workers in the Reich, including the newly annexed Austria. Beyond that, nothing changed. The penalties for failing to observe the directive remained the same. Employers who hired – or workers who let themselves be hired – without permission could be punished with a fee or up to three months' imprisonment.[143]

As Timothy Mason has documented thoroughly, the RfAA directives could neither suppress frequent job changes within the sector, nor stop the movement of labour from other sectors to the construction industry. When allotting labour, the labour offices were to consider a construction project's national political and economic importance. Workers of firms engaged in such crucial projects were not allowed to change their job. However, this did not solve anything, since almost all projects, apart from maybe residential building, could claim to be important.[144]

140 Document 66, Directive on the Employment of Masons and Carpenters, 6.10.37, in: Mason, *Arbeiterklasse und Volksgemeinschaft*, 501–2.

141 Document 38, Directive on the Employment of Workers in the Chemical Industry and Construction in the Districts of the Labour Offices Bitterfeld, Halle and Wittenberg, 27.4.37, in: Mason, 347–48.

142 Document 104, Excerpt from the monthly reports of the Trustees for Labour, March/April 1938, in: Mason, 637.

143 Document 80, Directive on the Employment of Workers and Technical Employees in Construction, 30.5.38, in: Mason, 550–51. See also the accompanying decree of 2 June: Document 81, Decree of the president of the RfAA to the labour offices, 2.6.38, in: Mason, 552–54.

144 Mason, *Arbeiterklasse und Volksgemeinschaft*, 553, fn. 3.

Furthermore, the threats of punishment seem to have had little effect on employers. The Reich Trustees continued to report "numerous" cases of companies hiring construction workers without labour offices' permission.[145] One striking example of the limits of the directive of 30 May 1938 stems from the *Reichsautobahnen* Corporation. In a meeting in 1939, it complained despairingly about the construction industry that lured away their officials with generous salaries. The heads of the regional OBRs reported consistently that the RfAA's directive had provided no relief.[146] Although there was little doubt about the national political importance of the *Reichsautobahnen* at the time, the directive of 30 May 1938 was obviously unable to prevent the construction industry draining a state agency of its officials.

By the summer of 1938, the labour shortage in the industries necessary for the development of the German war machine had reached new heights. Now, the construction of the *Westwall* suddenly placed massive, additional demand on the labour market. The RfAA's previous attempts at regulation had merely aimed to prevent job changes that would jeopardise rearmament. Now, Göring's Four Year Plan office intervened and took the opportunity to issue an ordinance that allowed the conscription of workers for tasks of special state importance.[147] The Ordinance for Securing Labour for Tasks of Special State Importance of 22 June 1938, often referred to as the *Dienstpflichtverordnung*, authorised labour offices to temporarily transfer workers to workplaces that were deemed crucial for the war effort.[148] It was not limited to the construction industry, although the labour demand of the *Westwall* construction sites was certainly huge in the summer of 1938. Details were laid down in a directive by RfAA president Syrup on 29 June.[149] Every German worker could be conscripted as long as he or she was physically fit for the new task. A compensation was paid if the worker had earned more in his or her former position. The latter clause caused some discontent among the workforce. Depending on their previous wage, some workers were able to earn more than others despite executing the same tasks in the same workplace. Therefore this regulation was later replaced and the pay scale of the new workplace applied to all workers alike.[150]

According to the RAM, approximately 1.3 million workers had been conscripted pursuant to the *Dienstpflichtverordnung* between 1 July 1938 – the day the Ordinance

---

145 Document 108, Excerpt from the monthly reports of the Trustees for Labour, May/June 1938, in: Mason, 656.

146 HHStAW, Abt. 485, 6, Conference with representatives of the *Reichsautobahnen-Direktion* and the OBRs on questions of personnel, finances, construction and operation in Salzburg, 1939, fol. 72.

147 Mason, *Arbeiterklasse und Volksgemeinschaft*, 251–53. See also: Tooze, *Wages*, 260–63.

148 RGBl. 1938, I, 652. From 13 February 1939, the labour offices were able to conscript workers for an unlimited period of time. See this chapter, fn. 527.

149 Document 111, Directive for the implementation of the Ordinance for Securing Labour for Tasks of Special State Importance, 29.6.38, in: Mason, *Arbeiterklasse und Volksgemeinschaft*, 670–74.

150 Document 119, Ordinance for Securing Labour for Tasks of Special State Importance, 13.2.39, in: Mason, 695–98. An example was the *Westwall* pay scale. See above, Table 9.

became effective – and the end of 1939. Despite these impressive figures and the Ordinance's broad scope, the Nazis were rather cautious to use it.[151] Most of the workers were conscripted on two occasions: the construction of the *Westwall* (400,000) and immediately after the outbreak of the war (500,000). Nevertheless, the Ordinance certainly was one of the most drastic interferences with labour relations during the National Socialist reign. It plays a central role in the history of the OT in particular, as it provided the Organisation with masses of conscripted workers on the *Westwall*. We might say that while the OT's relatively generous wages on the *Westwall* were the carrot, the *Dienstpflichtverordnung* was the stick.

The Ordinance interfered with entrepreneurial autonomy, too. Especially handwork firms in the construction sector complained that their businesses were threatened by labour offices sending their workers to the *Westwall*. It is hard to say how often entrepreneurs tried to prevent the conscription of their workers. Moreover, it is unclear whether they were brought to heel by fees or even imprisonment when they did so.[152]

It has been documented, however, that larger construction companies used the Ordinance as a means of draining the handwork sector. The companies' tactic was to take over large, important construction projects, but to deploy most of their regular workers on less important construction sites. They then banked on the labour offices conscripting the missing workers for the larger construction projects from smaller competitors. Handwork firms from Berlin complained bitterly about this behaviour.[153] Their opinion was not unfounded, as proven by a decree of the president of the RfAA of December 1938, which was confirmed by the RAM in April 1939. It enabled labour offices to interfere with the *internal* distribution of workers between a company's various building sites. The RAM was well aware that larger companies did not always use their regular workers on their most important building sites, but hoped for the allocation of additional labour by the authorities. Fritz Todt had raised concerns over this major interference into construction firms' internal affairs, albeit without success.[154] Although the two decrees alarmed Todt, it seems unlikely that the labour offices were able to effectively control the distribution of workers on companies' building sites, which were often scattered throughout large regions. A company could always claim that certain workers with special skills were urgently

151 Mason, 152–53; Michael Schneider, *Unterm Hakenkreuz: Arbeiter und Arbeiterbewegung 1933 bis 1939* (Bonn: Dietz, 1999), 320; Tooze, *Wages*, 261. Most of the conscriptions in late 1939 benefitted armaments producers, not the construction sector.

152 Mason, *Arbeiterklasse und Volksgemeinschaft*, 675, fn. 2.

153 Mason, 905. The large *Westwall* contractor Grün & Bilfinger, for example, gained 6,000 new workers on the *Westwall*, a tremendous number in times of labour shortage: Bilfinger Archive, A 4467, Report to the supervisory board, 1938.

154 Document 82, Decree by the President of the RfAA to the labour offices, 5.12.38 and Document 126, Decree of the Reich Minister for Labour to the labour offices, 30.4.39, in: Mason, 554–55, 733–35.

needed on another building site, which was nearly impossible for the authorities to disprove. We should also keep in mind that the Third Directive for the Implementation of the Four Year Plan of 7 November 1936 had required very similar controls, which had proven to be largely impracticable.

b The Call for a Strong Man in the Construction Sector

Beside labour, iron and steel became another limiting factor for the execution of construction projects. The Four Year Plan office thus amended its fourth directive in July 1937. The wage bill was no longer the decisive criterion. From now on, the labour offices had to authorise all projects that required more than 2 tonnes of steel.[155] The labour offices' influence in this field remained limited, however, because of the system of iron and steel rationing that the RWM had introduced in February 1937.[156] Initially, the RWM's Regulatory Office for Iron and Steel (*Überwachungsstelle für Eisen und Stahl*) established quotas for seven consumers of iron and steel, including the *Wehrmacht*, the Four Year Plan office, the GdSt and the National Railways. During the following months, this number was raised to 15.[157] The organisations in question could dispose freely over the iron and steel allocated to them. As long as they provided the materials from their quota, the labour offices had no right to stop the construction projects. The labour offices' influence was therefore limited to minor fields like residential building or municipal public construction projects. Hence, central steering of the construction sector through the allocation of iron and steel did not did not take place until the end of 1938.[158]

Over the course of 1938, the congestion of the construction sector became unbearable. The previous attempts of regulation had failed to yield the desired results, and thus calls for a central regulation authority for the construction sector were increasing. On 7 December 1937, the RWM had already come forward with the demand for a central body to register all public construction projects and rank them according to their national political and economic importance. Furthermore, the new authority was to draft consistent rules for all public procurement agencies in the sector. The RWM warned about the dangers of "random public procurement" and "the agglomeration of numerous public construction projects in certain regions of the Reich". This endangered both the stability of the wage structure and the

155 Hensler, *Stahlkontingentierung*, 75; Mason, *Arbeiterklasse und Volksgemeinschaft*, 228, fn. 3.
156 For a short overview of iron and steel rationing in the Third Reich in English, see Hensler, 'Iron and Steel Rationing'.
157 Hensler, *Stahlkontingentierung*, 53–54, 71.
158 Hensler, 75. Cf. Document 75, Letter from the Labour Deployment Business Group at the Administrator for the Four Year Plan to the Reich and Prussian Minister of Economics, 9.1.38, in: Mason, *Arbeiterklasse und Volksgemeinschaft*, 534–37.

production of building materials. The RWM's letter was sent to 16 ministries and public authorities. The ministry asked for comments on the proposal by 5 January 1938.[159]

In December 1937 likewise, Reich Price Commissioner Josef Wagner demanded that his administration be given the right to prioritise construction projects and control wages and salaries. This would have meant an enormous gain in power for the Price Commissioner.[160] However, Wagner's advance did not meet with the support of the Reich Chancellery, which preferred the RWM strategy (restriction of the construction volume) to Wagner's (fixing of wages and prices). It was of the opinion that the reduction of wages would increase the risk of discontent among the workforce, while price fixing was unfeasible as it required time-consuming pre-calculations by the procurement agencies.[161] Thus the Reich Chancellery arrived at a very similar conclusion to the RAM and the RWM in October 1936 when they had opposed the introduction of maximum wages.[162]

Reich Labour Minister Franz Seldte also renewed his call for a reduction of state demand by prioritising crucial – and cancelling less important – construction projects.[163] As drafts of *Wehrmacht's Inflationsdenkschrift* of autumn 1938 prove, experts within the war economy staff unit of the OKW came to the same conclusion, although a passage on the necessity of reducing public construction orders did not make it into the memorandum's final version.[164]

It was not only representatives of the state apparatus who expressed the desire for a strong, central authority in the construction sector. Smaller handwork firms in particular hoped for a limitation of large-scale public construction projects. The Reich Association of German Craftsmen (*Reichsstand des Deutschen Handwerks*) criticised the fact that building machines and vehicles were seized and distributed among larger firms working on major construction projects like the *Westwall*. While the handwork firms received a compensation or rental for their machines, many of them were no longer able to carry out their own projects. Furthermore, the large *Westwall* firms used the *Dienstpflichtverordnung* of 22 June 1938 to lure skilled workers away from handwork firms. Thus the Reich Association presented the RWM

---

159 Document 72, Letter from the RWM to several ministries and public agencies, 7.12.37, in: Mason, *Arbeiterklasse und Volksgemeinschaft*, 522–25. For unknown reasons, Hitler rejected the RWM's proposal in February 1938: Document 78, Note by the Reich Chancellery, 25.2.38, in: Mason, 546–47.

160 Document 128, Letter from the Reich Price Commissioner to the Administrator for the Four Year Plan, 12.12.37, in: Mason, *Arbeiterklasse und Volksgemeinschaft*, 747–54.

161 Document 74, Note by the Reich Chancellery, 5.1.38, in: Mason, 532–34. Document 129, Note by the Reich Chancellery, 5.1.38, in: Mason, 754–56.

162 See chapter III.2.a, fn. 81.

163 Document 105, Letter from the Reich and Prussian Labour Minister to the head of the Reich Chancellery, 23.5.38, in: Mason, *Arbeiterklasse und Volksgemeinschaft*, 648–49.

164 Geyer, 'Rüstungsbeschleunigung', 184, fn. 209.

with a list of claims, demanding among other things that awarding authorities place more orders with working combines of smaller firms. Furthermore, they requested that a "Reich Construction Commissioner" ("*Reichsbaukommissar*") be appointed. This man should be responsible for "all questions concerning the awarding as well as the deployment of labour, machines and building materials." The commissioner should also compile a priority list of construction projects, which would reduce insecurity for private businesses.[165]

In November, the Berlin branch of the Reich Guild Association of the Building Trades (*Reichsinnungsverband des Baugewerks*) complained about the *Westwall* works in particular. Firms were no longer able to calculate properly because their machines and workers could be conscripted at a moment's notice. They perceived the conscriptions as unjust and uncoordinated. Furthermore, smaller repair and maintenance works were postponed due to the construction of the *Westwall*. This hit craftsmen in particular, as those works were an important part of their business. Another problem was the allocation of building materials. Every material, like steel, timber or cement, was managed by different authorities. Hence, even if one material was allocated, construction projects could be delayed because another material was still missing.[166] Much of the building trades' critique was directed against "big business", but the complaints by member firms of Wigru Bau sounded very similar. Construction sites lay idle or worked inefficiently when skilled workers were ordered to the *Westwall*.[167] Therefore, also Wigru Bau called for some kind of construction commissioner.[168]

Finally, in a speech before the Reich Defence Council (*Reichsverteidigungsrat*) on 18 November 1938, Göring announced – albeit very vaguely and without using this particular term yet – the appointment of a General Commissioner for the Regulation of the Construction Industry.[169] As the previous examples have shown, there was a broad consensus within both the state apparatus and private industry on the need for a more centralised management of the German construction sector. There was agreement that state demand had to be reduced and ranked. Moreover, a central authority would have to untangle the confusing and inefficient rules on the allocation of building materials. The office's establishment was not an act of subjugation of private industry. Rather, it was universally welcomed.[170]

---

165 Document 187, Reich Association of German Craftsmen to the Reich Economic Minister, 18.8.38, in: *Die Regierung Hitler. Band V: 1938*, 617–21.

166 Document 151, Economic report for the fourth quarter of 1938 by the Mayor of Berlin, in: Mason, *Arbeiterklasse und Volksgemeinschaft*, 878–906.

167 BArch, R 43-II/1169, fol. 155–60, Wigru Bau to RWM, main department I, 5.8.38.

168 BArch, R 43-II/417b, fol. 33–40, Wigru Bau to RWM, 16.12.37.

169 Document 152, Speech by Göring at the first meeting of the Reich Defence Council, 18.11.38, in: Mason, *Arbeiterklasse und Volksgemeinschaft*, 908–33.

170 For positive reactions, see for example: AG für Beton- und Monierbau, business report 1938; Allgemeine Hoch- und Ingenieurbau AG, business reports 1938 and 1939; Boswau & Knauer AG,

c A "Minister for Construction": The GB Bau

On 9 December 1938, Göring, as head of the Four Year Plan, appointed Fritz Todt the General Commissioner for the Regulation of the Construction Industry. In his letter, Göring explicated:

> Due to the special situation in 1938, the demands on the construction market have risen more than in other branches of the economy. This has led to unwanted friction in the raw materials management, labour deployment and finally in the wage structure. Because of the plethora of urgent tasks, particularly the headhunting of workers and granting of lure wages have become common especially in construction. The result is that the performance enhancement on the construction market no longer corresponds to the amount of money expended: construction has become disproportionately expensive.[171]

Initially, it seems like the GB Bau office was not necessarily intended to be a permanent institution. When the war broke out, however, Göring confirmed that the GB Bau should remain active during wartime, and in 1941, plans were made for the post-war era, too.[172] The GB Bau's day-to-day affairs were mostly handled by Todt's right-hand man Günther Schulze-Fielitz and the chief of the office, Joachim Steffens.[173]

The main task of the GB Bau was to reconcile the available capacity of the German construction and building materials industry with the orders of institutions like the *Wehrmacht* and the OT.[174] The most important competence of the new office, located at Pariser Platz 3 in the shadow of the Brandenburg Gate, was the prioritisation of all construction projects in the German Reich. Todt's staff ranked projects according to their national political importance, which became the underlying criterion for all other measures of the GB Bau. Göring had already devised the order:

a. Fortifications, docks, locks and harbours for the defence of the Reich.
b. Production facilities for armament goods and facilities for the production of the necessary primary products for a. and b.
c. Securing the urgent demand for the maintenance of crucial factories.
d. "*Führer*" buildings, primarily in Berlin, Nuremberg and Munich.
e. Canals, motorways and railroads.

---

business report 1939; Julius Berger Tiefbau AG, business report 1938; Neue Baugesellschaft Wayss & Freytag AG, business report 1939.

171 Dittrich, *Vom Werden, Wesen und Wirken der Organisation Todt*, 422–23.

172 Botzet, 'Generalbevollmächtigter', 118.

173 Botzet, 119.

174 A detailed list of the GB Bau's responsibilities can be found in RAO, RAFA-2188/2/Hfa-L0007, file 27, Regelung des Bauschaffens in Norwegen, Undated document "Tasks of the General Commissioner for the Regulation of the Construction Industry and their execution by the Commissioner for the occupied Norwegian territories". The document refers to a Göring decree of 22 January 1941, which I was unable to locate.

f. Residential building, primarily for workers at Four Year Plan factories.
g. Other residential building.[175]

In the words of Joachim Steffens, the GB Bau's task in the months before the outbreak of the war was to support the industry in order to build as much as possible and meet the demands of the Four Year Plan. After 1939, activities had to be cut back to what was absolutely vital to the war effort. More than ever, speed became most important criterion in the execution of construction projects.[176] All other construction activities in the Reich were banned from August 1939 on. Provided the GB Bau approved, minor projects and vital maintenance and repair works were still permitted.[177] In November 1940, however, Todt limited the overall volume of these exceptions to 88.5 million RM: one reichsmark per citizen.[178]

The prime task of the GB Bau office was to gain an overview over the German construction firms' capacity. Todt's men also considered factors such as availability of building materials, machines and labour when setting the overall construction volume. Todt subsequently named 19 claimants (*Kontingentträger*), which received a quota of building materials.[179] Within their respective fields of responsibility, the claimants prepared lists of construction projects they wished to realise using their

175 Dittrich, *Vom Werden, Wesen und Wirken der Organisation Todt*, 422–23.
176 Steffens, 'Bauwesen', fol. 73–76.
177 BArch, R 3/1447, fol. 6, 6th decree of the GB Bau regarding the ban of new constructions, 4.8.39; ibid., fol. 14, 9th decree of the GB Bau regarding the ban of new constructions, 16.2.40. Only the 9th decree was published in: *Deutscher Reichsanzeiger und Preußischer Staatsanzeiger*, no. 44 (21.2.42).
178 RAO, RAFA-2188/1/E1a-L0002, file 2100, Exceptions from the ban of new constructions, 6.11.40.
179 In April 1939, the claimants were: the OKW for the three *Wehrmacht* branches, the Reichswerke Hermann Göring, the NSDAP, the Reich Centre for Economic Development (*Reichsstelle für Wirtschaftsausbau*), the Volkswagenwerke, the Reich Ministry for Food and Agriculture, the General Inspector for German Roads, the National Post, the DAF, the RAM, the General Building Inspector for the Reich Capital (*Generalbauinspektor für die Reichshauptstadt*) and the Reich Traffic Group for Tram Railways (*Reichsverkehrsgruppe Schienenbahnen*). The Reich Ministry of Transport (*Reichsverkehrsministerium*, RVM) administered the contingents of the National Railways and the Reich Waterways Administration (*Reichswasserstraßenverwaltung*). The RWM administered the contingents of three claimants: the mining industry, the iron- and steel-producing industry and the Reich Group for the Power Industry (*Reichsgruppe Energiewirtschaft*). The GB Bau administered two contingents itself: public construction of the Reich and the states, including the SS, and city development (*Ausbau der Gaustädte*). See Announcement by the GB Bau, 5.4.39, in: *Der deutsche Baumeister* 1, no. 4 (1939): 26. Cf. also Fritz Todt, 'Regelung der Bauwirtschaft', *Der Vierjahresplan* 3, no. 12 (1939): 762–64. The list of claimants changed constantly. For the contingents in June 1942, see *Deutscher Reichsanzeiger und Preußischer Staatsanzeiger*, no. 146 (25.6.42). For the contingents in March 1943, see Generalbevollmächtigter für die Regelung der Bauwirtschaft, ed., *Die Regelung der Bauwirtschaft. Übersicht über die für die Baugenehmigung und Kontingentierung wichtigsten Bestimmungen des GB Bau* (Berlin: E.S. Mittler & Sohn, 1943), 60–65.

quota, which they then discussed with the GB Bau. Based on these lists, building permissions were granted, and building materials and transport capacity allocated. Finally, the labour offices received the lists in order to allocate additional workers if necessary.[180]

In order to monitor construction activities locally, the GB Bau appointed regional representatives (*Gebietsbeauftragte*) in November 1939. They were to supervise and provide advice to the existing local administrations. Later, Albert Speer placed construction commissioners (*Baubevollmächtigte*) in every armaments inspectorate (*Rüstungsinspektion*) and representatives in every district. It has been pointed out that these men not only executed orders from Berlin, but felt obligated to take up the interests of local firms and authorities, too. This could undermine the GB Bau's attempts at central steering.[181] Construction commissioners were also appointed in the occupied territories of Northern and Western Europe, Serbia, the Protectorate of Bohemia and Moravia and the General Government.

After approving a construction project, the regional construction commissioners allocated building materials. After the outbreak of the war, the most important building materials were all allocated by one authority, the GB Bau. In particular, the control of iron and steel for construction purposes was separated from the general system of iron and steel rationing.[182] This was meant to remedy the problems of the preceding years, when the allocation of one material had not necessarily meant that all other materials could be provided, too.[183] Construction firms had repeatedly criticised the arrangement. The rationed materials were construction iron and steel, timber, cement, and fuel. From 1943 onwards, the GB Bau also distributed the materials for road construction (bitumen, tar, rock asphalt mastic), that previously had been under Todt's management in his function as General Inspector for German Roads.[184]

180 Joachim Steffens, 'Die deutsche Bauwirtschaft im Kriege', *Die Bauindustrie* 8, no. 6 (1940): 118–20.

181 Botzet, 'Generalbevollmächtigter', 120–21. For details on the tasks of the regional construction commissioners, see Steffens, 'Die deutsche Bauwirtschaft im Kriege'.

182 Hensler, *Stahlkontingentierung*, 71. See also: Josef Struck, 'Eisen und Stahl für die Bauwirtschaft im Kriege. Entwicklung der Lenkung des Eiseneinsatzes in der Bauwirtschaft von 1936 bis heute.', *Der deutsche Baumeister* 1, no. 12 (1939): 14–16.

183 Another problem, however, remained unsolved: the transport of the allocated materials to the construction sites. In 1940, Joachim Steffens was forced to admit that the lack of transport capacity had become the biggest obstacle to the construction process. Although the allocation of transport capacity was meant to be linked to the approved allocation of materials, in reality transport was mere improvisation: Steffens, 'Die deutsche Bauwirtschaft im Kriege', 120.

184 Details on the proceedings for the various materials are given in: Generalbevollmächtigter für die Regelung der Bauwirtschaft, *Regelung*, 80–102. For cement, see also: Anton Seeger, 'Kontingentierter Zement', *Der deutsche Baumeister* 1, no. 11 (1939): 17–21.

In 1939, the GB Bau allocated almost half of the German cement production and of iron and steel for construction purposes to the OKW. The *Westwall* construction sites were supplied from this source among others. The military also received one third of the rationed timber. Some timber was still available outside the quota system, and here, too, the *Wehrmacht* secured its share. Major amounts of iron and steel were given to private industry. Materials for residential construction and buildings of municipalities came from the quota of the Reich Labour Ministry, which explains its high share of timber. 12.9 percent of the German cement production was used on Fritz Todt's road construction sites (see Figure 8).

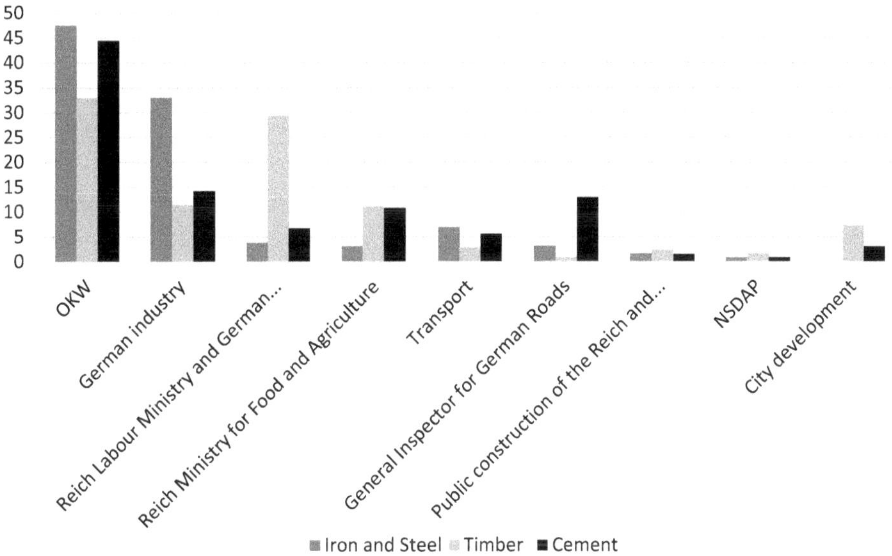

Figure 8: Allocations of building materials by the GB Bau, 1939, in % of the total allocated amount of the respective building material. Source: Steffens, 'Bauwesen', fol. 215–216. German industry includes the Reich Centre for Economic Development, the Reichswerke Hermann Göring, the Volkswagenwerke, the mining industry, the iron and steel producing industry and the Reich Group for the Power Industry. Transport includes the National Railways, the Reich Waterways Administration, the National Post and the Reich Traffic Group for Tram Railways. Public construction of the Reich and the states includes the SS and police forces.

Initially, the OT was no autonomous claimant with its own quota of building materials. The Organisation was building on behalf of other authorities, mostly *Wehrmacht* branches or the SS. Accordingly, building materials were provided by these institutions using their respective quotas. However, both the OT and the GB Bau office were headed by Fritz Todt and many officials held positions in both organisations simultaneously. Hence, the GB Bau officials were able to represent the OT's interests when negotiating on construction programmes. Only later in the war, when the Organisation carried out large projects in the Reich, it received its own quota of building materials.

Constant regroupings make it difficult to compare the values for 1939 with the quotas of later years. Some institutions lost their status as claimants entirely, like the General Building Inspector for the Reich Capital (*Generalbauinspektor für die Reichshauptstadt,* GBI) in September 1942,[185] while new programmes with their own quotas were set up, such as an air-raid protection programme (*LS-Führerprogramm*). Still, data from 1943 suggest that materials allocated to the *Wehrmacht* within Germany were reduced significantly, while the share of construction iron and steel allocated to industry rose to 43 percent. Of course this is a shift in relative terms, not a growth in absolute numbers, which can be explained primarily by the *Wehrmacht* relocating many of its construction activities to the occupied territories.[186]

To a certain degree, the distribution of power between the GB Bau and the claimants remained unresolved. Officially, Todt's office had the final say as it allocated the building materials.[187] On the other hand, Todt was forced to admit that his office had to rely on the lists compiled by the claimants. Even though his office was only controlling projects with a value of more than 1 million RM, this officially would have necessitated 13,000 checks per year. According to Todt, this would require a supervisory authority so large that it would not be in accordance with the *modus operandi* of the GB Bau.[188] The claimants thus retained some influence. The *Wehrmacht*, for example, initially even refused to send a list of their construction projects to the GB Bau, stating that the latter was not capable of judging the importance of military construction projects.[189] In 1939–40, the *Wehrmacht* also ignored Todt's authority when it came to the allocation of construction iron.[190] Similar resistance is documented for the SS.[191] It seems that the broadly shared support for a

185 BAK, N 1318/2, Speer Chronicle, 1942, fol. 88. Major regroupings happened for example in the field of residential building in December 1942: BArch, R 3901/21510, fol. 5, Reich Building Commissioner regarding the establishment of a building materials quota with the construction commissioners of the Speer ministry, 11.1.43.

186 Steffens, 'Bauwesen', fol. 215–216.

187 Generalbevollmächtigter für die Regelung der Bauwirtschaft, *Regelung*, 59.

188 Todt, 'Regelung'.

189 Botzet, 'Generalbevollmächtigter', 123.

190 Hensler, *Stahlkontingentierung*, 101.

191 Schweitzer, 'Plans and Markets', 107. In October 1941, Todt personally wrote to Heinrich Himmler and complained about the unauthorised construction project of an SS member who had erected a villa with a swimming pool in Berlin in the middle of the war. SS guards even had denied building inspectors access to the building site. Todt reminded Himmler of the case of a building entrepreneur who had been sent to a concentration camp after he had been caught building a swimming pool: BArch, NS 26/1188, fol. 381. Nevertheless, it seems that such drastic punishments happened rather rarely. In April 1944, the Speer ministry had to remind its construction commissioners to take action against illegal building activities. The preferred measure was fines rather than costly and lengthy criminal proceedings, however. That the same examples of violations of

central supervisory authority in the construction sector vanished very quickly as soon as cutbacks jeopardised the individual stakeholder's own projects.[192] However, according to Christiane Botzet, Todt managed to stabilise his position in the following months, not least since he made sure of Göring's backing in 1941.[193]

Nevertheless, allocations and building permits remained the object of tough negotiations between the GB Bau and the claimants. For negotiations to be successful, it was necessary to claim that a project was vital to the war effort. Furthermore, having the personal support of the "*Führer*" could be a crucial factor.[194] Being vital to the war effort, however, became an increasingly meaningless criterion. Almost everything could to some extent be regarded as *kriegswichtig*. Albert Speer annulled the existing classifications for construction projects in May 1942 as too many projects had been given high priority.[195] Speer introduced a more detailed classification system.[196]

Initially, the claimants were in charge of transporting the construction materials to their sites once the GB Bau had approved a project and allocated the materials. However, hoarding iron and steel in particular became such a significant problem that the GB Bau stepped in. In the middle of the war, the office took the right to distribute building materials out of the claimants' hands. After this, the claimants were only allowed to use the materials from their own quota once the regional construction commissioners had controlled the construction sites and registered that previously allocated materials really had been used.[197]

Through the prioritisation of construction projects and allocation of building materials, the GB Bau gained indirect influence over the distribution of labour. When a construction site was declared vital to the war effort and all materials had been allocated, the labour offices provided additional workers. Accordingly, construction firm workers were only safe from being drafted into the armed forces if

---

the construction ban are mentioned again and again in documents on the topic may indicate that harsh punishments were exemplary and rare. The notorious swimming pool appears once again in the document of April 1944: RAO, RAFA-2188/1/E1n-L0049, file 5360, Letter Stobbe-Dethleffsen, Reich Ministry for Armaments and War Production, to the construction commissioners, 10.4.44.

192 The minutes of a meeting between the head of the War Economy and Armaments Department at the OKW, General Georg Thomas, and representatives of the *Wehrmacht* branches on 3 February 1939 are evidence of the *Wehrmacht's* unwillingness to agree to any cutbacks, although building materials requests for 1939 exceeded the 1938 consumption by over 300 percent in the case of steel and 270 percent in the case of cement, for example. See BA-MA, RW 19/3080, fol. 2–7.

193 Botzet, 'Generalbevollmächtigter', 123–24. Seidler reaches the same conclusion: Seidler, *Fritz Todt*, 214.

194 Botzet, 'Generalbevollmächtigter', 123–24.

195 *Der Deutsche Baumeister* 4, no. 6/7 (1942): 14.

196 27th and 28th decree of the GB Bau, 27.5.42, in: *Der Deutsche Baumeister* 4, no. 6/7 (1942): 16–18; Steffens, 'Bauwesen', fol. 158–164.

197 Steffens, fol. 155–158.

the company was engaged in a vital project. This meant that there were no so-called *W-Betriebe* in the construction sector, that is, companies whose workforce was permanently protected from the grasp of the *Wehrmacht*. The status of a construction company always depended on the status of the site. So did the exemption from military service (*UK-Stellung*) for indispensable, often highly-skilled workers.[198] Thus it was crucial for the firms to be engaged in vital projects. However, as we will see below, when the Nazis increasingly relocated construction activities in the following years, firms more and more often had to go where the state's policy created demand: the occupied territories.[199]

Entrepreneurs were able to name workers who had been in their firm for at least three years "*Stammarbeiter*".[200] Firms had lost long-time workers to the *Wehrmacht* early in the war,[201] but it seems that the *Stammarbeiter* status gave workers a certain protection from being drafted into the *Wehrmacht* later in the war. Appointing *Stammarbeiter* thus became an important way for construction companies to keep their workforce together. The Philipp Holzmann AG, for example, had 2,585 *Stammarbeiter* in 1940, and gave this status to further 3,492 workers just in 1941.[202] Declaring this group of workers "untouchable" was one of the major concessions the GB Bau made to the construction industry.[203] After spring 1942, when *Wehrmacht* call-ups were decimating the workforce in construction, the GB Bau was once again keen to "[. . .] hold a protecting hand over the construction industry, so that it was able to meet the demands that German warfare was posing on it."[204] Speer confirmed the protected status of the *Stammarbeiter* once again in June 1943,[205] and still in late October 1944, Xaver Dorsch reminded his OT staff that any attempt to interfere with the firms' disposition of their workforce was strictly forbidden.[206] Overall, however, the average age of the German construction firms' workforce increased significantly during the war. Willi Henne, for example, estimated that the average age of EW's German workers was 56 years.[207]

198 Joachim Steffens, 'Die Bauwirtschaft im Kriege', *Der deutsche Baumeister* 1, no. 11 (1939): 9–11; Steffens, 'Die deutsche Bauwirtschaft im Kriege'; Steffens, 'Bauwesen', fol. 81–84.
199 On the GB Bau in occupied Norway, see below, chapter IV.4.
200 Reichsarbeitsblatt (RABl.) 1941, IV, 1592.
201 See for example Thormann & Stiefel AG, business report 1940.
202 Pohl, *Philipp Holzmann*, 233.
203 Steffens, 'Bauwesen', fol. 85–86.
204 Steffens, fol. 78–79. The image of GB Bau's "protecting hand" is also used by Schulze-Fielitz, *Die Bauwirtschaft im Kriege*, 1.
205 RAO, RAFA-2188/1/E1m-46, Joint decree by GB Bau Speer and GBA Sauckel (copy), 9.6.43.
206 Circular decree 494/44 of 31 October 1944, in: *Mitteilungsblatt der Organisation Todt-Zentrale* 3, no. 37 (1944), 350.
207 *OT-arkiv, Institutt for historiske studier, NTNU* (hereafter: IHS), Willi Henne, 'Die Organisation Todt in Norwegen: Aufgabe, Organisation, Art der Baudurchführung mit aufgetretenen Schwierigkeiten, Leistungen' (Unpublished manuscript, 1945), pt. III, 31–32.

Construction machinery became yet another bottleneck during those years. The overheated construction activities put an enormous strain on the industry's equipment. On many sites, machines were running almost non-stop, with little time for service and maintenance. While the GB Bau had gained control over the distribution of construction iron and steel, the material for new construction machines still came from the various claimants' steel quotas. Accordingly, the primary source of new construction equipment for a firm working for the *Wehrmacht* was *Wehrmacht's* steel quota.[208] As Jonas Scherner has shown, the amount of newly produced construction equipment and machinery in Germany remained at a relatively high and stable level throughout the war (Table 12). This highlights the enduring importance of the construction sector in the National Socialist war economy. As labour was scarce, rationalisation and mechanisation became particularly important in construction. In 1933, when construction machinery was almost banned on the *Autobahn* construction sites as it was regarded to put labourers out of work, the degree of mechanisation in construction was 0.8 horse powers per worker. By 1938, it had already risen to 1.4 horse powers.[209] In late 1947, it was estimated to be three horse powers per worker.[210]

Table 12: Domestic sales of newly produced construction equipment and machinery, million RM, constant prices.*

1928	1936	1937	1938	1939	1940	1941	1942	1943	1944
173	199	237	334	377	342	383	372	284	365

*Scherner, 'Preparation', 461. Equipment index used for deflation: Rolf Krengel, 'Die langfristige Entwicklung der Brutto-Anlage-Investitionen der westdeutschen Industrie von 1924 bis 1955/56', *Vierteljahrshefte zur Wirtschaftsforschung*, no. 2 (1957): 184.

Detailed data exist on Philipp Holzmann's transport and construction machinery (Figure 9). Germany's biggest construction company invested massively in machinery during the construction of the *Westwall*. These numbers dropped during the war as rapidly as they had risen up until 1939. Nevertheless, the firm managed to keep its machinery investments at a reasonable level throughout the war, still matching the investment figures of the mid-1930s in 1944. Only in 1944 and 1945 did the firm lose almost 2 million RM's worth of machinery while retreating and in bombings of their properties in Germany.

208 Konrad Weber, 'Baugeräte', in *Gelenkte Bauwirtschaft*, ed. Eitelfritz Kühne (Berlin: Otto Elsner, 1941).
209 *Die Bauindustrie* 7, no. 37 (1939): 1204.
210 IfS, W1-2/276, Handwritten calculations by Hans Meyer-Heinrich (Philipp Holzmann AG), February 1948.

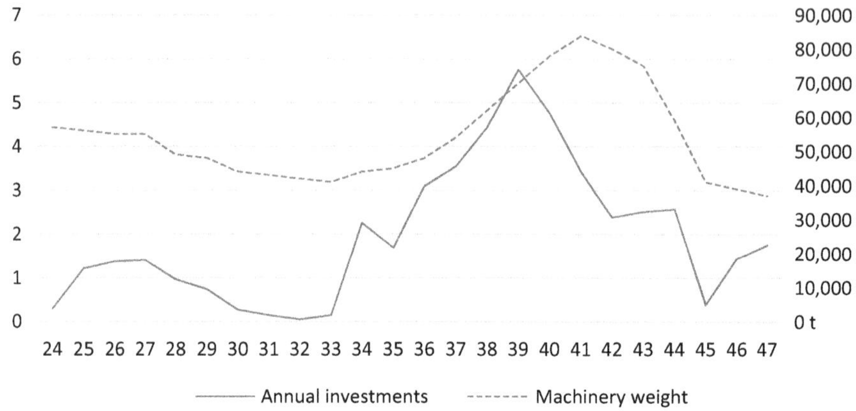

Figure 9: Annual machinery investments in million RM (current prices) and weight of Philipp Holzmann's transport and construction machinery, in tonnes, 1924–47. Source: IfS, W1-2/276, Statistics on Philipp Holzmann's transport and construction machinery, February 1948. For a possible explanation for the discrepancy between decreasing investments and the increase in machinery during the first half of the war, see below, chapter V.1.c.

As the data on the overall weight of Holzmann's machinery in Figure 9 show, the lost war and the expropriation of machinery of the East German branches in Halle, Leipzig and Nieder-Neuendorf threw the firm back onto the level of the Great Depression.

To support their member firms, Wigru Bau and the Reich Association of German Craftsmen received their own, small quotas of iron and steel for repairs and maintenance of machinery (so-called *U- und E-Kontingente*). Before 1 April 1940, companies were also allowed to buy spare parts on the free market as far as possible. Finally, as a last resort, the GB Bau allocated iron and steel from a special quota (*G.B.-Sonderkontingent*) that had been established in early 1939. To supply the *G.B.-Sonderkontingent*, the other iron and steel claimants had to provide 4 percent of their quota to the GB Bau.[211] However, as an official of the GB Bau admitted quite frankly, none of these channels was able to meet the demand of the industry. The GB Bau was managing a shortage. In 1939, the GB Bau allocated a singular appropriation of 200,000 tonnes of iron and steel to the construction industry to overhaul existing or buy new machinery. Until the end of 1942, the *G.B.-Sonderkontingent* provided the industry with 30,000 tonnes per quarter, and

211 Weber, 'Baugeräte', 79–80; 'Beschaffung von Baumaschinen und Geräten', *Technische und Wirtschaftliche Rundschau*, Supplement to Der deutsche Baumeister, 37 (1939): 207–8. When the war broke out in September 1939, the *G.B.-Sonderkontingent* had been cancelled, but it was already re-established in February 1940: *Die Bauindustrie* 8, no. 5 (1940), 91; *Die Bauindustrie* 8, no. 8/9 (1940), 161–62.

after a massive reduction to 3,000 tonnes per quarter in the first half of 1943, the *G.B.-Sonderkontingent* eventually ran dry in the summer of that year.[212]

The *G.B.-Sonderkontingent* allowed the GB Bau to gain some influence over the production of construction machinery, as the office only approved requests for certain machine types. The aim of this was to reduce the number of machine types used in construction, which again aimed to rationalise production and facilitate the supply of spare parts.[213] For the same reason, Todt initiated the construction of a "standard excavator" with only three different bucket types, for example, which was then manufactured by the major German producers.[214]

Beyond that, the GB Bau confined itself to registering and distributing the existing construction machinery. Here, more than in any other field within his purview, Todt relied on Wigru Bau and information provided by the companies themselves. Wigru Bau established a bureau that was supposed to act as an agent between firms that were looking for equipment and those who wished to rent out their idle machines. This initiative, set up in the spring of 1940, was a reaction to the situation in 1938–39, when confiscations benefitting *Westwall* contractors had aroused considerable concern and resentment within the sector.[215] Todt felt obliged to assure the industry of his support and forbid the confiscations of construction machines without his or the OKW's approval.[216]

Finally, the GB Bau tried to utilise the industry's existing capacities as well as possible. In countless appeals, Todt prompted the construction industry to rationalise work processes and increase the performance of their workforce through improved personnel management ("*Menschenführung*", in the parlance of the time).[217] Wigru Bau's periodical *Die Bauindustrie* in particular became an organ for the propagation of innovative construction methods and tips on how to save materials. Competitions were initiated in which construction workers and engineers from building sites all over the country could send in their tricks and ideas on how to make certain work steps more efficient. After the outbreak of the war, construction designs had to be simple, even "primitive", without any embellishment and using as few valuable materials as possible. Todt stated that it was sufficient if the constructions for the armaments sector barely outlasted the war.[218] But even in December 1941, two

212 Steffens, 'Bauwesen', fol. 106–107.
213 Steffens, fol. 106–107.
214 Todt, 'Regelung', 764.
215 The general contractor Grün & Bilfinger provides an example from its sector of the *Westwall.* Three quarters of construction machinery, more than two thirds of cars, and more than two thirds of sheds and huts on Grün & Bilfinger's sites were not their own: Bilfinger Archive, A 12, Report of the Kehl construction site, 1938–39.
216 Botzet, 'Generalbevollmächtigter', 129.
217 Seidler, *Fritz Todt*, 207–19.
218 Eichholtz, *Kriegswirtschaft II/2*, 303; Seidler, *Fritz Todt*, 215–16.

months before his death, it seems that he still observed that the claimants did not follow his saving measures often enough.[219]

It is difficult to assess the achievements of the GB Bau and the effects of the regulations implemented by the office. This is still the case even if one leaves aside – as we have done so far – GB Bau's role in the various occupied territories with their different institutional settings. Steering and regulating the German construction industry was one thing. An equally important task was to prevent other state agencies from constantly undermining GB Bau's regulatory work.[220] Historians have consistently pointed to the mixed results and the limits of Todt and Speer's policy as GB Bau, even when documenting its far-reaching competencies.[221] Joachim Steffens himself stated around 1943–44 that "[. . .] the larger claimants ruthlessly defied the decrees and regulations [. . .] in order to push through their own projects."[222] A popular tactic among awarding authorities was to start construction works with less material than necessary to finish the project. Their claimant was then constrained to allocate the missing materials so as not to abandon the project half-finished. Sometimes claimants allotted materials to unimportant rather than vital construction sites, assuming that it would be easier to enforce additional allocations for the latter. The GB Bau also reported on cases in which workers started their workday on a prioritised construction site but were sent to non-vital construction sites nearby in the afternoon.[223]

Ultimately, the GB Bau was unable to solve the underlying problems of the sector, that is, the discrepancy between supply and demand, the transport problems, and the obstinacy of the SS, *Wehrmacht* and other agencies. However, we do not know what the development would have looked like without the office's regulating measures. After all, the German construction industry did carry out gigantic projects within the Reich and all over Europe under conditions of war for more than five and a half years.

During the last two years of the war, the GB Bau's office and purview was subject to several reorganisations. In July 1943, Speer delegated certain tasks of the GB Bau to a newly established *Hauptausschuss Bau* led by the head of Wigru Bau,

219 BArch R 3/1445, fol. 74, Circular decree by the GB Bau regarding the saving of construction materials, 8.12.41.

220 Ironically enough, even Fritz Todt's own *Reichsautobahnen* Corporation disobeyed the GB Bau's regulations. In early 1941, Todt ordered that construction firms were to be deployed on motorway construction sites in the east, even though these construction sites were not ranked high enough on the GB Bau's priority lists. See BArch, R 13-VIII/69, Wigru Bau, head office, to all branch offices, 25.4.41.

221 Seidler, *Fritz Todt*, 201–19; Botzet, 'Generalbevollmächtigter'; Rolf-Dieter Müller, 'Albert Speer and Armaments Policy in Total War', in *Organization and Mobilization of the German Sphere of Power*, ed. Militärgeschichtliches Forschungsamt, Germany and the Second World War, V/II (Oxford: Clarendon Press, 2015), 483–90.

222 Steffens, 'Bauwesen', fol. 80.

223 *Die Bauindustrie* 9, no. 18 (1941): 743.

Bruno Gärtner.[224] Moreover, Carl Stobbe-Dethleffsen became Speer's special deputy for carrying out the tasks of the GB Bau.[225] Stobbe-Dethleffsen was a partner in the Dortmund-based construction company Wiemer & Trachte whom Speer apparently held in high esteem. After the war, Speer described Stobbe-Dethleffsen as a "very reliable and agreeable collaborator. Candid and upright. Suitable for a leading position."[226] In late October, an office in the Armaments Ministry, the *Amt Bau*, was established under Stobbe-Dethleffsen and took over all of the GB Bau's responsibilities. After "the industry declined to have this government authority telling it what to do[, p]lanning work was taken out of the hands of a 'dull' civil servant and entrusted to an industrialist [. . .]."[227]

On 28 October 1943, Speer gave far-reaching competencies to the *Hauptausschuss*.[228] The minister ordered that the deployment of construction firms was to be rationalised: instead of being scattered over various construction sites, companies' workforce and machinery were to be concentrated on one or few projects. Small and inefficient firms had to work either as subcontractors, in working combines, or – as a last resort – could be incorporated into a larger company. Although it was stressed that the workforce of the incorporated firm was to remain together and the incorporation was only a temporary measure for the time of the war, most owners of small handwork firms probably harboured no illusions that this meant the end of their business. What is remarkable in this respect is the fact that the decision on these concentration and rationalisation measures lay with the *Hauptausschuss* and its regional and local subdivisions, that is, with men from the larger industry firms. What is more, all clients such as the *Wehrmacht* and the OT were – at least theoretically – subject to the decisions of the *Hauptausschuss Bau* when it came to the deployment of construction firms. Within the borders of the Reich, industrial self-responsibility in the construction sector had reached a new high point.

There are no reliable data on the scope and effects of closures, concentration and rationalisation measures in the construction sector. Experience from closure

---

224 On the organisation and tasks of the new *Hauptausschuss Bau*, see Karl Knecht, 'Vorläufiger Organisationsaufbau der wirtschaftlichen Selbstverantwortung der Bauwirtschaft', *Die Bauindustrie* 11, no. 14 (1943): 373–77.

225 *Die Bauindustrie* 11, no. 13 (1943): 363. Cf. Reuss, *Verbände*, 94–95; Botzet, 'Generalbevollmächtigter', 133.

226 Office of the Publication Board, Department of Commerce, 'Interrogation of Albert Speer and Members of the Former Reich Ministry of Armaments and War Production [Technical Report No. PB 430]' (Washington, D.C., 1945), 78.

227 Müller, 'Armaments Policy', 486.

228 Directive on the Concentration and Simplification of the Deployment of Firms, 28.10.43, in: *Die Bauindustrie* 11, no. 18 (1943): 491. Cf. also Jecht, 'Vereinfachung des Firmeneinsatzes. Zur Anordnung des Reichsministers Speer vom 28.10.1943'.

programmes in other sectors showed that the results were not always overwhelming.[229] However, the decision on closures lay in the hands of the sectors' big players, who were represented in the various rings and committees. As Bernhard Kroener has put it, "the possibility that firms might have their own agenda when implementing industrial policy caused some observers to suspect that this was simply an easy way to get rid of unwelcome competition under the pretext of bowing to the demands of the war."[230]

The *Amt Bau* under Stobbe-Dethleffsen only enjoyed a short reign over the construction sector within Germany, however. In 1943, units of the OT had returned to the Reich for the first time after the outbreak of the war in order to restore the *Eder-* and *Möhnetalsperren*, two dams that were crucial to the water supply of the Ruhr area and that had been destroyed by the Royal Air Force's *Operation Chastise* in May 1943. During the following months, the OT under the ambitious Xaver Dorsch tried to gain control over the construction activities within the Reich, too. In particular, the OT disliked being subordinate to Stobbe-Dethleffsen's office. The conflict was a power struggle between Dorsch and his OT apparatus on the one side and Albert Speer and his supporters in the Armaments Ministry on the other. In spring 1944, Speer only managed to retain his power in the construction sector by making considerable concessions to Dorsch.[231] Speer had to sacrifice a pawn – Stobbe-Dethleffsen – and the *Amt Bau* was merged with the OTZ into an *Amt Bau-OT* under Dorsch. From the summer of 1944 onwards, all tasks of the GB Bau within the Reich were thus taken over by the OT. The GB Bau had ceased to exist as an independent institution.[232]

With regard to the question of entrepreneurial freedom of action, the establishment of the *Amt Bau-OT* has been interpreted as a noticeable change in the relation between state and construction industry. It has been argued that one dimension of the Dorsch-Speer conflict was that Dorsch tried to tighten the reins on the powerful construction companies, while Speer still thought that encouraging the industry to be self-responsible would produce the best results. Although representatives of the construction industry retained influence in the new office, too,[233] Dorsch's approach

---

229 Bernhard R. Kroener, 'Management of Human Resources, Deployment of the Population, and Manning the Armed Forces in the Second Half of the War (1942–1944)', in *Organization and Mobilization of the German Sphere of Power*, ed. Militärgeschichtliches Forschungsamt, Germany and the Second World War, V/II (Oxford: Clarendon Press, 2015), 949–50.

230 Kroener, 952.

231 On Xaver Dorsch and his relation to Speer, see below, chapter IV.1.a.

232 Botzet, 'Generalbevollmächtigter', 134–35.

233 With a decree of 8 August 1944, a task force (*Arbeitsstab Bauwirtschaft*) was established in the *Amt Bau-OT*, which was – among other things – to handle the deployment and relocation of firms and the distribution of newly produced construction machinery: Circular decree by the *Amt Bau-OT* regarding the collaboration of the construction industry, 8.8.44, in: *Ministerialblatt des Generalbevollmächtigten für die Bauwirtschaft, des Generalinspektors für das deutsche Straßenwesen, des Generalinspektors für Wasser und Energie und für den Rüstungsausbau, herausgegeben vom*

is said to have threatened construction companies' autonomy and in fact led to fierce complaints by Wigru Bau.[234]

Mark Buggeln has developed this argument further, stressing that the conflict was also a clash of interests between Germany's large construction companies and smaller businesses. According to Buggeln, who studied the contract negotiations of several military construction projects within the Reich in 1944, Wigru Bau and its large member companies feared that the OT would support the smaller businesses and enforce contracts unprofitable for the big players.

Buggeln's results are ambivalent, however. In the case of one construction project, the large companies were able push through their wishes in some important points as late as October 1944.[235] On another building site, the OT accepted the proposals of both large companies and their smaller subcontractors.[236] The question remains whether the fights and reorganisations on the administrative level and the merger of GB Bau and OTZ really had any tangible consequences for German construction companies' day-to-day work. In particular, it has not been investigated whether the changes in Germany had any consequences for the relation between the OT and its contractors in occupied Europe and whether the characterisation of the OT as a militarised organisation pursuing an anti-industry policy actually describes this relation accurately. Chapter V will touch upon these questions with regard to the EW and its contractors.

Between December 1938 and May 1944, the GB Bau certainly – and visibly – intervened with the private economy and gained far-reaching control over the industry's input factors in particular. However, its role was that of a moderator, rather than a "building dictator". In the maze of directives, rationing regulations and gargantuan construction projects, an authority like the GB Bau was genuinely appreciated. Wigru Bau retained a good deal of self-responsibility, and as we have seen, its responsibilities under the GB Bau office were even extended to more fields during wartime. As Günther Schulze-Fielitz recalled after the war, it was possible to keep

---

Reichsministerium Speer 2, no. 27 (15.9.44). The Arbeitstab was staffed with representatives of the industry and led by Hermann Malsy of the construction firm Peter Bauwens. Cf. Reuss, *Verbände*, 96; Botzet, 'Generalbevollmächtigter', 134.

234 Müller, 'Armaments Policy', 483–91; Botzet, 'Generalbevollmächtigter', 133–35; Seidler, *Die Organisation Todt*, 238–54; Lemmes, 'Arbeiten für das Reich', 382, 404–5; Dietrich Eichholtz, *Geschichte der deutschen Kriegswirtschaft 1939–1945. Band III/1: 1943–1945* (1996; repr., München: K. G. Saur, 2003), 29–32; Marc Buggeln, *Das System der KZ-Außenlager: Krieg, Sklavenarbeit und Massengewalt* (Bonn: Friedrich-Ebert-Stiftung, 2012), 103–5.

235 Marc Buggeln, '"Menschenhandel" als Vorwurf im Nationalsozialismus. Der Streit um den Gewinn aus den militärischen Großbaustellen am Kriegsende (1944/45)', in *Rüstung, Kriegswirtschaft und Zwangsarbeit im 'Dritten Reich'*, ed. Andreas Heusler, Mark Spoerer, and Helmuth Trischler (München: Oldenbourg, 2010), 217.

236 Buggeln, 213–14. However, Buggeln's study of the protracted and tough contract negotiations provides another argument against the notion of a "command economy" as late as 1944.

the construction industry “on a long leash”. He actually identified this policy as one reason for his post-war career in the private construction industry.[237] During the war, Joachim Steffens revealed how he envisaged the relation between state and construction industry developing in the long run:

> At the moment, entrepreneurs are not yet ready to justify such far-reaching customer trust that supervision could be dispensed with, but education has to go in the direction that the construction industry delivers branded quality goods [“*Markenware*”] like the rest of the industry. Just as you buy a Mercedes without monitoring the construction of the vehicle and know that you are getting a vehicle of a particular quality, it should be possible to know that you get a certain quality when construction works are executed by firm X, even if not every work step is monitored.[238]

4 Limiting Profits

The obvious alternative to measures taken before or during construction was to limit profits to a permissible level, audit the firms and skim off excess profits. As the struggle for wage and price controls yielded only meagre results, attempts of agencies such as the Price Commissioner and the Court of Audit to monitor and limit profits gained importance in the course of the war.

The term “excess profits” that will be used frequently in this chapter was of course no objective variable. A profit by itself, as an indicator of entrepreneurial success, cannot be “excessive”. Moreover, it would *per se* be a rather unspectacular finding that firms in a booming sector were realising very healthy profits. However, in the National Socialist (war) economy, profits above a certain level could be regarded as excessive compared to the level that was politically defined as acceptable. What is more, realising excess profits could – at least potentially – be heavily punished. Accordingly, the extent to which construction firms were able to realise excess profits can be understood as a measure of entrepreneurial room for manoeuvre under the Third Reich.

Two different indicators of profit will be used in this chapter, the first being the rate of return on equity (RoE).[239] The RoE was disclosed in the companies’ balance sheets and was regarded as a significant indicator of entrepreneurial success both by contemporaries and present-day analysts. Secondly, I will use the firms’ rate of return on sales (RoS).[240] It has been pointed out that little speaks for using the RoS

237 Botzet, ‘Generalbevollmächtigter’, 136–37.
238 Steffens, ‘Bauwesen’, 136–37.
239 For the considerations and arguments in this paragraph, see Spoerer, *Eigenkapitalrentabilität*, 26–40.
240 Note that RoS and RoE are not directly comparable.

for measuring entrepreneurial success from a theoretical point of view.[241] There are practical reasons, however. On the one hand, it has been possible to compile a particularly large sample of annual results disclosing the RoS of German construction firms. This gives an unusually broad picture of the development in the sector, especially because it includes small and mid-sized firms that were not joint-stock corporations. On the other hand, contemporary discussions on the development in the sector, on the limitation and collection of excess profits as well as all contract negotiations between state agencies and the construction industry all used the RoS as an indicator of entrepreneurial success. Hence, in the historical context, the RoS was highly relevant.

Part a of this chapter tries to assess the development of profits in the German construction industry during the Third Reich, especially in comparison to other sectors and the sector's own performance during earlier economic booms. Furthermore, a sample of more than 1,000 annual results from more than 600 firms has been compiled to gain an idea of the profit level in construction under the National Socialist rule.

Part b investigates the topic from another angle. Here, the focus lays on the capital adjustments of German joint-stock companies after June 1941. They revealed the considerable liquidity that German companies had accumulated over the course of the 1930s. Again, the extent of these capital adjustments in construction will be compared to the rest of the German industry.

Finally, part c will analyse the state's attempts to limit and collect the excess profits in the construction sector. I argue that the efforts' success was limited mostly due to two reason. First, the state lacked the capacity to overcome the information asymmetries that prohibited a comprehensive registration of excess profits. Second, awarding authorities used healthy profits as the main mechanism to mobilise private construction businesses. Limiting and collecting these profits too rigorously would have destroyed this incentive structure and jeopardised the state's own projects.

a Assessing Entrepreneurial Profits in Construction

Rising profits in construction were no accident. It has been repeatedly documented how the National Socialists exploited the entrepreneurial goal of profit maximisation in order to reach their political goals of rearmament and autarky. It is hardly surprising that the construction industry was among the winners of this armaments boom. On the one hand, construction traditionally reacted more sensitively than other sectors to economic ups and downs, while on the other hand, the regime's

241 It should be noted that the RoS is hardly suitable for inter-sectoral comparison as it discriminates against companies with a high turnover. Therefore, I have only used the RoS for comparing construction companies with each other.

specific wishes, like the build-up of armaments capacities or the construction of military facilities, placed an enormous demand on construction companies. Although German firms' tax burden of German firms rose significantly from the mid-1930s on,[242] the level of profits became a source of deep resentment in the ranks of the RFM, the Court of Audit and the office of the Reich Price Commissioner.

An indication of the level of entrepreneurial profits, here understood as RoE, is given by the disclosed commercial balance sheets of German industrial stock corporations, which were collected by the SRA up until 1939. In the case of construction, 43 companies met the criteria for SRA registration in 1939.[243] The data can be compared with the rest of the German industry and with earlier years to gain a first impression of the level of profitability in construction during the 1930s. The point of reference, especially when it comes to methodological questions, is Mark Spoerer's pathbreaking study on German industrial profits between 1925 and 1941.[244] Spoerer has compiled data for most sectors of the German industry, but did not include construction in his study. Therefore, data for the construction sector have been collected by using the same sources, methods and selection criteria as Spoerer. It is presented alongside his figures in Figure 10.

Figure 10 shows the profitability (RoE) of joint-stock companies in the German industry in general and in the construction sector in particular, as published in the commercial balance sheets, as far as they met the criteria of the SRA.[245] However, the figures immediately suggest that caution is required: Is it really plausible that German businesses earned less even in the strongest years of the industrial boom of the 1930s than in an average year of the period before the First World War?[246] Indeed, Mark Spoerer has stressed that the figures do not reveal the absolute and real level of entrepreneurial profits during the National Socialist years. Simply put, German accounting regulations allowed firms to undervalue their assets during the boom. By doing so, they created considerable hidden reserves in order to conceal their rising profits, a strategy often described as "window-dressing".[247] Therefore, Spoerer has evaluated

---

242 Ralf Banken, *Hitlers Steuerstaat: Die Steuerpolitik im Dritten Reich*, Das Reichsfinanzministerium im Nationalsozialismus 2 (München, Wien: De Gruyter Oldenbourg, 2018), 404–56.

243 On the criteria, see chapter III.4.a, fn. 245.

244 Spoerer, *Eigenkapitalrentabilität*. For a concise overview in English over his methods, sources and findings, see Mark Spoerer, 'Industrial Profitability in the Nazi Economy', in *After the Slump. Industry and Politics in 1930s Britain and Germany*, ed. Christoph Buchheim and Redvers Garside (Frankfurt am Main: Peter Lang, 2000), 53–80.

245 The SRA gathered information on all joint-stock companies up to 1924–25, from 1925–26 on all joint-stock companies that were listed on the stock market or had a share capital of more than 1 million RM, and from 1933–34 on all joint-stock companies that were listed on the stock market or had a share capital of more than 500,000 RM.

246 Spoerer, *Eigenkapitalrentabilität*, 61.

247 Spoerer, 51–61; Spoerer, 'Profitability', 58–59.

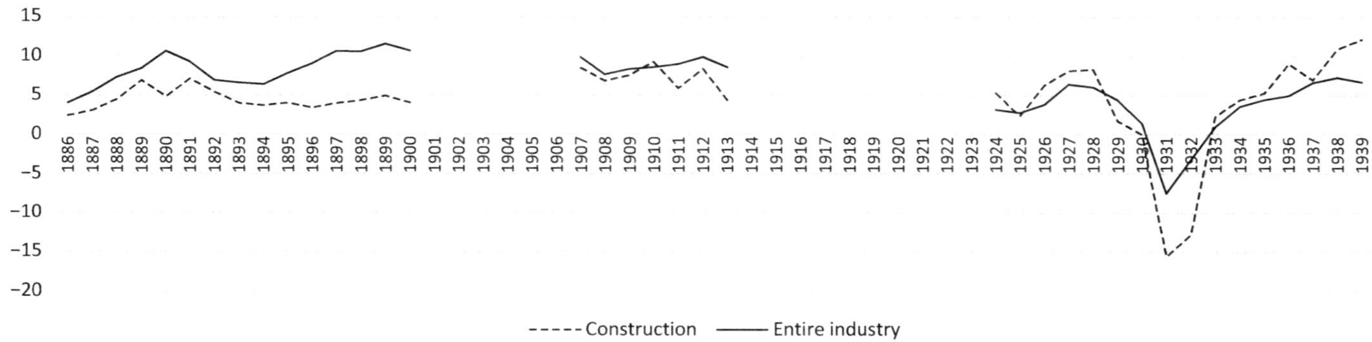

Figure 10: Rate of return on equity of German joint-stock companies in construction and in the entire industry, as reported in their published commercial balance sheets, 1886–1900, 1907–13, 1924–39. Source: For entire industry, see Spoerer, 172–79. For construction, see appendix, Table A5.1.

more than 700 reports from government tax audits as well as corporate income tax data, which give a much more realistic picture of the German industry's actual profits. According to Spoerer, German joint-stock companies' average rate of RoE was almost 15 percent between 1936 and 1941. He assumes that even this figure still underestimates the real profitability to some extent, and there are indications that profitability remained high in 1942 and 1943, too.[248] From 1935 onwards, the capital goods sector in particular experienced a veritable profit explosion.[249]

Nevertheless, the easily accessible information from disclosed commercial balance sheets can be used for comparisons between different sectors of the economy and for comparison over time within one sector. The inter-sectoral comparison reveals that during the Nazi period, entrepreneurial profits in construction were extraordinarily high. As early as 1933, profitability in construction was exceeding that of the rest of the German industry and was almost twice as high in 1939. The average disclosed RoE in construction was 7.31 percent between 1933 and 1939. This was a rate only few other sectors at the core of the armaments and autarky industry were able to match, like the petroleum industry (7.07 percent) or vehicle manufacturing (6.76 percent). Only rubber and asbestos producers reported higher profits (8.50 percent) than construction did during the peace years of the Nazi rule.[250] The figures also allow for an intra-sectoral comparison, showing that in 1936, corporate profits in construction for the first time surpassed the excellent results of the boom years 1927 and 1928. The comparison also indicates that construction companies earned better between 1933 and 1939 than they had done both before the turn of the century and in the Weimar years. In particular, construction companies exceeded the average profit rates of the very strong years between 1907 and 1913. Interestingly enough, profit rates in construction had been below the industrial average during the economic upswings before the First World War. Under the Nazi rule, construction was spearheading the development (Table 13).

Table 13: Average rate of return on equity of German joint-stock companies in the construction sector and industry in general as reported in the published commercial balance sheets.*

	1886–1894	1895–1900	1907–1913	1924–1929	1930–1932	1933–1939
Construction	4,67	4,12	7,27	5,37	−9,41	7,31
German industry	7,24	10,04	8,85	4,47	−3,10	4,96

*For the values for the German industry, see Spoerer, *Eigenkapitalrentabilität*, 172. The values for the construction industry are calculated based on the figures in the appendix, Table A5.1.

248 Spoerer, *Eigenkapitalrentabilität*, 146–61.
249 Spoerer, 169.
250 Spoerer, 179.

Parallel to the findings from Spoerer's sample, which did not include construction companies, it is reasonable to assume that the actual profit rates in construction during the 1930s exceeded the disclosed rates considerably. The German construction industry profited massively from the National Socialist drive for rearmament, autarky and, ultimately, war.

The development in the sector was already noticed by contemporaries like the economic journalist Gustav Plum in 1940. In a manuscript that never saw publication, Plum evaluated the published balance sheets of 45 German joint-stock companies in the sector. Plum's figures on sales and (published) after-tax profits make it possible to calculate the RoS ratio of these firms for the years 1933–39.

Table 14 distinguishes between a group of 20 firms with more than one million RM share capital and a second group of 25 firms with less than one million RM share capital. Turnovers in constant prices in the two groups of firms had risen by 582 percent and 511 percent respectively between 1933 and 1938. The respective rate for the entire German construction sector was 276 percent.[251] The rate of RoS fell over the course of the 1930s. Table 14 also suggests that the larger joint-stock companies coped with the aftermath of the slump (1933) and the lack of workforce, transport capacity and building materials (1937) better than the smaller joint-stock companies.

Table 14: Average rate of return on sales of 45 German construction joint-stock companies according to their disclosed balance sheets, 1933–39, in percent.*

	1933	**1934**	**1935**	**1936**	**1937**	**1938**	**1939**
Group 1: Companies with more than 1 million RM stock capital	4.39	2.82	2.65	2.74	3.19	2.84	2.09
Group 2: Companies with less than 1 million RM stock capital	3.39	3.05	2.44	2.45	1.54	2.56	2.11

*See the appendix in: Plum, 'Abschlüsse'.

However, the data in Table 14 are also based on the published commercial balance sheets of the joint-stock companies, in which they tried to conceal their real profits to some extent. Behind the curtain, that is, in the tax balance sheets and government audit reports, the situation looked quite different. In his sample of firms, Plum noticed a suspicious discrepancy between the disclosed after-tax profits, which had risen by 219 percent in the group of larger firms and 188 percent in the group of smaller firms, and the tax sums that the firms had paid on their earnings and assets. The latter had risen by 1,300 percent in the first group and 1,700 percent

251 Plum. For the rate of the entire construction sector, see appendix, Table A1.2.

in the second group. The taxes were paid on basis of the more accurate, but confidential tax balance sheets, not the sugar-coated published balance sheets. Plum noted that tax rates had not been rising that much and that hence the figures revealed a considerable, hidden accumulation of capital in the sector.[252]

For many construction firms, the reality thus looked quite different from what Table 14 suggests. From experience, the Court of Audit expected a firm's RoS rate to fall when turnover was rising. But despite rising turnovers, RoS rates were not falling as the official picture showed. Instead, they were growing! The Court of Audit noticed this anomaly in the case of Dyckerhoff & Widmann, for example, where audits had shown that turnover between 1934 and 1939 had risen by 445 percent, the RoS, however, by 1,045 percent: "This result of one of the biggest German construction businesses raises considerable doubt about the correctness of the pricing [. . .]."[253] In 1939, Dyckerhoff & Widmann, as one of the main contractors on the *Westwall*, realised a RoS of astonishing 17 percent.

Dyckerhoff & Widmann was not an isolated case. The regime was very much aware of what was really going on in the sectors crucial to the war effort. In the early 1940s, department XI 1 of the German Court of Audit in Potsdam, led by Friedrich Wilhelm Elfert, tried to gather information on the profits that construction companies realised on public construction sites. The abovementioned special reports of the tax authorities all over the country formed the basis for Elfert's work. They had been requested either by the Court of Audit or *Wehrmacht* auditors. Each of these special reports was concerned with one construction firm and could cover several years, mostly from the period 1936–40.[254] In contrast to the disclosed commercial balance sheets, the special reports were based on the tax balance sheets, which the authorities had corrected for excessive tax deductions, disguised dividends, and the like. This source therefore provided a much more truthful picture of the firms' profitability. The profit rates calculated by the tax auditors often lay several hundred percent (!) above those disclosed in the published balance sheets.

Figures 11–12 are based on the special reports for 603 construction firms from all parts of Germany and all fields of construction activity.[255] Altogether, 1,094 annual results could be evaluated.[256] I was able to locate these annual results in five lists found among the documents of the German Court of Audit at the Federal

252 Plum, 112–15.
253 BArch, R 2301/7172, fol. 486–88, Letter by the German Court of Audit to the Reich Price Commissioner regarding excess profits of the firm Dyckerhoff & Widmann KG and other large companies in construction, 31.1.41. From 1937 onwards, despite being one of the biggest German construction firms, Dyckerhoff & Widmann was a limited partnership (*Kommanditgesellschaft*), not a joint-stock company, and is thus not included in Figure 10, Table 14 and Table 15.
254 BArch, R 2301/7183–7186 and 9950–9951.
255 After 1938, some Austrian companies, too, were included in the sample.
256 For more details on the sample, see appendix A6.

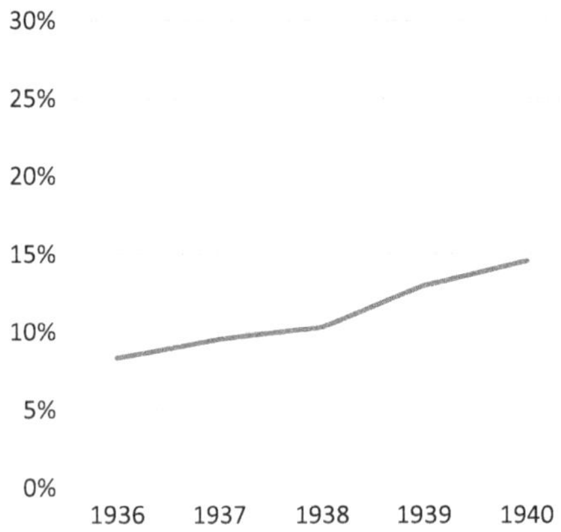

Figure 11: Average taxable return on sales (arithmetic mean) of German construction businesses according to the special reports of German tax authorities, 1936–40. Source: appendix, Table A6.2.

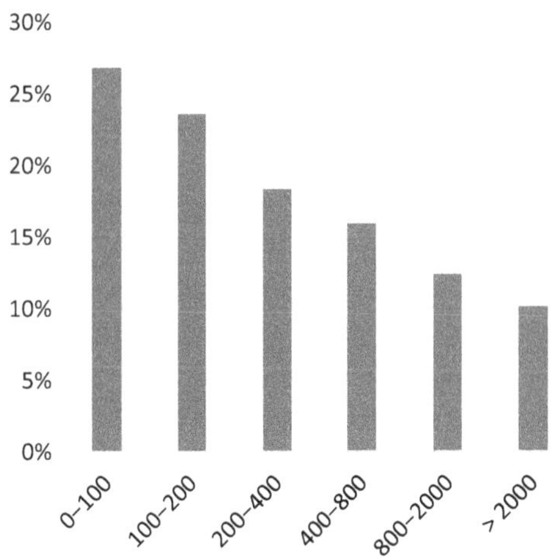

Figure 12: Average taxable returns on sales (geometric mean) of German construction businesses in the period 1936–40, grouped according to their annual turnover in 1,000 RM. Source: appendix, Table A6.2.

Archives in Berlin.[257] As the lists partly overlap and contain cross references, they have been carefully evaluated to avoid double counts before merging them into one

257 BArch, R 2301/7187, fol. 27–41, list Gesamtverzeichnis (1), 1936–38; ibid., fol. 42–43, list Gesamtverzeichnis (2), August 1937 – February 1939; ibid., fol. 138–48, list I, 1939 – May 1941; ibid., fol. 149–86, second version of list I, 9.11.42; ibid., fol. 187–225, list II, July 1942.

sample. As of today, this is the biggest sample of annual results that has been compiled of any sector of the National Socialist economy.

Two trends are clearly identifiable. First, Figure 11 shows the increasing profits during the late 1930s, measured by return on sales. On average, all sizes of construction companies were able to increase their profits during the 1930s. The rate continued to rise in 1940, showing that the outbreak of the war had no immediate negative effect on profitability. The returns on sales were not only high compared to the lower level of the early 1930s, they were so in absolute terms, too. At the time, a RoS of approximately 4–7 percent was regarded as decent in construction.[258] The profits granted in public contracts also usually lay within this range: in 1942, in his commentary on the revised version of the LSBÖ, the Reich Price Commissioner assessed a RoS of 4.3 percent under cost-plus contracts and 6.1 percent under the *Selbstkostenfestpreisvertrag* as tight but decent.[259] Figure 11 shows that the reports uncovered far higher rates, between 8.3 and 14.45 percent on average, in the period 1936–40. Almost two thirds of the annual results showed a RoS of more than 10 percent.[260] Second, Figure 12 shows that the returns on sales of smaller companies were significantly higher than those of larger businesses. This is partly due to the fact that the owners of small handwork businesses still had to take the income for themselves and their families from the profits. Large firms, on the other hand, paid their managers a salary, which is already subtracted as expenses. Nevertheless, nothing indicates that small handwork firms were run over by "big business" and were unable to participate in or profit from the National Socialist building boom. In 1943, the president of the Court of Audit identified the profits of mid-sized firms as particularly inflated.[261]

The special reports did not reveal a specific geographic pattern, which is due to a peculiarity of the construction sector: the location of a construction firm's headquarters allows no conclusions to be drawn about the locations of its construction sites. The most prominent example in this respect was the *Westwall*, where firms from all over Germany were working and earning money. Furthermore, the sample gives no indication that any awarding authorities stood out as allowing particularly high profits. Excess profits, sometimes up to 30, 40 or even 50 percent of turnover, were identified all over the Reich on construction sites of the *Wehrmacht*, the OT,

---

258 Matis and Stiefel judge the 4 percent profit that the OT granted the Viennese Allgemeine Baugesellschaft – A. Porr AG to be "respectable": Matis and Stiefel, *Porr*, 411. During the negotiations on the cost-plus contract introduced by *Einsatzgruppe Wiking* in 1942, Wigru Bau's pricing expert Gerhard Opitz calculated with 4–5 percent: BArch R 13-VIII/264, Draft by Gerhard Opitz on mark-ups for profit and operating costs in the new *Wiking-Bauvertrag*, 6.6.42. Rates lay in this range in other countries as well, like in Denmark, where the average RoS in construction had been 6.05 percent before the war: Andersen, *De gjorde Danmark større*, 213.

259 Herbert Prange, 'Die neuen LSBÖ-Vorschriften', *Die Bauindustrie* 10, no. 7 (1942): 171–74.

260 Appendix, Table A6.3.

261 BArch, R 2301/3007, Memorandum of the President of the German Court of Audit on the Reich budget account for the budget year 1939, 19.4.43, fol. 260.

the *Reichsautobahnen* and other state agencies alike. However, due to the dimensions of the project and the sheer number of firms engaged in it, the *Westwall* certainly formed the Potsdam auditors' centre of the attention. The Court of Audit complained in 1942 that some OBLs of the OT and the fortification pioneers had been "particularly 'accommodating' when controlling the prices" and found "such a far-reaching failure of the awarding authorities [. . .] that it seems advisable to consider legal measures against the responsible officers."[262] Above all, the Court of Audit was concerned that the same officers were now responsible for other construction projects where "they pursued their 'Westwall customs' maybe even with the same firms elsewhere to the detriment of the treasury."[263] These misgivings were not unjustified. In November 1942, the Court of Audit found that a large number of firms that had realised extraordinary profits at the *Westwall* had been transferred to Bremen, where they continued to work for the OT. The office of the General Inspector in Wiesbaden, however, defended the profits: they had been the result of negotiations between the firms, the General Inspector and Wigru Bau's self-regulating committee and could not be rejected.[264] There are no indications that firms that had been caught demanding excess prices were later banned from public construction projects. And indeed, as we will see below, the *Westwall* firms even became sought-after partners in the following years, as the OT appreciated the expertise and experience they had gained from the large-scale project.[265]

The statistics presented in Figure 11 and Figure 12 are assumed to give a correct impression of the major trends in the construction sector of the time. Some comments on the source material's validity and reliability have to be made, though. The special reports on the companies' profits were prepared by tax inspectors of the regional tax authorities. They visited the firms in order to control their books and accounts. The reliability of the reports is sometimes reduced because of the sketchy book-keeping particularly of small handwork businesses. In such cases, the auditors sometimes had to estimate turnovers or profits. In general, however, it seems that the auditors were able to gain a very detailed overview over the company results.

When it comes to the material's validity, that is, its explanatory power with regard to the entire German construction sector, two aspects should be kept in mind. First, the special reports were concerned with public construction projects only. Hence, the question is to what extent the results display the situation in the entire construction sector. However, the share of public construction in the overall German construction volume of the late 1930s and early 1940s was extraordinarily high.

262 BArch R2301/7187, fol. 85–89, Note regarding the evaluation of the results of the special reports of the tax offices, 29.7.42.
263 Ibid.
264 BArch, R 2301/7172, fol. 281–82, Letter by Elfert, 4.11.42.
265 Cf. chapter IV.2.a.

Moreover, we can assume that private customers had to match the prices paid by public awarding authorities in order to attract construction firms. Thus, the results of the special reports are probably representative for the entire German construction sector. A more serious reservation is that the results in the sample stem from a possibly biased selection of firms. It cannot be entirely ruled out that the auditors requested information particularly on those firms they already suspected of excess profits. This holds true particularly for smaller firms. The share of annual results with a RoS of more than 10 percent was higher the smaller the firm was.[266] This thesis has argued that smaller firms realised higher profits because they were harder to monitor, their book-keeping was often sketchy and because the profit still contained the "salary" of the owner. However, it is of course possible that auditors focused primarily on larger firms and bothered to control small firms only when they had strong indications for excess profits. As long as we have no further information on how the auditors of the Court of Audit and the *Wehrmacht* selected the firms, readers are well advised to read the results with care and regard the presented figures as upper estimates.

b Construction Companies' Capital Adjustments after 12 June 1941

By the early 1940s, German firms had become so solvent and distributed dividends so high that the government saw itself compelled to revise the law on the distribution of profits. The Ordinance on the Limitation of Dividend Distributions (*Verordnung zur Begrenzung von Gewinnausschüttungen (Dividendenabgabeverordnung)*, DAV) of 12 June 1941 limited the dividend that joint-stock companies were allowed to distribute to 6 percent of their share capital. Dividends between 6 and 8 percent were heavily taxed, and distributions of more than 8 percent entirely forbidden. The intention behind the DAV, however, was not primarily to put industry on a short leash. Rather, two-digit dividends revealed an extent of war profiteering that could easily arouse discontent among a population forced to tighten their belts in the second year of the war. Thus the DAV included the possibility of massive capital adjustments: by raising the assessment base – the share capital –, firms could push the dividend under the required 6 percent without actually distributing less money to their shareholders.[267] Grün & Bilfinger, for example, already decided on 12 July 1941 to triple their share capital from 4.41 million RM to 13.23 million RM. The company could now pay a dividend of 5 percent, which in absolute numbers exactly matched the value of the previous dividend of 15 percent.[268] Another reason why the state accepted this behaviour was that

266 Cf. appendix, Table A6.3.
267 Curt Fischer, *Die Gesellschaft mit beschränkter Haftung* (Berlin, Göttingen, Heidelberg: Springer-Verlag, 1948), 15–17; Eichholtz, *Kriegswirtschaft II/2*, 533–38; Spoerer, *Eigenkapitalrentabilität*, 87–89; Banken, *Steuerpolitik*, 414–15.
268 Grün & Bilfinger AG, business report 1940.

the state was interested in companies able to finance their investments from their own equity, instead of competing with the hard-up state on the capital market.[269]

The majority of companies in the late 1930s was quite solvent and had created significant hidden reserves. By doing so, the industry had partly avoided the taxation of their rising profits under the increasing corporate taxes. The capital adjustments can serve as an indicator of the extent to which construction firms had accumulated hidden reserves and undervalued their equipment in the published balance sheets before 1941. For this purpose, I compiled a survey of 76 construction joint-stock companies.[270] It can be compared to the capital adjustments of joint-stock companies in the entire German industry, which earlier research has assessed (Table 15).

Table 15: Capital adjustments of German joint-stock companies according to Fischer[271] and of construction joint-stock companies (own calculation) as a consequence of the DAV of 12 June 1941.

	All joint-stock companies	Construction joint-stock companies
Total number, end of 1940	5,397	76
. . . and their capital, end of 1940, in million RM	*21,494*	*138*
Number of companies adjusting their capital	1,272	35
. . . and their capital before adjustment in million RM	*9,020*	*76*
. . . and their capital after adjustment in million RM	*13,397*	*133*
Adjustments in million RM	4,377	57
Adjustments in % of capital before adjustments	48.5	75
Number of companies adjusting their capital in % of all companies	23.6	40.1
Adjusted capital in % of total capital at the end of 1940	20.4	41.3

269 Banken, *Steuerpolitik*, 415.
270 The sample was compiled from: *Handbuch der deutschen Aktiengesellschaften 1943: Das Spezial-Archiv der deutschen Wirtschaft*, vol. 48 (Berlin: Hoppenstedt, 1943). This handbook covers more or less all of the more than 5,000 German joint-stock companies in 1943. All firms that are member of *Wirtschaftsgruppe Bauindustrie* or can be clearly identified as construction firms based on their company profile have been included in the sample. Real-estate companies, which earn their money mostly with the purchase and sale of estate and buildings, have been excluded. Thus, the survey of 76 construction firms represents practically all construction joint-stock companies in the German Reich as of 1943. Not included: Deutsche Asphalt- und Tiefbau-Aktiengesellschaft, Braunschweig, which was founded on 30 June 1942 and almost immediately raised its share capital by 50 percent.
271 Fischer, *GmbH*, 160. Other calculations of the capital adjustments of all German joint-stock companies arrive at very similar figures. Cf. Kurt Pritzkoleit, *Gott erhält die Mächtigen: Rück- und Rundblick auf den deutschen Wohlstand* (Düsseldorf: Karl Rauch Verlag, 1963), 59 ff. Eichholtz, *Kriegswirtschaft II/2*, 535.

Among the 76 construction companies in the survey, 35 (40.1 percent) increased their share capital as a consequence of the DAV of 12 June 1941. This was a share almost twice as high as in the rest of the industry. On average, the construction firms raised their share capital by 75 percent (against 48.5 percent in the case of all joint-stock companies). Overall, the adjusted capital made up 41.3 percent of the capital that all firms in the survey possessed by the end 1940. The corresponding figure for all German joint-stock companies was less than half of that, only 20.4 percent. In other words: not only were construction firms more likely than the average German joint-stock company to raise their capital between 1941 and 1943, the average value of adjustments was higher, too.[272]

Excess profits remained a problem after the outbreak of the war and the introduction of the KWVO.[273] According to the Court of Audit, special reports had shown that "a large number of construction firms" continued to realise excess profits in 1940 and 1941.[274] In March 1943, the Reich Price Commissioner (Hans Fischböck, who had succeeded Josef Wagner) still noted that "building costs have been rising in the last 6 years at a progressive rate".[275] Fischböck felt impelled to issue guidelines for the officers of the awarding authorities, reminding them of the basics of contracting and monitoring.[276] The outbreak of the war, which had caused a decline of construction activity within the Reich, does not seem to have reduced the profit opportunities of the German construction industry. Thus the regime began to pursue a two-pronged strategy, introducing laws and regulations while at the same time increasingly resorting to public threats. References were made to the First World War, when, according to the well-known stab-in-the-back myth, irresponsible war profiteers had contributed to the German defeat. This time, no one at home was to profit unduly from the war while German soldiers were sacrificing their lives for "*Führer*" and fatherland.[277]

---

272 It is worth noting that Table 15 does not even display all capital adjustments in the late 1930s and early 1940s. Most Austrian construction companies did not raise their share capital after June 1941, because they had done so already in 1938–39 during the conversion from schilling to reichsmark. The Viennese Bau-Aktiengesellschaft "Negrelli", for example, raised their capital by 800 percent after the conversion in 1939. Furthermore, many other firms had likewise adjusted their capital between the outbreak of the war and the introduction of the DAV, like the Huta Hoch- und Tiefbau AG, Breslau (+ 66 percent), Held & Francke Bau-Aktiengesellschaft, Munich (+ 100 percent), or the Elsässische Tief- und Hochbau AG, Strassburg (+ 300 percent).

273 Hermann Krauthause, 'Der Lohnfaktor bei der Preissenkung und Gewinnabführung', *Die Verwaltung* 18, no. 21 (1941): 398–401.

274 BArch, R 2301/7172, fol. 281–82, Letter by Elfert, 4.11.42.

275 Ibid., fol. 467–68, Letter by Reich Price Commissioner Fischböck, 8.3.43.

276 Ibid., fol. 469–70.

277 Rauh-Kühne, 'Hitlers Hehler', 9–15.

c Limiting and Collecting Excess Profits

In early 1939, a new tax was already introduced: firms had to pay a 15 percent tax on the differential between their profits in 1938 and 1937, that is, only on the share of the profit that exceeded the 1937 level.[278] After the outbreak of the war, section IV of the KWVO of 4 September 1939 provided the legal basis for further measures against excess industry profits. The KWVO had prohibited to demanding excess prices and compelled firms to reduce their prices when production costs decreased.[279] But this could not entirely prevent excess profits occurring in the overheated war economy. Hence, in a circular decree of 11 March 1941, Reich Price Commissioner Wagner now declared that firms that had realised excess profits after 1 September 1939 were obliged to report them. Wagner highlighted the responsibility of private business:

> The German people are waging war for the national existence of their *völkische* community and for every individual's physical and economic existence. This fact places an absolute duty of war upon every member of the German people's community. The German economy's war duty consists of securing the victory by achieving the highest performance. This includes considering the financial foundation of the community, especially the protection of the currency. The war has to be made as cheap as possible.[280]

To determine the level above which a profit was too high, the Price Commissioner defined certain reference values (*Richtpunkte*) for various sectors. The reference values for the construction sector did not constitute an undue hardship for the firms. The Price Commissioner allowed 12 percent RoS for the first million RM of a company's turnover, 11 percent between one and five million RM, 10 percent for the part of the turnover between five and ten million RM and 8 percent above ten million.[281] Firms were not allowed to lower their profits by deducting overly high contributions to social funds, wages above the legal level, or expenses for research and development that exceeded the customary level. The collected excess profits were to be frozen in special accounts of the Price Commissioner until the end of the war; officially, they were withheld only temporarily. The excess profits could be set off against losses in future years, should the situation arise.[282]

However, there were no systematic controls by the Price Commissioner or other institutions. Often, different agencies only monitored a single construction site of a

---

278 Spoerer, *Eigenkapitalrentabilität*, 91.
279 RGBl. 1939, I, 1609–13.
280 Circular decree no. 28/41 by the Reich Price Commissioner on the proceedings for the implementation of section IV of the War Economy Ordinance, 11.3.41, in: *Mitteilungsblatt des Reichskommissars für die Preisbildung* 5, part I, no. 12 (1941): 159–62.
281 BArch, R 13-VIII/188, Information by the Reich Price Commissioner for Wigru Bau regarding reference values of the KWVO, 27.5.41.
282 Spoerer, *Eigenkapitalrentabilität*, 91. See also: Rauh-Kühne, 'Hitlers Hehler', 22.

firm or certain aspects of book-keeping, not the prices and profits of the entire firm.[283] First of all, the firms themselves were supposed to report and repay excess profits to their local tax offices. This, of course, amounted to little more than wishful thinking. Normally, the tax offices detected excess profits. They sent their special reports to the Court of Audit, which again could inform the Price Commissioner. If the auditors detected that firms had not discharged their responsibility, they could initiate renegotiations on the prices and fine them.[284] In some blatant cases, criminal proceedings were initiated, as in the case of a salesman in Paderborn who had delivered overpriced camouflage paint.[285]

To begin with, the Reich Price Commissioner attempted to skim off the excess profits made in 1939 and 1940. This was extremely complex and time-consuming, and the number of auditors was hardly sufficient to maintain a system as intricate as that implemented by Wagner.[286] The process of detecting a firm's excess profits, renegotiating prices and imposing penalties could easily take a year. In 1943, the audits for 1940 were still not finished. By February 1942, only 232 million RM had been collected from the entire German industry: "[. . .] the fiscal result", as Cornelia Rauh-Kühne has rightly concluded, "was out of all proportion to the effort expended."[287] Tellingly enough, after the war a former head of department at the office of the Reich Price Commissioner, Wilhelm Rentrop, referred to this policy as "psychological warfare", rather than effective economic measures.[288]

In March 1942, the collection of excess profits realised in 1941 and 1942 was put on a new basis. Göring's Ordinance on the Registration of Exceptional Increases in Profits during Wartime (*Verordnung über die Erfassung außergewöhnlicher Gewinnsteigerungen während des Krieges,* GAV), of 31 March 1942 was now the relevant provision.[289] The first decree on the implementation of the GAV, issued on the same day, set out that every profit of 1941 that was more than one and a half times higher than the value in 1938 would be regarded as an extraordinary increase in profits. Of the sum exceeding the 1938 value, unincorporated businesses (*Personengesellschaften*) would have to pay 25 percent, corporations (*Körperschaften*) 30 percent to their local tax offices. The profit of Austrian firms could exceed the value of 1939 by one and three quarters, while firms

283 See for example the audits at Dyckerhoff & Widmann in the second half of the 1930s: BArch, R 2301/7172, fol. 486–88.
284 Streb, 'Negotiating Contract Types', 376.
285 BArch R2301/7187, fol. 88.
286 The fact that the system had overburdened the administration during wartime was the main reason for its subsequent abolishment: Franz Ott, 'Bauwirtschaftliche Fragen bei der Gewinnabführung 1943', *Die Bauindustrie* 12, no. 10 (1944): 217.
287 Rauh-Kühne, 'Hitlers Hehler', 25.
288 BArch, R 26-II ANH./2, Wilhelm Rentrop, 'Zu den Materialien zur Geschichte des Reichskommissars für die Preisbildung' (Unpublished manuscript, circa 1976), fol. 32.
289 Printed, together with the first decree on its implementation, in: *Die Bauindustrie* 10, no. 9 (1942): 245–47. See also Banken, *Steuerpolitik*, 311.

from the Sudetenland were allowed to earn twice as much as in 1939. Firms with a profit of less than 30,000 RM in 1941 were generally exempted from the payments. As the RFM specified on 20 May 1942, firms could also choose the average profit of the years 1936–38 as a reference value instead of 1938. Several exceptions existed for firms in the border regions of the Reich.[290]

What had been introduced with the GAV was *de facto* a tax. The tax offices could now detect the excess profits in the tax returns, which made their registration *potentially* much more comprehensive. Although tax authorities had constantly tightened their grip on businesses during the previous years,[291] it seems that the financial authorities still lacked the capacity to audit all tax returns of construction firms comprehensively. In July 1943, the office of the Price Commissioner insinuated that excess profits of the years 1941 and 1942 were "in the majority of cases only insufficiently registered."[292] Moreover, two very important aspects of the GAV should be kept in mind. The GAV also covered profits that German firms realised in the occupied territories, for example on *Wehrmacht* and OT construction sites.[293] However, as shown below, the results of the remote construction sites all over Europe were particularly difficult to monitor. Secondly, in choosing 1938, the state had selected a base year for the determination of excess profits when construction activity in the Reich had reached fever pitch and firms were already realising extraordinary profits, not least those working on the *Westwall*. Accordingly, in his memorandum on the Reich budget of 1939, the president of the Court of Audit came to the conclusion that the GAV had had no effect in large parts of the construction sector.[294] Therefore, when the GAV for the year 1943 (*Verordnung über die Gewinnabführung im Kalenderjahr 1943*)[295] was issued on 15 May 1944, the tax authorities no longer used 1938 as the reference year. Instead, they allowed firms a certain standard profit. The sum consisted of two elements, the first being 4–6 percent of the firm's capital, the second being dependent on the firm's turnover in relation to its assessed value (between 10 percent for a turnover up to 20 percent of the firm's assessed value and 1 percent for a turnover of 1,000 percent or more of the assessed value). Unincorporated businesses, where the owner made his or her living from the

290 *Die Bauindustrie* 10, no. 10 (1942): 333–35.
291 Banken, *Steuerpolitik*, 314–45.
292 BArch, R 13-VIII 178, Minutes of a meeting between representatives of the Reich Price Commissioner and Wigru Bau in Berlin, 7.7.43.
293 Elisabeth Meierkord, 'Die Abführung von Übergewinnen für das Wirtschaftsjahr 1941', *Die Bauindustrie* 10, no. 12 (1942): 312–15.
294 BArch, R 2301/3007, Memorandum of the president of the German Court of Audit on the Reich budget account for the budget year 1939, 19.4.43, fol. 260. See also: Ott, 'Bauwirtschaftliche Fragen bei der Gewinnabführung 1943', 217.
295 RGBl. 1944, I, 120–23, Decree on the collection of excess profits in 1943, 15.5.44.

firm's profits, could add an additional sum. All profits exceeding the permitted profit rate were taxed at 30 percent for unincorporated businesses and 35 percent for corporations.[296]

Despite these efforts, the Price Commissioner admitted at the beginning of the last year of the war that "[. . .] the regulations on the reduction of excess prices and the formation of decent profits are still not sufficiently observed by large parts of the construction industry."[297] The inability to truly effectively limit excess profits can be partly explained by the personal links between Wigru Bau and the agency of the Reich Price Commissioner. The latter was highly dependent on industry expertise when it came to calculating what constituted a "fair price" in construction. Some of the leading experts on pricing in the ranks of the Price Commissioner were actually men from the construction industry, like Herbert Prange of Siemens-Bauunion. For the draft of the Construction Price Ordinance of 16 June 1939, the Price Commissioner called in two experts, Gerhard Opitz of Siemens-Bauunion and Carl Baresel of C. Baresel AG. Opitz had finished a book manuscript on pricing in construction in 1938, which became the basis of the Construction Price Ordinance and further agency guidelines.[298] As it will be shown in chapter V.2, Opitz was also behind important parts of the contracts that EW introduced in occupied Norway from 1942 on. In the second half of the war, Reinhard Jecht of Dyckerhoff & Widmann replaced Opitz in his position as Wigru Bau's delegate to the Price Commissioner, while another pricing expert, Max Lütze of Wayss & Freytag, was the association's delegate to the OT.[299] Last but not least, with Hans Fischböck, a former board member of Universale Hoch- & Tiefbau AG, Vienna, had taken office as Reich Price Commissioner in 1942. When looking back on its history in 1966, the association of the construction industry proudly claimed that it had not been a mere heel-clicking stooge of the National Socialist regime, but had in fact been able to represent the industry's interests against the state.[300]

Nevertheless, a serious confrontation between state and construction industry emerged in the last year of the war, when the office of the Price Commissioner, in a last attack on excess profits in construction, initiated a retroactive limitation of the profits of the year 1942. The Price Commissioner's regional offices analysed the mandatory declarations on turnover and profits that the firms had to deliver according to the KWVO. Depending on the size of the firm, profits of 6.5 to 10 percent of turnover were allowed. The Price Commissioner tolerated profits exceeding these

296 Max Henze, 'Die gesetzlichen Grundlagen der Gewinnabführung für 1943', *Die Bauindustrie* 12, no. 10 (1944): 221–25; Banken, *Steuerpolitik*, 312.

297 BArch, R 13-VIII/188, Circular decree 6/44 by the Reich Price Commissioner regarding profit formation and price reduction in construction, 14.3.44.

298 Reuss, *Verbände*, 78–81.

299 Reuss, 81.

300 Reuss, 53.

rates by up to three percentage points, ultimately allowing firms to realise RoS rates of 9.5 to 13 percent. Profits between this level and up to 14.5 to 18 percent were confiscated and a caution issued. The fee for the caution was 3 RM. Only when firms realised profits above 14.5 to 18 percent, did they have to pay a severe fine in addition to repaying the excess profit.[301] The campaign was flanked by another circular decree, issued on 25 March 1944, which tightened regulations for accounting and invoicing and rendered them more concrete.[302]

This retroactive limitation of profits was accompanied by a lot of public sabre-rattling. The head of the office within the Price Commissioner agency concerned with the construction sector, Hess, launched articles in the press that contained fierce attacks on war profiteers. The tenor was that profits had risen to an extent that could no longer be explained by increased productivity, that large parts of the sector were ignoring the KWVO and that "generous" price calculations had become a habit even in reputable firms.[303]

The campaign provoked vigorous responses from entrepreneurs, who felt publicly pilloried and seemed genuinely surprised at this new tone.[304] Gerhard Opitz made a quasi-official response on behalf of the association in May 1944. Opitz's main argument was that it was not unusual for a construction firm's RoS to be subject to great fluctuations. High profits in boom years needed to compensate for losses in other years or single construction sites that did not break even.[305] Opitz' article caused Hess' office to inform all regional offices of the Price Commissioner that he did not approve of the article, as it "[. . .] could yield results that would run counter to the principles of a war economy and by that to §§ 22 ff. of the KWVO."[306] This statement had to be published in the Price Commissioner's own bulletin, since Wigru Bau refused to print it in its publication *Die Bauindustrie.*[307] At the same time, Wigru Bau asked the director of Universale Hoch- & Tiefbau, Heinrich Goldemund, to meet with Price Commissioner Fischböck in person in Vienna to discuss the matter

---

301 BArch, R 13-VIII/188, Circular decree 6/44 by the Reich Price Commissioner regarding profit formation and price reduction in construction, 14.3.44.
302 Circular decree 9/44 by the Reich Price Commissioner regarding the implementation of the Construction Price Ordinance, 25.3.44, in: *Mitteilungsblatt des Reichskommissars für die Preisbildung* 8, part I, no. 16 (1944): 149–56.
303 Two of the articles were: Kurt Enderlein, 'Probleme der Preispolitik in der Bauwirtschaft', *Die Bauindustrie* 11, no. 19 (1943): 509–19; 'Preisdisziplin auch für die Bauwirtschaft', *Neuer Wirtschaftsdienst* 11, no. 7 (1944): 1–2.
304 See for example: BArch, R 13-VIII/178, Letter by the construction firm Heinrich Butzer to Wigru Bau, 29.1.44.
305 Gerhard Opitz, 'Die Gewinne in der Bauindustrie', *Die Bauindustrie* 12, no. 7 (1944): 147–50.
306 *Mitteilungsblatt des Reichskommissars für die Preisbildung* 8, part I, no. 30 (1944): 322.
307 BArch, R 13-VIII/178, Letter by the leader of Wigru Bau, Gärtner, to the Reich Price Commissioner, 28.6.44.

and possibly arrange a meeting with Wigru Bau's leader Bruno Gärtner. Fischböck, however, did not seem to be in the picture about Hess' approach.[308]

We do not know whether a meeting between Gärtner and Fischböck took place in the end. It may have soon become meaningless, anyway, as the tangible economic consequences of the retroactive limitation of profits were less dramatic for the sector than what might have been expected in light of the public slugfest between Wigru Bau and the Price Commissioner. First, as the latter admitted, the allowed profit rates of 9.5 to 13 percent were quite generous and met the level of RoS in the second half of the 1930s when the industry had been realising very healthy profits.[309] Second, since the administrative effort would have been too high, the circular decree of 14 March 1944 had excluded all firms with an annual turnover of less than 100,000 RM from the campaign. This example supports the hypothesis recently formulated by Jonas Scherner, according to which small businesses were harder to monitor and hence were able to capitalise on information asymmetries to a higher degree.[310] Furthermore, in numerous meetings with representatives of the Price Commissioner during the summer of 1944, the construction firms and Wigru Bau were able to negotiate some exceptions and modifications in order to prevent cases of undue hardship. By and large, Wigru Bau was satisfied with the results of these negotiations.[311] Finally, a survey of the association among its regional divisions revealed that there was little reason for serious concern. Although this was only based on first, preliminary reports from the regional divisions, Wigru Bau noted that offices of the Price Commissioner acted rather reluctantly, showing "[. . .] greater or lesser courtesy [. . .]" when deciding on fines and penalties.[312] Last but not least, the Reich Price Commissioner's activities of were massively reduced after the summer of 1944. The *Wehrmacht* was combing out the administration for the last men able to stand straight and hold a rifle. The increasing lack of personnel forced Fischböck to issue a "simplification decree" on 17 August 1944. Price controls were reduced and the formation of prices completely suspended. Excess profits detected during the evaluation of the mandatory declarations on turnover and profits according to §§ 22 ff. of the KWVO were no longer to be collected.[313] In desperate need for money, the state also closed the accounts in which the collected excess profits from 1939–40 had

308 Ibid., Letter by Goldemund, Universale Hoch- & Tiefbau AG, to Gärtner, Wigru Bau, 11.7.44.

309 BArch, R 13-VIII/188, Circular decree 6/44 by the Reich Price Commissioner regarding profit formation and price reduction in construction, 14.3.44.

310 Scherner, 'Keine primäre moralische Verantwortung?', 166. The *Wehrmacht* had already expressed the same opinion in its *Inflationsdenkschrift* of autumn 1938: Geyer, 'Rüstungsbeschleunigung', 158.

311 See the circulars of Wigru Bau on this topic from summer and autumn 1944 in: BArch, R 13-VIII /188.

312 BArch, R 13-VIII/188, Letter Neubacher, Wigru Bau, to Knecht, Wigru Bau, 14.8.44.

313 Simplification decree, 17.8.44, in: *Mitteilungsblatt des Reichskommissars für die Preisbildung* 8, part I, no. 38 (1944): 416.

been frozen.[314] If the firms left half of the sum to the state, they would get back the other half of their excess profits.[315]

The discrepancy between the public campaign against war profiteering and the rather lenient policy actually pursued by the auditors reveals the dilemma that the Nazi regime was facing in public procurement. Rising profits burdened the state budget and were hardly compatible with Nazi propaganda. But at least in part, they were the result of the transition to fixed-price contracts, that is, of the state's own policy. Fixed-price contracts were to stimulate higher efficiency as the companies were able to keep the money saved as additional earnings. When these additional profits were later confiscated, "the fixed-price contracts that had been in use in 1942 were retroactively transformed into something more akin to cost-plus contracts."[316] The retroactive limitation of profits destroyed the main incentive mechanism in fixed-price contracts. The representative of one construction firm noted in September 1944:

> Under the current circumstances, the economic incentive to reduce costs by high effort fades away because the premium contractors sometimes receive from their clients are taxed away the next year. The state gives with the one hand and takes away with the other.[317]

As the regime collapsed shortly after, we can only speculate about the consequences of this breach of trust for the relation between the construction industry and the state, but it certainly was counterproductive to Todt and Speer's goal of increasing the construction industry's efficiency by exploiting its objective of profit maximisation.

As this chapter has shown, the German construction industry realised extraordinary profits from the mid-1930s on. As in other parts of the industry, their profit level was far higher than the disclosed commercial balance sheets suggested and remained on a healthy level far into the war. The conflict in 1944 shows that the regime remained responsive to the pleas of the industry to a certain degree even in the last year of the Nazi rule. What is more, the retroactive collection of profits ran counter to the interests not only of industry, but also of the experts in Fritz Todt's Armaments Ministry. In their view, it could only be counterproductive when companies had to fear that their cost savings from efficient work would be skimmed off later on.[318]

Finally, the rising inflation was no reason for the construction firms to shy away from a serious confrontation with the state on prices and profits. This

314 See above, chapter III.4.c, fn. 282.
315 Rauh-Kühne, 'Hitlers Hehler', 45, fn. 130.
316 Streb, 'Negotiating Contract Types', 377. For this thought, cf. Scherner, *Logik der Industriepolitik*, 43.
317 BArch, R 13-VIII/178, Letter Martens to Wigru Bau, 7.9.44. Here quoted in the translation of Streb, 'Negotiating Contract Types', 377.
318 Rauh-Kühne, 'Hitlers Hehler', 23.

confirms the findings of Marc Buggeln, who has suggested that even in a disintegrating economy, profit rates were something the industry still considered worth fighting for. In the light of the destruction of industrial facilities and houses all over Germany, construction firms knew that big tasks would be waiting for them after the war, and that they would thus sooner or later re-enter negotiations with German state authorities. Thus, firms did not want to agree on low profit rates because they feared that they could be pinned down to these rates after the war.[319]

319 Buggeln, 'Menschenhandel', 2010, 217–18.

IV German Construction Activities in Occupied Europe: The Case of Norway

1 The Organisation Todt in Wartime Europe

This chapter focuses on the OT's activities in occupied Europe. First, part a briefly sketches the OT's history and *modus operandi* between 1939 and 1945. Part b then discusses the legal status of the German OT firms in occupied Europe and analyses the increasing militarisation of work relations under the Organisation. Although existing literature has often pointed out this fact, so far it has remained unclear what it actually meant for the day-to-day work on the construction sites when large parts of the OT were given the status of a "*Wehrmacht* auxiliary force", and to what degree this impacted the OT contractors' scope of action. The central question is whether OT firms lost their identity as independent economic entities when they followed the Organisation to the occupied territories. In my opinion, answering this question is indispensable if we are to understand what kind of organisation the OT actually was. In part b, I also stress the need for a more systematic analysis of the OT's role in the Nazis' genocidal policy, an aspect this thesis can touch upon only briefly. Finally, part c gives an overview over the major reorganisation of the OT apparatus that took place after Fritz Todt's death in 1942.

a Expansion into Occupied Europe

Wherever German soldiers crossed a border in the following months and years to subject yet another country to German rule, OT units followed on their heels. After the invasion of Poland, the OT sent units for the reconstruction of roads, bridges and other infrastructure. When the *Wehrmacht* attacked France, Belgium, Luxembourg and the Netherlands in the spring of 1940, the main contractors of the *Westwall* formed mobile construction units, which often bore the firms' names. Germany's largest construction firm, the Philipp Holzmann AG, formed several of these units.[1] Like in Poland, their prime task was to rebuild destroyed infrastructure.[2] During the invasion of the Soviet Union, too, the OT worked with mobile and flexible units behind the front line.[3] It was during these military campaigns when the line became blurred between the OT as a mere construction organisation and as an auxiliary force with military tasks. To support the *Wehrmacht's* rapid advance in the West, the OT

1 Pohl, *Philipp Holzmann*, 248–50; Lemmes, 'Arbeiten für das Reich', 117.
2 Seidler, *Die Organisation Todt*, 26–31.
3 Seidler, 87 ff.

https://doi.org/10.1515/9783110694291-004

transported troops, ammunition and military equipment to the front lines.[4] In the Soviet Union, the OT Front Aid (*OT-Fronthilfe*) helped to evacuate wounded soldiers from the combat zone.[5]

However, missions of that kind remained the exception. Infrastructure projects continued to be a main field of activity, while the OT more and more often took over long-term projects. Among these, massive fortification lines were the most evident reference to the Organisation's roots. First and foremost, the Atlantic Wall was intended to protect the European coast from the Spanish-French border to the North Cape from an Allied invasion. From February 1943 on, the French Mediterranean coast was also fortified ("*Südwall*").[6] In Italy, several fortification lines along the coasts and west-east were built to impede the Allied advance.[7] The construction of airfields and submarine bunkers served military purposes, too. Other projects were related to the occupied territories' economic exploitation, such as the expansion of the Bor copper ore mines in Serbia[8] or the construction of facilities for shale oil extraction for the Baltische Öl GmbH in Estonia,[9] to name just two of the more prominent projects. The OT's widespread activities soon made it necessary to establish a central OT headquarters, the OTZ in Berlin, which subsequently also absorbed the apparatus of the *Reichsautobahnen.*

OT's headquarters was led by one of Todt's closest co-workers, Franz Xaver Dorsch (1899–1986).[10] Born in Illertissen, Dorsch had studied civil engineering in Stuttgart and Munich and had worked for the National Railways, before he entered Munich-based Sager & Woerner in 1929, where he met Fritz Todt. At the road construction company, Dorsch became "thoroughly acquainted with all of the details of the planning and projecting of the autobahns."[11] In January 1934, Fritz Todt asked Dorsch to join the *Reichsautobahnen* Corporation. Like many GdSt and OT officials would do after him, Dorsch had changed sides – from working for a

4 Seidler, 30.
5 Seidler, 96–97.
6 Lemmes, 'Arbeiten für das Reich', 133.
7 Lemmes, 168–71.
8 Rutar, 'Bor'.
9 On the background of the project, see Rainer Karlsch and Raymond G. Stokes, *'Faktor Öl': Die Mineralölwirtschaft in Deutschland, 1859–1974* (München: C.H. Beck, 2003), 213 ff. Karlsch and Stokes do not mention the contribution of the OT, though. Cf. also: Seidler, *Die Organisation Todt,* 100–101.
10 The following biographical sketch is based on Xaver Dorsch, 'The Organisation Todt in France and in Germany [MS B-670]', in *World War II German Military Studies: Part X. Special Topics, Continued*, ed. Donald Detwiler, vol. 24, World War II German Military Studies: A Collection of 213 Special Reports on the Second World War Prepared by Former Officers of the Wehrmacht for the United States Army (1946; repr., New York, London: Garland Publishing, 1979), 9–12; Seidler, *Die Organisation Todt*, 17–18.
11 Dorsch, 'OT in France and Germany', 10.

Reichsautobahnen contractor to supervising contractors. It would remain a cornerstone of the OT's *modus operandi*, especially during the war, that employees of construction firms took crucial positions at every level within Todt's and Speer's sphere of power.

In the course of the 1930s, Xaver Dorsch advanced to be the number two in the OT behind Todt and Speer respectively. At the *Westwall*, he was responsible for all questions regarding the workforce, before he led OT's mission in occupied France and became head of the OTZ in 1942. Since Albert Speer was only little involved in the day-to-day work at the OTZ, Xaver Dorsch emerged as the central figure in the German construction sector during the second half of the war.

Dorsch's political vita reads like the prime example of a National Socialist biography. After returning from the First World War, Dorsch had fought in one of the right-wing *Freikorps* that helped to put down the Communist uprisings in Bavaria in 1919. In 1922, he became a member of the SA, participated in the 1923 Beer Hall Putsch and joined the NSDAP in 1929. As Political Representative of the NDSAP (*Politischer Beauftragter der NSDAP*) in Todt's apparatus, Dorsch was certainly one of the "ideologists" among Todt's engineers and architects. His relation to Albert Speer, however, was stressed. Dorsch's loyalty was primarily directed towards the late Todt and at the same time, he was self-willed and highly ambitious. He managed to establish a permanent, personal access to Hitler, which he used to attack Speer in the serious power struggle of spring 1944. It took Speer all of his diplomatic and tactical skills to contain the conflict and maintain control over the OT and the Reich's construction sector.[12] It does not come as a surprise that Speer characterised Dorsch with blatant antipathy as "reprehensible" and "ruthless", when he was interrogated by Allied officers in 1945.[13]

One year before the EW was established, the OT dispatched an independent mission to Norway, too. In April 1941, OBL Trondheim, often referred to as *OT-Nord*, was commissioned with the construction of two massive submarine bunkers in Trondheim, *Dora I* and *Dora II*. The leading German construction firm on the site was Sager & Woerner, which had previously employed both Todt and Dorsch.[14] A small office in Oslo, *Leitstelle Oslo*, was set up to coordinate the supplies of building materials and machines to Trondheim. Within 27 months, *Dora I* was finished.[15]

12 On the conflict between Speer and Dorsch, see above, chapter III.3.c, as well as Müller, 'Armaments Policy', 413–15, 425–28, 485–89; Buggeln, 'Menschenhandel', 2010, 200–204; Brechtken, *Speer*, 240–51.

13 Office of the Publication Board, Department of Commerce, 'Interrogation of Albert Speer and Members of the Former Reich Ministry of Armaments and War Production [Technical Report No. PB 430]', 77.

14 Sönke Neitzel, *Die deutschen Ubootbunker und Bunkerwerften. Bau, Verwendung und Bedeutung verbunkerter Ubootstützpunkte in beiden Weltkriegen* (Koblenz: Bernard & Graefe Verlag, 1991), 120–30.

15 *Dora II* could not be finished.

Almost all supplies – cement, machines, construction iron and coal, as well as trucks and ships for their transportation – had to come from abroad. According to the head of OBL Trondheim, Fritz Altinger, the construction site was provided with 600,000 tonnes of material. Transports from Germany took 4 to 16 weeks.[16] The unit coordinated its construction works closely with the RK's Technical and Transport Department (*Abteilung Technik und Verkehr*), as the bunker was supposed to fit in with the plans for a whole new city: Neu-Drontheim.[17] The Nazis envisaged making Neu-Drontheim an important naval base and the northernmost metropolis of a German-dominated continent. After Todt visited Trondheim, Xaver Dorsch ordered OBL Trondheim to conduct drilling works in order to test the suitability of two potential building sites for the future city.[18] Hardly surprisingly, the city was to remain a National Socialist pipe dream.

This list of the OT's tasks in the occupied territories is far from complete. Activities were so extensive that Albert Speer's chronicler Rudolf Wolters estimated that in 1943, half of the German construction workers were working outside the Reich under the OT.[19] The Planning Office in Speer's Armaments Ministry noted in December 1943 that the aggregate consumption of cement in Greater Germany and of German authorities in the occupied territories matched the German pre-war consumption.[20] The OT's construction volume in reichsmark alone almost matched the German domestic volume in 1943.[21] In other words, through their engagement in occupied Europe, German construction firms were able to compensate to a considerable degree for the reduced construction volume within the borders of the Reich. There was probably no sector of the German economy that relocated its activities to the occupied territories to such an extent as construction.

During this time, the shift from domestic projects to OT projects in occupied Europe was considerable. The *Tiefbau* branch of Philipp Holzmann's office in Frankfurt am Main can serve as an example of what this could mean for a construction firm (Figure 13). The company was involved in OT projects in occupied France and Italy, which became the main pillar of the office's activities from 1942 on. The figure also shows that the firm relied on local subcontractors to a large extent.

---

16 Dorsch, *Die Organisation Todt*, 541.

17 RAO, RAFA-2174/Ef-L0003, file 10, Letters by Dorsch to Klein and Altinger, 22.9.41.

18 Ibid. On the Neu-Drontheim project, see also Sæveraas, 'Beton macht Geschichte', 91–92.

19 BAK, N 1318/5, Speer Chronicle, 1944, fol. 58.

20 BArch, R 3/1957, fol. 15–16, 2. weekly report of the Planning Office, completed on 12 December 1943, 20.12.43.

21 BArch, R 3/1808, fol. 285–86. In 1943, the OT was active in the occupied territories, apart from first deployments for the clearance of air raid damage in the Reich.

Figure 13: Turnover of the Tiefbau branch of Philipp Holzmann, Frankfurt am Main office, 1924–47, in million RM, current prices. Source: IfS, W1-2/269, Table and hand-drawn chart on annual turnover of the *Tiefbau* branch of Philipp Holzmann's office in Frankfurt am Main.

b Legal Status of the OT Firms and Militarisation of the Workforce

Compared to other sectors of the German economy, construction's involvement in occupied Europe stood out – and not only in terms of quantity. Private companies were embedded into a military structure in a new and unique way and had to make their decisions within an increasingly militarised framework. Their workers and employees were subject to militarisation and constant ideological indoctrination. The following part of the chapter will investigate two questions that are crucial for understanding the relation between the OT and its contractors. First: did the OT firms lose their identity as private businesses when working under the Organisation and accordingly, did they act increasingly out of a military, rather than economic, logic? And second: what did the militarisation of work relations under the OT mean for day-to-day work in the construction sites and the contractors' German workers? Were they *de facto* soldiers or still primarily civilian employees of private construction firms?

The question of the OT firms' legal status already concerned Allied intelligence and interrogators in and after 1945. During the post-war interrogations and trials, the former OT leaders Dorsch and Speer were of course keen to highlight the OT's civilian character. Albert Speer's defender in Nuremberg, Hans Flächsner, for instance, asserted in July 1946 that the Organisation "[. . .] was in no way a paramilitary organization as has been falsely asserted. Apparently this false assumption has been strengthened by the fact that the German members of the administration of the Todt Organization abroad wore a uniform. These people were considered as Armed Forces followers; but on the other hand the labor engaged by the firms and the construction workers of the firms, as well as the technical personnel, stood in no such relation [. . .]."[22] The British experts that compiled the *Handbook of the Organisation Todt*, on the other hand, argued that "the status of a firm in the OT is fundamentally similar to that of a unit or individual in the Armed Forces of the Reich. Whether such a firm applied of its own accord for enrolment as an OT-Firm or whether it was 'abgestellt für OT' (conscripted for the OT), it contracted itself to carry out construction under OT administration, just as a soldier, by taking the oath, contracts himself to Army Service." As will be shown, while the *Handbook's* comparison between an OT firm and a soldier is misleading, Flächsner's statement is demonstrably wrong. Discussions on the firms' legal status and relation to the military can be also be found in recent research literature. Most pointedly, Sæveraas has argued that firms were "conscripted" into the OT and that a differentiation between the OT and its contractors is "misleading".[23]

The OT's personnel can be divided into three groups. The OT organic personnel was the OT's own administrative and technical staff. The firm personnel consisted of the firms' salaried employees such as engineers, site managers, highly-skilled workers in a supervisory capacity and clerical personnel. Finally, there was the large group of recruited, conscripted and forced labourers, which were either civilians, POWs or inmates. All of these groups contained both Germans and foreigners. The share of foreigners was of course highest in the latter group.[24]

During the Second World War, the OT declared zones susceptible to enemy action "front zones" and those parts of the Organisation that were employed in these zones as *Front-OT*". Front zones were the *Wehrmacht's* operational zones along the western and eastern front, heavily bombed areas within the Reich, and the entire area of the *Einsatzgruppen Italien* and *Wiking*.[25] Within these zones, the firms' regular German construction workers, but not the abovementioned salaried employees, were declared front workers (*OT-Frontarbeiter*). Like the OT organic personnel, these front workers were first sent to the OT camps in Frankfurt am Main or Berlin-Wannsee. In these camps, they were given OT uniforms and received their pay and identity books as well as their identification badges. As all of occupied Norway was considered a front zone, all German construction workers under the EW were considered front workers. Salaried employees like the firms' management and clerical

22 International Military Tribunal, *International Military Tribunal (1945–1946 Nürnberg) Trial of the Major War Criminals before the International Military Tribunal Nuremberg: The Tribunal, 1947–1949*, vol. 19: Proceedings: 19 July 1946–29 July 1946 (Nürnberg: The Tribunal, 1948), 206.

23 Torgeir Ekerholt Sæveraas, 'OT, Wehrmacht og byggingen av Festung Norwegen', *Historisk Tidsskrift* 97, no. 3 (2018): 197–98.

24 Lemmes, 'Arbeiten für das Reich', 79–80.

25 *Handbook of the Organisation Todt*, 21–22.

personnel seem to have received neither uniforms, nor identification badges and pay and identity books.[26]

Only within the front zone, the OT was declared a *Wehrmacht* auxiliary force (*Wehrmachtgefolge*).[27] This status applied to all groups of OT personnel, regardless of whether they were wearing a uniform or not, or whether they were German or not.[28] The German personnel underwent basic military training and alarm drills.[29] In combat situations, the OT was unconditionally subordinate to *Wehrmacht* command.[30] At least in Norway, OT units normally did not participate in military activities, although the Allied forces received reports that members of EW had taken part in anti-invasion manoeuvres near Haugesund and Kristiansand.[31] In the light of what has been said here, it becomes clear that Flächsner's characterisation of the OT in Nuremberg had little to do with reality.

As members of the *Wehrmacht* auxiliary forces, OT members were protected by the Hague Convention just like regular soldiers. In general, they were subject to the military penal code (*Militärstrafgesetzbuch*, MStGB) and the *Wehrmacht* disciplinary code (*Disziplinarstrafordnung*, WDStO) and were thus tried by *Wehrmacht* courts if they committed an offence.[32] However, the OT workers never became soldiers in a

---

26 *Handbook of the Organisation Todt*, 138–45, 411. Cf. ThHStAW, Straßenbau AG Weimar Nr. 107, fol. 298–99.

27 On EW's status as *Wehrmacht* auxiliary force, cf. also Henne, 'Organisation Todt in Norwegen', pt. II, 9.

28 Lemmes incorrectly assumes that the private German construction firms' workers and employees were not declared *Wehrmacht* auxiliary forces: Lemmes, 'Arbeiten für das Reich', 79–80. Cf. Sæveraas, 'Festung Norwegen', 197.

29 As a training programme of spring 1944 from the Stavanger region shows, all OT members had to gather for training every Sunday from 9 to 11.30 a.m. This training included the handling of rifles and hand granades, basic combat tactics and instructions for the erection of temporary bridges, anti-tank obstacles and similar constructions: ThHStAW, Straßenbau AG Weimar Nr. 147, fol. 763, training programme by pioneer battalion 274 for all parts of the OT deployed around Stavanger, 1.2.44.

30 BA-MA, RW 39/39, fol. 75–76, Order by WBN von Falkenhorst regarding OT units in combat situations, 24.11.42.

31 RAO, RAFA-2188/2/Hh-L0003, Report on the set-up and main activities of Organisation Todt in Norway by Lt.Colonel A. Roscher Lund in the High Command of the Royal Norwegian Forces, 10.1.44, fol. 1.

32 This statement is certainly an oversimplification. It is very difficult to determine where and when various groups of OT members became subject to *Wehrmacht* jurisdiction. Likewise, what kind of offenses should be tried by *Wehrmacht* courts seems to have been a matter of dispute also among contemporaries. More research is necessary on this topic. See Peter Kalmbach, *Wehrmachtjustiz* (Berlin: Metropol Verlag, 2012), 98–102; Kerstin Theis, *Wehrmachtjustiz an der 'Heimatfront': Die Militärgerichte des Ersatzheeres im Zweiten Weltkrieg* (Berlin, Boston: De Gruyter Oldenbourg, 2016), 371–80.

legal sense. In principle, the labour laws preceded the MStGB and the WDStO.[33] Kerstin Theis has studied *Wehrmacht* court trials in the area of the Reserve Army. She found that the judges handed cases over to civilian courts more frequently in the case of members of the *Wehrmacht* auxiliary forces than in the case of regular soldiers. Moreover, sentences against the former were normally milder. Theis also assumes that many cases remained unreported by the OT and its contractors, because neither of them was interested in losing a worker in times of scarce labour.[34]

Also in occupied Norway, it seems that OT members were tried by *Wehrmacht* courts only seldom.[35] Rather than proceedings against OT men before *Wehrmacht* courts, the example of a German OT front worker from September 1943 was probably more representative for the situation in Norway: the man, working for Arge Stallmann under OBL Mo i Rana, had refused to work, which was a severe offence under the National Socialist regime. After having consulted the OBL, however, the firm only warned the worker and subtracted 30 kroners from his next pay check as a fine.[36] Court proceedings were always directed against individuals, mostly concerning property and currency offences, personal misconduct and disobedience. *Wehrmacht* court proceedings were not an instrument to monitor and penalise OT firms.

It seems that the decision to declare the OT in the front zone an auxiliary force of the *Wehrmacht* had very practical reasons, as this meant that the Organisation was able to rely on the military infrastructure in the occupied countries. The OT could use the military's postal service, railway tickets and telephone system.[37] Most importantly, it could send its construction materials on *Wehrmacht* transports. After the war, Xaver Dorsch emphasised that it was these practical reasons that lay behind the decision to declare the OT *Wehrmachtgefolge*: "The OT was not directly under the command of the *Wehrmacht*, but was treated as a subsidiary organisation (*Wehrmachtsgefolge*) because, lacking the facilities thus granted in the war zone and the occupied territories, it would have encountered extreme difficulties."[38] Furthermore, OT members received their provisions from the same sources as *Wehrmacht* soldiers.

---

33 Rudolf Absolon, *Die Wehrmacht im Dritten Reich*, vol. V (Boppard am Rhein: Harald Boldt Verlag, 1988), 239–40.

34 Theis, *Wehrmachtjustiz an der 'Heimatfront'*, 371–80.

35 Information provided by Professor Maria Fritsche, NTNU, who is currently researching the activities of *Wehrmacht* courts in occupied Norway.

36 Note by Arbeitsgemeinschaft Stallmann, 24.9.43. The document is in private hands and was made available to me by Björn Stemler.

37 Dorsch, 'OT in France and Germany', 24.

38 Franz Xaver Dorsch, 'Organization Todt – Operations in the West [MS B-671]', in *World War II German Military Studies: Part X. Special Topics, Continued*, ed. Donald Detwiler, vol. 24, World War II German Military Studies: A Collection of 213 Special Reports on the Second World War Prepared by Former Officers of the Wehrmacht for the United States Army (1947; repr., New York, London: Garland Publishing, 1979).

Beyond such benefits, however, the construction workers employed by OT contractors were exposed to a growing militarisation of work relations. The OT had been a driving force in this development during the previous years.[39] Fritz Todt had given new orders to the remaining workers on the *Westwall* on 4 September 1939, leaving no doubt that the Organisation more than ever intended to spearhead the militarisation of German society. “In the state of war,” Todt wrote, “every single one, worker, businessman, or OBL engineer, even if he does not wear a uniform, is no longer a civilian, but a soldier in his attitude.”[40] The firms’ foremen were to guide and supervise the daily routines before and after work. Workers had to gather, were counted and marched to the construction site together. The weekly payment of wages was also staged as a muster, at which the firm’s site manager should also announced new directives and read the latest *Wehrmacht* reports.[41] Throughout the war, the OT’s propaganda flamboyantly highlighted the military character of the Organisation’s work, for example in its magazine *Der Frontarbeiter*.

The OT’s front area personnel section was an important institution with regard to the militarisation of work relations and the National Socialist “education” of workers. Named the *OT-Frontführung*, it was led by so-called *Frontführer*. The task of these *Frontführer*, who were based with every OBL, was to educate the workforce on a conduct appropriate for wartime, establish routines and facilities on the construction sites with regard to safety, camouflage and alarms, conduct controls in the living quarters and keep an eye on the workers’ clothing and equipment.[42] Over the course of the war, the *Frontführer* below the OBL level were increasingly recruited from the site managers or *Stammarbeiter* of the private construction firms.[43] These men were supposed to distinguish themselves by a “soldierly spirit and commitment, social understanding and passion for people, National Socialist thinking and innate leadership qualities.”[44] Thus the OT directly entrusted its contractors with tasks of National Socialist indoctrination and security.

Last but not least, the OT equipped some regular employees of private construction firms with weapons. Initially, only the OT’s own security guard units (*Schutzkommandos*) carried arms to supervise forced labourers on the construction sites, while *Wehrmacht* soldiers guarded the POWs. However, the units were poorly manned, which made it necessary to assign supervisory tasks to the members of private construction firms, too. What is more, the men needed to be able to

---

39 Lemmes, ‘Arbeiten für das Reich’, 58–62.
40 BArch, NS 26/1188, fol. 221–22.
41 Ibid.
42 Seidler, *Die Organisation Todt*, 148.
43 Henne, ‘Organisation Todt in Norwegen’, pt. III, 32.
44 BArch, R 50-I/226, Guidelines for the *Frontführer* in the OT construction units (construction firms) of OT, *Einsatzgruppe West*, 24.12.43.

defend themselves in case of attacks by Allied forces or partisans.[45] For obvious reasons, Xaver Dorsch emphasised that the arming of OT members was defensive in nature when he provided a summary of the OT's activities to his interrogators after the war. The OT first armed workers during the campaign in the Soviet Union and on the Balkans, where partisan activity was particularly high.[46] However, Dorsch failed to mention that the armed supervision of forced labourers by German construction workers was a key element of the OT's mode of operation. In what has been described as the OT's "hinge function" (*Scharnierfunktion*) between military, industrial and racial-ideological aims, the Organisation integrated private businesses into the National Socialist system of oppressing and exploiting millions of forced labourers.[47] The OT promised to solve one of the big dilemmas of the National Socialist war economy: the dilemma between the need for workers and the security concerns about a mass deployment of foreigners in German-controlled areas.

"A new task", the OT propaganda announced, "has accrued to the front worker: he must show that his technical competence and his maturity, which have grown through wartime experiences and enormous commitment, qualifies him to be accepted as leader by the foreign [*fremdvölkische*] workers."[48] The relation between the German and the foreign workers, however, was based far more on force than on "acceptance". Guidelines by *Einsatzgruppe West* in France illustrate what the Organisation expected from its German construction workers. If a forced labourer tried to escape, attacked a German or defied orders in any way, guards should shoot if other measures failed. In case of a coordinated attempt to escape, guns should be used immediately and ruthlessly. Every guard who was lenient was to be punished himself. The guidelines explicitly applied to all armed Germans on construction sites, including the companies' foremen and workers.[49] The vast literature on National Socialist forced labour had already documented that German workers had significant personal room for manoeuvre in their contact with forced labourers.[50] The members of construction firms under the OT were in a particularly powerful position. Their superior position as Germans, at the top of the Nazi racial hierarchy, was enhanced by their supervisory function within the OT. In June 1944, *Einsatzgruppe West* once more hammered home the role of German construction workers: "No German must be unarmed in his quarters and on the construction site. [. . .] Every [German, S.G.] OT man is more than ever responsible for holding

45 Sæveraas, 'Beton macht Geschichte', 193–95; Hatlehol, 'Norwegeneinsatz', 221–22.
46 Dorsch, 'OT in France and Germany', 24–25.
47 Herbert, *Fremdarbeiter*, 419.
48 Theodor von Kohary, 'Die Organisation Todt am Jahresbeginn 1941', *Die Bauindustrie* 9, no. 5 (1941): 1–3.
49 BArch, R 50-I/224, Special instructions for guards by OT Einsatzgruppe West, 15.3.44.
50 Spoerer, *Zwangsarbeit unter dem Hakenkreuz*, 190 ff.; Herbert, *Fremdarbeiter*.

the foreign workers together. [. . .] The rounding up of foreign workers is also necessary for reasons of security. Every runaway [foreign, S.G.] OT man can defect to the terrorists and reappear as an enemy."[51]

Not least, Albert Speer linked the wartime deployment of the OT workers to the large-scale aims of the National Socialist *Lebensraum* ideology. On Christmas Eve 1942, Speer held a speech in front of OT workers at the Atlantic Wall: "He knew full well", the minister assured his workers, "how difficult it was to execute this task when often, a single German foreman was responsible for 20 to 50 foreigners. But this leadership functions would prepare the German worker for the great tasks of colonisation during peacetime."[52] In Speer's logic, being a member of the future "master race" in conquered Eastern Europe would reward the OT worker for the hardships of his wartime duty.

Fragmentary reports and sporadic references in the literature mention war crimes committed by OT men. Several cases are documented for the Channel Islands. The situation was worst on the island of Alderney, where "[. . .] the combination of their haste to achieve construction targets [. . .], the corruption which flourished on this outpost of the Reich, and their disregard for the safety of the slave workers ensured a terrible cost in human life."[53] A former forced labourer recalled that "it wasn't just the fascists who beat us but the civilians [OT personnel] as well."[54] Other witnesses reported murders by OT men on Alderney and two OT members were sentenced to prison by a French military tribunal in 1949.[55] The death rate among the forced labourers was as high during the time when OT and *Wehrmacht* held control over the islands as under the rule of the SS.[56] In January 1945, approximately 150 OT men – Germans, but also men from Ukraine, the Baltic states, Belgium and France – helped to guard a death march of 7,000 Jewish inmates from the Stutthof concentration camp to Eastern Prussia. Only 200 of the inmates survived the brutal march and the following massacre by SS units.[57]

In Ukraine, OT *Einsatzgruppe Rußland-Süd*and its contractors built the *Durchgangsstraße IV* on behalf of the SS, using more than 100,000 forced labourers. As a result of the National Socialist policy of "extermination through labour" or by shootings, approximately 25,000 Jews perished during the road's construction. OT members' personal scope of action was immense and their cooperation with the SS

51 BArch, R 50-I/304, Political Information no. 4 of the front area personnel section of *Einsatzgruppe West*, 25.6.44.
52 BAK, R 1318/2, Speer Chronicle, 1942, fol. 126–27.
53 Bunting, *Model Occupation*, 158.
54 Bunting, 173.
55 Bunting, 283, 294.
56 Karola Fings, *Krieg, Gesellschaft und KZ: Himmlers SS-Baubrigaden* (Paderborn: Ferdinand Schöningh, 2005), 197–214.
57 Ian Kershaw, *Das Ende: Kampf bis in den Untergang, NS-Deutschland 1944/45* (München: Deutsche Verlags-Anstalt, 2011), 266–69.

men close. Hermann Kaienburg has pointed out the central role that road construction played in the National Socialist plans for occupying, punishing and finally killing the European Jews between 1933 and 1945.[58] The GdSt and the OT were the two agencies that became almost synonymous with road construction in Nazi Germany and later occupied Europe. It is thus remarkable how seldom the literature has established a connection between them and the Nazi genocidal policy, considering that the probably most infamous document on the Shoah, the Wannsee Protocol, stated that

> in the course of the Final Solution and under appropriate leadership, the Jews should be put to work in the East. In large, single-sex labour columns, Jews fit to work will work their way eastwards *constructing roads*. Doubtless the large majority will be eliminated by natural causes. And doubtless any final remnant that survives will consist of the most resistant elements. They will have to be dealt with appropriately, because otherwise, by natural selection, they would form the germ cell of a new Jewish revival.[59]

It was at the *Durchgangsstraße IV* building site in 1943 that a local military court convicted Johann Meißlein, an overseer of the construction firm Eras KG from Nuremberg. He had ordered the shooting of two enfeebled Jewish women. Meißlein was sentenced to three months in prison, but neither for murder nor manslaughter. In the eyes of the court, Meißlein's crime had been *Amtsanmaßung*, the unauthorised assumption of powers he did not have. Ordering the execution of the two women, the court clarified, would have been a matter for the police or SS.[60]

These are quite well-documented cases, but in general, direct evidence of mistreatment by German workers is seldom in the documents available to today's historians. In particular, it is often hard to distinguish whether crimes were committed by SS or *Wehrmacht* guards, by the OT's own security guard units or by employees of construction firms. A letter from EW's OBL Kristiansand to Strabag's site manager in Stavanger is quite revealing, however. The OBL obviously saw "[. . .] reason to

58 Hermann Kaienburg, 'Jüdische Arbeitslager an der "Straße der SS"', *1999. Zeitschrift für Sozialgeschichte des 20. und 21. Jahrhunderts* 11, no. 1 (1996): 13–39. Fritz Todt also played a pioneering role in employing non-German Jewish forced labourers within the borders of the Reich. From autumn 1940 onwards, the *Reichsautobahnen* corporation deployed Polish Jews on its construction sites in Brandenburg and Silesia: Wolf Gruner, 'Juden bauen die "Straßen des Führers". Zwangsarbeit und Zwangsarbeitslager für nichtdeutsche Juden im Altreich 1940 bis 1943/44', *Zeitschrift für Geschichtswissenschaft* 44 (1996): 789–808.

59 Mark Roseman, *The Villa, the Lake, the Meeting: Wannsee and the Final Solution* (London: Penguin Books, 2002), 113. [My own emphasis, S.G.]. Ulrich Herbert has rightly emphasised that the references to work projects in the east at the Wannsee Conference were mostly a fiction: Ulrich Herbert, 'Labour and Extermination: Economic Interest and the Primacy of Weltanschauung in National Socialism', *Past & Present* 138 (1993): 170–71. The example of *Durchgangsstraße IV* shows, however, that this fiction sometimes materialised.

60 Konrad Kwiet, 'Judenmord als Amtsanmaßung. Das Feldurteil vom 12. März 1943 gegen Johann Meißlein', *Dachauer Hefte* 16 (2000): 125–35.

emphasise that the mistreatment of prisoners of war is forbidden and moreover beneath the dignity of a German man and front zone worker." Of course, OBL Kristiansand needed to make sure this would not be misunderstood. The directive should by no means have the effect of decreasing the efficiency of the POWs: "Sluggish performance has to be punished." The guards of the POW camps would be able to "punish the person in question in a way that a sufficient work performance is still guaranteed", OBL Kristiansand explained.[61] A letter from the head of EW's road construction department, Hermann Hesse, to the leader of EW, Willi Henne, of June 1943 leaves no doubt about EW's responsibility. Hesse writes that he had been informed by the construction firm Hans Röllinger KG (Fürth) that they had recently stopped the beating of Yugoslavian inmates on their road construction site in the hope that this would raise the men's performance.[62]

In 1946, several Germans were tried in Belgrade for war crimes against Yugoslavian forced labourers around Trondheim.[63] As the witness accounts and trial documents reveal, some of the defendants were employees of the firms Müller-Altvatter (Stuttgart) and Eschweiler Tiefbau – J. Pellini (Eschweiler). They forced the foreign workers to work until exhausted, and beat them with their bare hands, spades and picks. Dozens of Yugoslavian workers died in this region on the OT's construction sites or in the camps.[64]

Rolf-Dieter Müller rightly concludes that "[. . .] it became more and more obvious that the OT were not just 'soldier-workers', but part of the Nazi system of terror and annihilation. Its members were committed in anti-partisan campaigns and supervised teams of Jewish slave-workers and Soviet POWs. They were technicians, slave-drivers, and in some cases murderers."[65] One of these "anti-partisan campaigns", a common code for mass murders during Germany's war of annihilation in the East,[66] is mentioned in the Speer chronicle. In the area of *Einsatzgruppe Rußland-Mitte*, the chronicler Rudolf Wolters notes in July 1943 that "a further rise of the danger of partisans could be prevented temporarily by means of purges ("*Säuberungsaktionen*") conducted conjointly by the Wehrmacht, the SS and the OT."[67] Research on OT members' participation in war crimes is still a desideratum

61 ThHStAW, Straßenbau AG Weimar Nr. 139, fol. 174, Letter OBL Kristiansand to site manager Korffmacher, Strabag Weimar, 10.12.43.
62 RAO, RAFA-2188/1/E2b-L0014, file B.V.c. Rognan-Langset, Letter Hesse to Henne, 15.6.43.
63 Hatlehol, 'Norwegeneinsatz', 278–80.
64 For detailed references to these sources, mostly from the Military Archives in Belgrade (*Vojni Arhiv*), see Hatlehol, 279–80. I would like to thank Michael Stokke for providing me with these documents.
65 Rolf-Dieter Müller, 'Todt Organization', in *The Oxford Companion to World War II*, ed. Ian Dear (Oxford: Oxford University Press, 2001), 870.
66 Peter Longerich, *Holocaust: The Nazi Persecution and Murder of the Jews* (Oxford: Oxford University Press, 2010), 214.
67 BAK, N 1318/4, Speer Chronicle, 1943, fol. 20.

in the scholarship on the Second World War. Analysis is needed that goes beyond the documentation of single cases and that systematically investigates the role and tasks of the OT during the German retreat in particular.[68]

As has been shown, German construction firms and their workers acted under the OT in an increasingly militarised environment during the Second World War. They were part of a society in which moral standards had been razed for years, particularly when it came to the treatment of people defined as "racially inferior". When working under the OT all over Europe, the firms became part of the occupying power, insofar as they took part in supervising forced labourers and implementing the National Socialist ideology in day-to-day work. In this respect, Sæveraas is right when stating that "a contract with the OT meant something more than just an economic relation between a state agency and a private business."[69]

The central question in this thesis, however, is whether construction firms lost their identity and – so to speak – dissolved in the OT apparatus and whether their motives and strategies can no longer be understood as being based on an economic rationale. In this respect, comparing German OT contractors to conscripted soldiers appears rather misleading. Contrary to a soldier, firms entered contracts with the Organisation for a limited period of time and negotiated on contract conditions and prices. Moreover, for most firms, the OT was only one of several customers. The management of a German firm that had one of its branches working under the EW in Norway, for example, was neither subordinate to the EW nor part of the *Wehrmachtgefolge*. In the sources from private construction firms that have been analysed for this thesis, the distinction between "us" and "the OT" is always clearly identifiable. The interests of the OT and its contractors often overlapped, but they were never identical.

Thus, chapter V will show in detail that economic rationality remained the guiding principle for OT firms and that they did not lose their identity as independent economic entities under the OT.[70] This, however, does in no way rule out that the firms' employees were influenced by National Socialist ideology, became part of criminal occupation regimes and committed war crimes.

---

68 Magnus Brechtken emphasised this recently: Brechtken, *Speer*, 668, fn. 47. Franz W. Seidler's description of the OT's tasks during the campaign in the Soviet Union and on the retreat is completely inadequate in this respect. The chapter in Seidler's book is mostly based on propaganda articles, without any noticeable critical assessment of the material: Seidler, *Die Organisation Todt*, 88–108.

69 Sæveraas, 'Festung Norwegen', 196.

70 For certain regions of Europe, a reservation must probably be made for the last months of the war when the established administrative and economic structures collapsed in the face of the rapid Allied advance.

c Reorganisation under Albert Speer

By February 1942, the OT's construction firms were working from Trondheim and Rovaniemi in the north to the Balkan peninsula in the south, from Brittany in the west to Minsk and Kiev in the east. The Organisation had doubtlessly become a major power in the war economy of German-occupied Europe, when Fritz Todt died in a plane crash at the "*Führer*" headquarters near Rastenburg (Kętrzyn) in East Prussia on 8 February. The accident's circumstances were never entirely clarified and thus left room for speculation. In all likelihood, however, the accident was the result of a pilot error, technical fault, bad weather conditions or a combination of the three. Only few hours after the accident, Hitler appointed Albert Speer Todt's successor in all his positions. At not even 37 years old, Speer was at the height of his power as Reich Minister for Armaments and Ammunition, head of the OT, General Inspector for German Roads and GB Bau – to name only some of his offices.

Albert Speer was born on 19 March 1905 in Mannheim into a wealthy upper-middle-class family, his father a successful architect, his mother from a family of merchants.[71] This background allowed him not only to grow up with seven servants, but spared him financial worries during his studies in the 1920s. Speer studied architecture in Karlsruhe, Munich and Berlin. He gained his diploma from the Technical University in Berlin in November 1927 and continued to work there as a faculty assistant. The university was known as a stronghold of the Nazi movement and Albert Speer became a member of the NSDAP and the SA on 1 March 1931, after having attended several Nazi campaign events over the course of the winter, including speeches by Hitler and Goebbels. In the autumn of 1932, Speer became a member of the SS. His career as an architect gained momentum after Hitler's seizure of power when he first took over smaller construction projects for the party and then, in 1934, was commissioned to design the architecture of the Nazi party rally grounds in Nuremberg. Speer managed to establish himself in Hitler's inner circle, and on 30 January 1937 he was appointed GBI, responsible for making Hitler's dream of a metropolis named "Germania" reality. Closely related to his duties as GBI were several transport and construction units that he established around this time and that later became a part of the OT empire. After taking over Fritz Todt's posts, Albert Speer became the *de facto* number two in the Nazi state between 1942 and 1945.

Of the innermost circle of the Nazi regime, Albert Speer certainly was the most prominent figure to survive the war and the post-war trials. At Nuremberg, he was sentenced to 20 years in prison, which he served in full in Spandau. From his

---

71 Brechtken, *Speer*, 19. Brechtken provides the latest and most comprehensive critical biography on Speer. This short biographical sketch is based on Brechtken as well as Matthias Schmidt, *Albert Speer. Das Ende eines Mythos* (1982; repr., Berlin: NZ Netzeitung, 2005), 46–77.

release in 1966 to his death in 1981, Albert Speer shaped our picture of the Third Reich like no other representative of the Nazi state. Historians and journalists alike paid court to this unique source – without paying too much attention to the inconsistencies and plain lies in his enthralling accounts. This holds true not least for historians from the English-speaking world, such as Hugh Trevor-Roper or Alan Bullock.[72] For them, the eloquent and well-mannered Speer was the opposite of the simple-minded SA thug. Speer represented those Germans who had unfortunately become entrapped in the Nazis' crimes, but were about to become a reliable part of the Western world again. Moreover, Speer willingly shared his expertise on the German war economy with his post-war interrogators.[73] In his *Erinnerungen* (1969), the *Spandauer Tagebücher* (1975), *Der Sklavenstaat* (1981) as well as in numerous interviews and home stories, Speer knew perfectly how to regale the public with voyeuristic details from inside the regime.[74] At the same time, Speer embellished the defence strategy that had already saved his neck during the Nuremberg trials and now fell on fertile ground in German post-war society. To be sure, he claimed to be the organiser and mastermind behind the alleged German "armaments miracle" that enabled Germany to wage war for more than five years. Nevertheless, he presented himself as a politically naive technocrat, a common man among the "real Nazis", his advancement in the Nazi hierarchy being little more than a sequence of coincidences. Speer cleverly played the remorseful Nazi by taking on an impersonal, abstract responsibility for the National Socialists' crimes. He knew full well that he could not deny having been a member of a criminal government. But he strictly avoided admitting to any personal guilt. According to his account, he neither knew what had been happening in the concentration and extermination camps, nor did he have knowledge of the conditions in armament plants and on construction sites where forced labourers worked themselves to death.

Speer managed to maintain this public image until his death in 1981, and both historians and the public only slowly freed themselves from the spell of this historic figure. It took decades, to name just one example, to refute one of Speer's biggest and most absurd lies, according to which he had had no knowledge of the genocide of the European Jews.[75] The first critical study, Matthias Schmidt's doctoral thesis

72 Trommer, *Rechtfertigung und Entlastung*, 53–56.

73 Brechtken, *Speer*, 295–98, 421–36.

74 Albert Speer, *Erinnerungen* (Frankfurt am Main, Berlin: Ullstein, 1969); Albert Speer, *Spandauer Tagebücher* (Berlin: Propyläen, 1975); Albert Speer, *Der Sklavenstaat. Meine Auseinandersetzung mit der SS* (Berlin: Ullstein, 1981). English editions: Albert Speer, *Inside the Third Reich* (New York: Macmillan, 1970); Albert Speer, *Spandau: The Secret Diaries* (New York: Macmillan, 1976); Albert Speer, *Infiltration: How Heinrich Himmler Schemed to Build an SS Industrial Empire* (New York: Macmillan, 1981).

75 In practice, the debate boiled down to the question whether Speer had been present at the town hall of Posen (Poznań) in 1943, when the leader of the SS, Heinrich Himmler, spoke openly about the extermination of the European Jewry. Speer himself had spoken in the morning and claimed

of 1982, remained largely unnoticed, before Gitta Sereny and Dan van der Vat presented two major biographies in the 1990s.[76] In 2002, Susanne Willems was able to identify Speer as one of the main driving forces behind the deportation of the Jews of Berlin, as the "*Führer's*" architect needed their homes for the Berliners who had to make room for his ambitious reshaping of the capital.[77] In recent years, economic historians have debunked the myth that the German "armaments miracle" after 1942 primarily originated in the minister's brilliant mind, although Speer tried to take credit for it as long as he lived, wrote and gave interviews.[78] Most recently, Isabell Trommer's doctoral thesis and Magnus Brechtken's comprehensive biography have put the focus on Speer's self-staging after 1945 and not least on how the German and international public wanted to believe the stories of the "good Nazi".

As the new head of the OT, Albert Speer initiated some major changes at the organisational level after his predecessor's death in February 1942. On 18 February 1942, Speer already announced the details of the reorganisation.[79] The OT's tasks in occupied Europe were now executed by *Einsatzgruppen*. *Einsatzgruppe West* in Paris was responsible for the construction projects in France, Belgium, the Netherlands and on the Channel Islands, *Einsatzgruppe Wiking* in Oslo for Norway, Denmark and a small part of Northern Finland. *Einsatzgruppe Südost* had its headquarters in Belgrade. Four *Einsatzgruppen* were established in the occupied Soviet Union. Three of them were attached to the army groups of the *Wehrmacht* (*Rußland-Nord*, *Rußland-Mitte* and *Rußland-Süd*), while *Einsatzgruppe Jakob* was a mobile unit operating directly behind the frontline. In 1943, *Einsatzgruppe Italien* was established in Sirmione and for a short time, an *Einsatzgruppe* also existed in the Caucasus, the easternmost region under German occupation.[80] The *Einsatzgruppen* were able to work relatively autonomously and were structured in the familiar way, with *Oberbauleitungen* (OBLs) on the regional level, *Bauleitungen* (BLs) and *Abschnittsbauleitungen* (ABLs) on the local level, and

---

that he had left Posen before Himmler's speech. However, in 1971, a transcript of the speech surfaced which showed that Himmler had personally addressed Speer during the speech. Speer therefore asked old confidants to provide him with an alibi and sign declarations in lieu of oath that Speer had pre-formulated for them. See Erich Goldhagen, 'Albert Speer, Himmler and the Secrecy of the Final Solution', *Midstream* 17 (1971): 43–50; Gitta Sereny, *Das Ringen mit der Wahrheit: Albert Speer und das deutsche Trauma* (München: Kindler, 1995), 451–68; Heinrich Breloer, *Unterwegs zur Familie Speer: Begegnungen, Gespräche, Interviews* (Berlin: Propyläen, 2005), 578–79; Brechtken, *Speer*, 458–64.

76 Schmidt, *Speer*; Sereny, *Speer*; van der Vat, *The Good Nazi*.

77 Susanne Willems, *Der entsiedelte Jude: Albert Speers Wohnungsmarktpolitik für den Berliner Hauptstadtbau* (Berlin: Edition Hentrich, 2002).

78 Tooze, 'No Room'; Scherner and Streb, 'Ende eines Mythos?'; Tooze, *Wages*, 552–89; Budraß, Scherner, and Streb, 'Fixed-Price Contracts, Learning and Outsourcing'.

79 BArch, R 50-I/3a, Letter by Speer regarding the reorganisation of the OT, 18.2.42.

80 The *Einsatzgruppen* were subject to constant reorganisation. *Einsatzgruppe Jakob* seems to have been merged with the *Einsatzgruppen Rußland Nord/Mitte/Süd* in the course of 1942. *Einsatzgruppe Rußland-Nord* moved its headquarters to Königsberg (Kaliningrad) in 1944 and became

finally the actual construction sites. Sometimes, so-called *Einsätze* functioned as coordinating entities between several OBLs and the headquarters of an *Einsatzgruppe*.

In the following years, Albert Speer seems to have not bothered too much about the details of the OT's everyday work. Xaver Dorsch, head of the OTZ, thus became the *de facto* leader of the OT for the rest of the war. It is not entirely clear whether the reorganisation of the OT and establishment of the *Einsatzgruppen* was a result of the change in leadership from Todt to Speer. However, only ten days lay between Todt's death and Speer's reorganisation decree. It is very likely that the OT already had had some plans in the pipeline.

By and by, Speer integrated several construction and transport units from his own sphere of power into the OT. Speer had been GBI since 1937 and had founded the Construction Staff Speer (*Baustab Speer*) to carry out his construction projects in Berlin. After the invasion of the Soviet Union, the *Baustab Speer* was deployed to the occupied East under Speer's co-worker Walter Brugmann. It was this unit that formed the OT's *Einsatzgruppe Rußland-Süd* in spring 1942; after Brugmann's death in May 1944, it was renamed *Einsatzgruppe Brugmann*.[81] In order to supply the *Baustab Speer's* "Germania" construction sites with materials, the *NSKK Baustab Speer* was founded in 1938 (later renamed *NSKK-Transportstandarte Speer*) before following the *Wehrmacht* to the Soviet Union in 1941 as *NSKK-Transport-Brigade Speer*.[82] This unit was merged with other OT transport organisations in 1942.[83] The GBI had also established a transport fleet (*Transportflotte Speer*) for seaborn transport, primarily to deliver granite from Scandinavia to GBI's construction sites in Berlin. In June 1942, the fleet became part of the OT and shipped building machines and construction materials to EW's construction sites in Norway and Denmark. Primarily Dutch sailors, but also Norwegians worked in the *Transportflotte Speer*, which had its wartime headquarters in the Netherlands and ran a school in the Southern Norwegian town Sandefjord.[84]

Only very rough estimates exist on the total number of workers under the OT in occupied Europe. The OT missions were constantly reorganised and there was a high fluctuation among the workforce. Moreover, the different *Einsatzgruppen* lacked a consistent system of counting their workers. Figures from German wartime

---

Einsatzgruppe Tannenberg. *Einsatzgruppe Brugmann* was relocated to the General Government in 1944. In the last year of the war, the OT finally established several *Einsatzgruppen* within the Reich.

81 Brechtken, *Speer*, 626, fn. 20.

82 *Handbook of the Organisation Todt*, 77–84; Seidler, 'NSKK', 632–36.

83 The new organisation was called *NSKK-Transportgruppe Todt*, and also absorbed the *NSKK-Transport-Brigade Todt* that had been established at the *Westwall* and the *Legion Speer*, which consisted of non-German personnel. To render the terminological confusion perfect, the *NSKK-Transportgruppe Todt* was renamed *Transportkorps Speer* in mid-1944.

84 Seidler, 'NSKK', 635–36. The archive (38 linear metres) of *Transportflotte Speer*, *Einsatz Norwegen*, is kept at the National Archives in Oslo: RAO, RAFA-2189. Finding aids that will make the archive accessible are currently under production.

sources range from little more than 700,000 to 1.4 million men and women.[85] In its 1945 study on the OT, British intelligence assumed that the Organisation reached its greatest extent between May 1942 and May 1943, when it allegedly deployed more than 2 million workers on its construction sites.[86]

During the first half of the war, the OT worked almost exclusively in the occupied territories. Only in 1943 did the OT deploy larger units in the Rhein-Ruhr area to restore facilities that had been destroyed by Allied air strikes on Germany's industrial heartland. Until the end of the war, the OT executed numerous special tasks on German soil, mostly to protect the armaments production from air raids. To this end, eight *Einsatzgruppen* were established within Germany in the last year of the war, which recruited their personnel mostly from dissolved or downsized *Einsatzgruppen* in occupied Europe.[87]

2 The Establishment of Einsatzgruppe Wiking

The following chapter is about EW's activities in occupied Norway and especially the process that led to its establishment in 1942. Part a presents the German construction activities under the *Wehrmacht* and the Reich Commissariat during the first two years of the occupation, before giving an overview over the tasks that the newly established *Einsatzgruppe* assumed after 1 April 1942. Part b then focuses on EW's establishment. The reasons behind EW's entry into Norway in spring 1942 have been subject to debate in recent Norwegian historiography. I provide a novel interpretation of this process, as I show that the OT was actually rather reluctant to take over some of the huge tasks in Northern Europe. Part c investigates EW's organisational structure. I argue that the Organisation was under permanent stress. I demonstrate the close integration of private business into the OT under the slogan of "industrial self-responsibility", which was both a necessity, given the lack of administrative personnel during the war, and a deliberate organisational strategy to effectively utilise the industry's potential.

a The Wiking Order of 13 May 1942

Einsatzgruppe Wiking was the northernmost of the *Einsatzgruppen* in occupied Europe formed after the OT's reorganisation in 1942. EW became operative on

85 Seidler, *Die Organisation Todt*, 146–47.
86 *Handbook of the Organisation Todt*, 3.
87 Seidler, *Die Organisation Todt*, 114–27.

1 April 1942 and Albert Speer chose Willi Henne as its head.[88] Henne's appointment shows how crucial the minister must have deemed the mission in Norway. As one of Fritz Todt's closest co-workers, Henne had led the construction of the *Westwall* and organised the deployment of OT units after the attack on the Soviet Union in 1941. Speer could hardly have found a man higher up in the OT hierarchy other than sending the head of OTZ, Xaver Dorsch, to Norway. Within few months, Henne rose to be one of the most powerful figures in occupied Norway. As leader of the EW, he only answered to Xaver Dorsch and Albert Speer. In addition, Henne became head of the Main Department for Technology (*Hauptabteilung Technik*), the department that had been responsible for the Reich Commissariat's construction activities in the first years of the occupation.[89] Later, Speer also made him *Bevollmächtigter für die Regelung der Bauwirtschaft*, the GB Bau's representative in Norway.

In spring 1942, Norway had already seen two years of rampant building activity.[90] The German building spree put an enormous strain on the country's economy with its mere 3 million inhabitants. The undertakings usurped construction materials, transport capacity and labour. By the autumn of 1941, unemployment had been practically wiped out in Norway.[91]

Since 1940, construction activities in occupied Norway had been controlled mostly by the *Wehrmacht*. The driving force behind the military aggression had

88 Willi Henne (1907–1977) had been a high-ranking official under Fritz Todt in the 1930s. After the war, Henne was arrested and sent to the Soviet Union, where he was sentenced to 25 years in prison. He was back in Germany in 1955 as one of the last German POWs returning from Soviet captivity. From 1957 to his retirement in 1972, he was head of the Hessian Office for Road Construction. Cf. Hatlehol, 'Norwegeneinsatz', 1–2. Albert Speer obviously thought highly of Henne. When interrogated in June 1945, Speer described Henne as one of his best building experts, "very energetic and too ruthless. Tough and determined good organiser": Office of the Publication Board, Department of Commerce, 'Interrogation of Albert Speer and Members of the Former Reich Ministry of Armaments and War Production [Technical Report No. PB 430]', 77.

89 During the first weeks of the Reich Commissariat, the name of the department seems not to have been clear yet. In some early documents, the department is called "*Abteilung Bauwesen und Verkehr*" before "*Abteilung Technik und Verkehr*" became accepted in mid-June. In summer 1941, the department was in practice split up into a Department for Technology (*Abteilung Technik*) and a Department for Transport (*Abteilung Verkehr*). In letters of the department from early summer 1941, the word "*Verkehr*" in the letterhead is crossed out (cf. RAO, RAFA-2174/Ef-L0002). Shortly after, "*Technik und Verkehr*" is no longer used. When EW was established in spring 1942, the Department for Technology was upgraded to a Main Department for Technology on 1 April: RAO, RAFA-2174/Ef-L0002, file 9, Letter Neyer (Main Department for Technology) to Main Department for Propaganda (*Hauptabteilung für Volksaufklärung und Propaganda*), 11.4.42. The new main department, EW (and the GB Bau office) *de facto* functioned as one unit in following years. The less important Department for Transport remained under the Main Department for Economy.

90 Cf. chapter IV.3.

91 Harald Espeli, 'Economic Consequences of the German Occupation of Norway, 1940–1945', *Scandinavian Journal of History* 38, no. 4 (2013): 506.

been the navy, which feared that an Allied blockade would trap the German fleet in the North Sea and therefore had already been drafting plans for an invasion since 1938. The Norwegian harbours, first and foremost Bergen and Trondheim, promised free access to the Atlantic.[92] And just as the main reason for the attack itself had been strategic, the lion's share of construction works in the summer of 1940 was rooted in the needs of the *Wehrmacht*. Accordingly, the three *Wehrmacht* branches had established their own construction units. The air force had its *Bauleitung der Luftwaffe*, while the navy established a central *Marinebaudirektion* under the Marine Group Command in Norway (*Marineoberkommando Norwegen*), in addition to three *Marineoberbauämter* under the regional offices of the *Marineoberkommando* in Bergen, Trondheim and Tromsø. Finally, the army commanded special pioneer and engineer troops, like the railway pioneers or the fortification pioneers, and used units of the RAD.[93]

After the invasion, the air force immediately insisted on the construction of new or enlargement of existing airfields. They were to serve as bases for the fights in Northern Norway and later England.[94] In Northern Norway, airfields such as Bardufoss were needed to fight the Allied convoys heading for Arkhangelsk and Murmansk after August 1941. The army became involved in the expansion of road and railway infrastructure. In 1940, neither the country's south-west nor its northern two thirds were connected to the railway network. Neither did the road network meet the requirements of heavy military traffic, especially in the north, which was intended to serve as concentration area for the northern attack on the Soviet Union. Hence, the army started the extension of national road 50 from Oslo to Kirkenes and of railway lines between Kristiansand and Sira (*Sørlandsbanen*) and between Mosjøen and Bodø (*Nordlandsbanen*) .[95] Massive fortifications along the Norwegian coast were to protect the country from a possible Allied invasion by turning it into a huge stronghold, *Festung Norwegen*. It was no secret that the tortuous Norwegian coastline was hard to defend and therefore had to be fortified immediately after the invasion. But construction works gained momentum especially after Operation Claymore, the British raid on Svolvær in March 1941.[96]

---

92 Carl-Axel Gemzell, *Raeder, Hitler und Skandinavien. Der Kampf für einen maritimen Operationsplan* (Lund: C.W.K. Gleerup, 1965); Hans-Martin Ottmer, 'Skandinavien in der deutschen marinestrategischen Planung der Reichs- bzw. Kriegsmarine', in *Neutralität und totalitäre Aggression: Nordeuropa und die Großmächte im Zweiten Weltkrieg*, ed. Robert Bohn et al. (Stuttgart: Franz Steiner Verlag, 1991), 49–72. Economic motives, such as access to Norwegian resources and Swedish ore, which was shipped from Narvik, always played a role, but were never the prime factor in the decision to invade Norway.

93 Hatlehol, 'Norwegeneinsatz', 77–78; Frøland and Hatlehol, 'Forced Labour', 48.

94 Bohn, *Reichskommissariat Norwegen*, 357.

95 Hatlehol, 'Norwegeneinsatz', 79; Frøland and Hatlehol, 'Forced Labour', 48.

96 On the different stages of the construction of *Festung Norwegen*, see Sæveraas, 'Beton macht Geschichte', 88–117.

The Reich Commissariat also became an important player in the Norwegian construction sector. The RK's Technical and Transport Department had been established in early June 1940 as *Abteilung Technik und Verkehr*, led by Heinz Klein. He, too, was a close co-worker of Fritz Todt's who had already held a leading position at the *Westwall*.[97] Although there was no institutional connection to the OT, the department was staffed with many former OT members. Its structure, with seven *Oberbauleitungen* around the country, was also clearly adopted from the OT.[98] Klein's department was responsible for a multitude of infrastructural, industrial and military construction works during the first two years of the occupation. Although part of the civilian administration, the department closely coordinated its construction projects with the *Wehrmacht*. The improvement of the infrastructure, which in 1940 hardly met West and Central European standards, was a precondition for the economic utilisation of occupied Norway as well as serving military needs. Hence, road construction was a major field of activity, with 474 km of new roads and 106 reconstructed bridges in 1940 and 1941 alone. In higher regions and the north, snow clearance and the erection of 155 km of snow fences formed an important part of the task. Moreover, the department tried to repair the damage resulting from the military campaign in spring 1940. Large parts of cities like Kristiansund had to be rebuilt, as did the harbour of Narvik. On behalf of the *Wehrmacht*, the Technical and Transport Department built 61 coastal fortifications, 199 anti-aircraft positions, 70 piers, 12 communication facilities, eight ammunition recycling facilities and 64 bunkers during the two first years of the occupation. 767 barracks and 605 wooden houses were built to house the men serving in the fortifications. In addition, almost 2,000 barracks were erected for the SS, police, navy and other agencies.[99] The Technical and Transport Department mostly relied on local Norwegian construction firms, accomplishing its tasks with very few administrative staff. In early 1942, the southern OBLs consisted of 10–15 people only, including auxiliary staff, typists, and Norwegian drivers and interpreters. Farther north, the situation

---

97 Helge Paulsen, 'Reichskommissariat og "motytelsene" under riksrådsforhandlingene', in *Norge og den 2. verdenskrig: 1940 – fra nøytralt til okkupert*, ed. Samtidshistorisk Forskningsgruppe (Oslo: Universitetsforlaget, 1969), 344, fn. 32. Klein was already a civil servant in Munich before the National Socialists seized power. His field of activity was bridge construction. See *Der Bauingenieur* 10, no. 33 (1929): 588.

98 The *Oberbauleitungen* were located in Kristiansand, Bergen, Trondheim, Narvik, Tromsø, Hammerfest and Kirkenes. An eighth OBL was established on 1 February 1942 in Alta: RAO, RA-2174/Ef-L0002, file 9, Organisational chart of the Department for Technology for the time before 1 April 1942, 11.4.42. However, it seems that OBL Alta only became operative when *Einsatzgruppe Wiking* took over construction activities in Norway: HHStAW, Abt. 485, 480, fol. 36.

99 RAO, RAFA-2174/Ef-L0002, file 9, Draft of/information for a "*Führer*" report on the accomplishments of the Main Department for Technology in the second year of the Reich Commissariat, 3.3.42 and 9.3.42.

was even more problematic. Organisation plans for OBL Hammerfest, for example, in fact reveal only six persons. OBL Kirkenes consisted of four men in spring 1942.[100]

In the winter of 1941–42, however, Hitler was convinced that Norway had become the war's "zone of destiny", as he expressed it in a meeting with navy leaders on 22 January.[101] Accordingly, plans began to take shape that would dwarf every undertaking of the previous years. Among these was the demand for an intensified extension of Norway's coastal fortifications and the vision of a railway line through Northern Norway, heading for Kirkenes. It was clear that the fortifications and the railway project would require an administration, labour and logistics of completely different dimensions than was available in Norway at that point in time.

In May, Hitler set out in writing which construction projects were to be prioritised during the next years. His Wiking order of 13 May 1942 was the epitome of the National Socialist will and ambition to quite literally move mountains. The document specified six tasks: (1) The construction of permanent fortifications for the formation of infantry and artillery posts. (2) A winter-proof extension of the national road 50 to Lakself and the construction of a winter-proof connection to the *Eismeerstraße*,[102] in addition to a general expansion of national road 50. (3) The construction of airfields. (4) A single-track line from Mo via Fauske – Narvik to Kirkenes. Most urgently the line to Narvik and the railway connection between the Nordreisa – Alta area and the Kirkenes area. (5) The erection of submarine bases in Drontheim and Bergen with shellproof docks together with the construction of a dry dock for the biggest types of vessels. (6) The construction programme for the increased production of aluminium and magnesium.[103]

Hitler gave the responsibility for the fortifications programme, the roadworks in Northern Norway, the railway to Kirkenes, the submarine base in Trondheim and the construction programme of the light metals programme to Albert Speer in his position as GB Bau and by that *de facto* to the EW. The air force and the navy remained in charge of their construction activities after EW's establishment. The air force continued with the construction of airfields, while the navy's main undertaking was the dry dock and submarine bunker in Bergen. Only over the course of 1944

---

100 HHStAW, Abt. 485, 480, fol. 36–37.

101 Percy E. Schramm, ed., *Kriegstagebuch des Oberkommandos der Wehrmacht (Wehrmachtführungsstab) 1940–1945. Band 2: 1. Januar 1942–31. Dezember 1942. Zusammengestellt und erläutert von Andreas Hillgruber. Erster Halbband (= KTB/OKW II/1)*, Kriegstagebuch des Oberkommandos der Wehrmacht (Wehrmachtführungsstab) 1940–1945. Geführt von Helmuth Greiner und Percy Ernst Schramm (Bonn: Bernard & Graefe Verlag, 2000), 123–24.

102 A connection to the *Eismeerstraße*, which connected the Petsamo region with the rest of Finland, would considerably enhance the mobility of the German troops in the High North.

103 Martin Moll, ed., *'Führer-Erlasse' 1939–1945. Edition sämtlicher überlieferter, nicht im Reichsgesetzblatt abgedruckter, von Hitler während des Zweiten Weltkrieges schriftlich erteilter Direktiven aus den Bereichen Staat, Partei, Wirtschaft, Besatzungspolitik und Militärverwaltung* (Stuttgart: Franz Steiner Verlag, 1997), 249–50.

were the construction units of the navy and air force integrated into EW's apparatus. By this point, however, the major projects were already more or less finished or no longer a priority.[104]

With the establishment of the EW, *OT-Nord*, that is, the *Dora* project in Trondheim, became part of EW as OBL Trondheim. National road 50 was also given over to the OT at a very early point.[105] After all, road construction was the core competence of Todt's former co-workers. At the beginning of March, around 80 men of the OT unit *Edelweiß* from OBR Villach (Austria) already travelled via Finland to Northern Norway, where they formed the core of the future *Linienchef für die Reichsstraße 50* office in Alta.[106] In September 1942, EW also became responsible for the remaining road construction sites in Norway.[107] The construction works of the aluminium and light metals programme under the companies A/S Nordag and A/S Nordisk Lettmetall never gained major importance for Willi Henne and his staff. The two companies worked quite autonomously and EW contented itself with a coordinating role.[108] Matters were more complicated, however, in the case of the two largest projects on Hitler's agenda: the fortifications programme and the railway in Northern Norway.

b The Reluctant Organisation: OT's Railway Project in Northern Norway

The Wiking order of 13 May 1942 was not the starting point for the establishment of *Einsatzgruppe Wiking*. Rather, it was the result of almost half a year of power struggles on its future tasks and its position within the occupation regime. In particular, for a long time it remained unclear to what degree Willi Henne's responsibilities would overlap with those of the highly ambitious Heinz Klein, the head of the Main Department for Technology. Would EW work alongside the RK's engineers and architects, would they take up a subordinate role or, on the contrary, assume control of the department? As it will be shown, the rise of EW and its head Willi Henne is a prime example of the polycratic infightings in the National Socialist administration.

It is noteworthy that Adolf Hitler issued the Wiking order on 13 May 1942, one and a half months after the EW had been established in Oslo. To date, EW is the only *Einsatzgruppe* we know about that received a written order specifying its tasks.

104 Sæveraas, 'Beton macht Geschichte', 79.

105 Today's European route E6.

106 HHStAW, Abt. 485, 480, fol. 114–22, Report on the activities of Supply Staff Wiking for the time from 1 March 1942 to 30 June 1942, 15.7.42.

107 Frøland and Hatlehol, 'Forced Labour', 55.

108 Frøland and Lervold, 'Light Metals Program'. When the programmes were downsized in the following years, however, Henne as GB Bau made sure to take over materials and tools to avoid them being sent to Germany.

In the following, I will argue that this might be explained by the Organisation's reluctance to assume responsibility for the extremely challenging railway project in Northern Norway.

Recent Norwegian historiography has debated the question why the OT actually came to Norway. Was the railway or the fortifications programme the initial reason behind EW's establishment? Both Ketil Gjølme Andersen and Frøland and Hatlehol have recently suggested that the Polar Line was EW's entrance ticket to Norway after Hitler had commissioned Albert Speer with the project's execution in mid-February 1942. Accordingly, they argue that the Organisation only gradually assumed more responsibilities in other fields, too. However, they focus mostly on Hitler's will and his orders to the OT's new head, Albert Speer. They do not investigate how the OT apparatus actually reacted to and processed Hitler's wishes.[109]

In contrast, I will provide an in-depth analysis of the process of mobilisation of EW's administrative apparatus in early 1942, which leads to a novel interpretation. I argue that the building of *Festung Norwegen* was EW's top priority. It was the initial motive behind the establishment of EW, while the *Einsatzgruppe* was remarkably reluctant to take over the railway construction project demanded by Hitler.

Concerned by his troops' halting advance in the Russian winter and the United States' entry into the war only three days earlier, Adolf Hitler had ordered the fortification of the western coasts of Europe on 14 December 1941.[110] Resembling a "new *Westwall*" stretching from the Bay of Biscay to the Polar Sea, this fortification was to enable the defence of Western Europe with as few troops as possible. Norway was particularly vulnerable in Hitler's eyes. Opinions were split among the *Wehrmacht* leaders on how likely an Allied attack on the Norwegian coast was. But the scenario was certainly more than just one of the "*Führer's*" private obsessions. Accordingly, during the following weeks preparations began for a massive step-up of construction activities along the coast.[111]

When Albert Speer announced the establishment of *Einsatzgruppe Wiking* few days after Todt's death, a logistics unit, Supply Staff (*Nachschubstab*) Mitte-Nord-Wiking, was established in the Berlin headquarters of the *Reichsautobahnen* Corporation.[112] The Supply Staff were to initiate the mobilisation of both administrative staff and construction firms and organise the transport of building materials and machines to Norway.

---

109 Hatlehol, 'Norwegeneinsatz', 83; Frøland and Hatlehol, 'Forced Labour', 53; Ketil G. Andersen, '"Teknisk æresoppdrag av høyeste orden". Organisation Todt og byggingen av Hitlers polarjernbane', *Historisk Tidsskrift* 97, no. 3 (2018): 213.

110 Percy E. Schramm, ed., *Kriegstagebuch des Oberkommandos der Wehrmacht (Wehrmachtführungsstab) 1940–1945. Band 2: 1. Januar 1942–31. Dezember 1942. Zusammengestellt und erläutert von Andreas Hillgruber. Zweiter Halbband (= KTB/OKW II/2)*, Kriegstagebuch des Oberkommandos der Wehrmacht (Wehrmachtführungsstab) 1940–1945. Geführt von Helmuth Greiner und Percy Ernst Schramm (Bonn: Bernard & Graefe Verlag, 2000), 1262–64.

111 Sæveraas, 'Beton macht Geschichte', 94–111.

112 BArch, R 50-I/3a, Decree by Speer on the reorganisation of the OT, 18.2.42.

Eduard Schönleben, the late Todt's right-hand man in the *Reichsautobahnen* Corporation, officially headed the unit, but it was Schönleben's representative Hugo Koester who was in charge of everyday affairs.[113] Koester chose OBR Frankfurt am Main to provide the administrative personnel for the mission in Norway. Five to six units, each with 25–30 persons, were to form the future OBLs in Norway.

In early March, Koester had envisaged sending the OBRs Essen and Cologne to Denmark, where they were to organise the construction of fortifications.[114] However, it soon became clear that Frankfurt alone was unable to provide the cadre for the mission in Norway. Hence, the Supply Staff ordered that the OBRs Essen and Cologne should each send two units to Norway instead of Denmark. Frankfurt provided the personnel for the future OBLs Kirkenes, Narvik and Alta, Cologne for OBL Nordwest in Trondheim, and Essen for the OBLs Bergen and Kristiansand.[115] The units were merged with the RK's OBLs that already existed in these places.[116] All of them were concerned with fortifications. In addition, the OBL in Trondheim in charge of the submarine bunker project (the former *OT-Nord*) also continued to exist under EW, together with an OBL dealing with hydraulic engineering in Kirkenes and the aforementioned *Linienchef* for national road 50, located in Alta.[117]

Like at the *Westwall*, the OT received their orders from the Wehrmacht regarding what and where to build. Between spring 1942 and the end of 1944, EW built more than 2,000 structures on behalf of Wehrmacht's Inspector of the Northern Fortifications (*Inspekteur der Landesbefestigung Nord*, *IdLN*). The construction sites for bunkers, torpedo batteries, and gun emplacements were scattered along the long-stretched coast, especially in the country's southern part. More than 555,000 m^3 of reinforced concrete were poured until the end of 1944. The construction of *Festung Norwegen* became the EW's most important undertaking in occupied Norway, both financially and in terms of labour used.[118]

Compared to other German fortifications projects, however, the amount of reinforced concrete poured in Norway was rather modest. Until the end of 1944, more

113 Cf. the organisational chart of Supply Staff Mitte-Nord-Wiking in HHStAW, Abt. 485, 480, fol. 129.

114 Ibid.

115 HHStAW, Abt. 485, 480, fol. 41–42, Express letter Supply Staff Wiking to OBRs Frankfurt am Main, Cologne and Essen, 24.3.42.

116 Where possible, the heads of the existing OBLs remained in office after EW took over: HHStAW, Abt. 485, 480, fol. 33–39, Minutes of a meeting concerning the Wiking mission on 17.3.42 in Berlin, 18.3.42.

117 BArch, R 50-I/3a, Undated organisational chart for *Einsatzgruppe Wiking*, probably from mid-April 1942. See also: Sæveraas, 'Beton macht Geschichte', 73.

118 RAO, RAFA-2188/2/Hh-L0002, file 91-IV, Report on the construction of fortifications by the OT in Norway and Denmark 1942–1944, 1.5.45. Cf. Sæveraas, 84–86.

than 1.4 million m^3 had been poured in Denmark[119] and until July 1944, more than 10 million m^3 on the Atlantic Wall in France, Belgium and the Netherlands, including 613,000 m^3 on the Channel Islands alone.[120] At the Westwall, the output had been 6.65 million m^3 of reinforced concrete between June 1938 and 1 September 1939.[121] The reason for the comparably low figures for Norway was the extremely challenging geographic conditions in Norway, which hampered efficient transports in particular. Moreover, many installations in Norway could be blasted directly into the rock, a construction method that required less concrete.[122] However, when we take into account that Norway had barely three million inhabitants when the war broke out, the figures are quite impressive.[123]

EW's fortifications programme also included the construction of roads, tunnels, piers and airstrips in order to provide access to these remote sites. After mid-1944, however, fuel shortages became increasingly problematic. On 30 September 1944, the OKW stopped the supply of cement to the construction sites in Norway and provided no more tonnage, which spelled the end of the undertakings. From then until the German capitulation, EW only carried out minor remaining works when it came to fortifications.[124]

In contrast to the fortifications programme, there is nothing in the available sources that points to a large-scale railway construction project under EW before late April 1942. Even in mid-April, EW's organisational chart does not show any institutional structures intended for railway construction.[125] In the documents of the Supply Staff, the possibility of additional road and railway works is sometimes mentioned, but in March and April, the unit was clearly projecting the construction of bunkers.[126] Accordingly, the works were to be executed by firms that had gained experience at the *Westwall*.[127] Moreover, the three regional liaison

119 RAO, RAFA-2188/2/Hh-L0002, file 91-IV, Report on the construction of fortifications by the OT in Norway and Denmark 1942–1944, 1.5.45.

120 Colin Partridge, *Hitler's Atlantic Wall* (Guernsey: D.I. Publications, 1976), 66–67.

121 BA-MA, RH 11-III/170 K-4.

122 Cf. Sæveraas, 'Beton macht Geschichte', 123.

123 Norway: 0.19 m^3 per inhabitant; France, Netherlands, Belgium: 0.19 m^3 per inhabitant; Denmark: 0.36 m^3 per inhabitant. Population figures according to Hans Umbreit, 'German Rule in the Occupied Territories 1942–1945', in *Organization and Mobilization of the German Sphere of Power*, ed. Militärgeschichtliches Forschungsamt, Germany and the Second World War, V/II (Oxford: Clarendon Press, 2015), 11.

124 RAO, RAFA-2188/2/Hh-L0002, file 91-IV, Report on the construction of fortifications by the OT in Norway and Denmark 1942–1944, 1.5.45.

125 BArch, R 50-I/3a, Undated organisational chart for *Einsatzgruppe Wiking*, probably from mid-April 1942.

126 HHStAW, Abt. 485, 480, fol 9–11, Supply Staff Wiking to OBR Frankfurt am Main concerning the use of OBR Frankfurt am Main under EW – Construction of fortifications in Norway, 9.3.42.

127 BArch, R 3901/20215, fol. 13–15, Reich Labour Minister to the presidents of the labour offices, 14.3.42.

officers (*Bezirksreferenten*) ensuring the contact between the OBLs and EW's headquarter in Oslo could "only be people with experience in the construction of fortifications". The lists of materials and machinery destined for Norway clearly reveal that the mission's purpose was not railway construction.[128] The documents of the Supply Staff hence support the position of Torgeir E. Sæveraas, who has identified the construction of fortifications as the initial motive behind the OT's mission in Norway. In particular, Sæveraas has shown how the *Wehrmacht's* fortification pioneers anticipated the OT's arrival and already discussed the Organisation's tasks in March 1942.[129]

In September 1941, Hitler had commissioned Fritz Todt to examine on site the prospects of a railway to Petsamo (Pechenga) in order to make troop reinforcements possible and to improve the supply situation of the forces already fighting in the area.[130] Todt explored and dismissed several possible projects, especially that of a narrow-gauge railway on Finnish territory from Rovaniemi to Petsamo. But after inspecting the terrain personally by plane – Todt had been a pilot during World War I – Todt was convinced that the construction works would overburden the area's infrastructure. Todt was similarly sceptical about a railway in Northern Norway, which he seems to have regarded as a long-term undertaking for the time after the war.[131] After presenting his views to Hitler, the Finnish railway project was dropped, and Todt was against the Norwegian line, too.[132]

Thus, in the winter of 1941–42, the railway in Northern Norway was a venture of Reich Commissioner Josef Terboven and of the Department for Technology under Heinz Klein.[133] Todt's opposition, and especially the cancellation of the Finnish alternative, offered an opportunity for Terboven to attract his "*Führer's*" attention. It was undoubtedly the most spectacular and ambitious undertaking in occupied Norway. Several versions of the idea were promoted during the early phase of the occupation, the most ambitious of which envisaged a line heading for Kirkenes. Since the Norwegian railway network ended a few kilometres north of Trondheim, extending it to the town of Kirkenes would have meant laying more than 1,200

128 HHStAW, Abt. 485, 480, fol. 33–39, Minutes of a meeting concerning the Wiking mission on 17.3.42 in Berlin, 18.3.42.

129 Sæveraas, 'Beton macht Geschichte', 101–6.

130 *Norges Hjemmefrontmuseum*, Oslo (hereafter NHM), FO.II/div. 24, Report and interrogation of Colonel General von Falkenhorst, Oslo, Sept./Oct. 1945, 42–43.

131 BA-MA, RH 20-20/203, Memorandum by Todt on securing supplies and the connection Northern Norway – Finland, 9.10.41.

132 Andersen, 'Æresoppdrag', 210–11.

133 Most likely, Terboven seized the railway project in October or November 1941. On 2 October, Klein still indicated to the navy that the RK's OBLs Kirkenes, Hammerfest and Tromsø would be liquidated by the end of the year due to lack of labour, trucks, working materials and tonnage: RAO, RAFA-2188/2/He-L0005, file 20 (Dr. Klein Mappe IV), Letter Klein to *Admiral Norwegen*, 2.10.41.

kilometres of railway tracks through a sub-polar region that was partly mountainous, partly swampy. The idea was initially promoted by the *Wehrmacht*, but Hitler himself had also shown interest in the Polar Line since Norway had become part of the German sphere of control.[134] The railway had the potential to bring together strategic (increased mobility of troops) and economic considerations (access to resources in Northern Europe) with the alluring vision of a continuous railway line from Oslo to Murmansk and even St. Petersburg.[135]

Terboven proposed expanding the *Nordlandsbane* to Narvik, excluding the extremely mountainous terrain between Narvik and Nordreisa from the project. Ships would be able to transport troops and materials in this region. Between Nordreisa and Kirkenes, a railway would be built through Finnish territory. Terboven's confidence and, probably even more, his promise to finish this masterpiece of engineering within two years (!) convinced Hitler.[136]

In RK's Department for Technology, Heinz Klein immediately realised that the railway project would be his chance to emerge as the central figure in the Norwegian construction sector. Klein was an ambitious man – apparently too ambitious for most of his superiors. His uncontrollable hunger for power alienated him from almost all important protagonists during his time in Norway. Although he worked in the Reich Commissariat, Klein had come to Norway in 1940 as Todt's "man of confidence".[137] In the following months, however, he tried to expand his sphere of power by taking over numerous construction works for the *Wehrmacht*, tasks for which he requested additional civil servants in Berlin. This seems to have upset the OT's Xaver Dorsch and even resulted in a personal letter from Todt. The minister accused Klein of dissipating his energies ("*sich verzetteln*").[138] Berlin was rather reluctant to pump money and resources into projects in occupied Norway.[139] In spring 1941, the navy ignored Klein's department when it awarded the construction of the *Dora* bunkers in Trondheim directly to the OT. Klein speculated that this was simply a misunderstanding, because his men had sometimes described themselves as OT members. Klein begged Todt not to ignore him, but it was obvious that he had lost the backing of his former colleagues.[140]

134 Bohn, *Reichskommissariat Norwegen*, 363–70.
135 Milward, *Fascist Economy*, 272. Andersen, 'Æresoppdrag', 207.
136 Willi A. Boelcke, *Deutschlands Rüstung im Zweiten Weltkrieg. Hitlers Konferenzen mit Albert Speer 1942–1945* (Frankfurt am Main: Akad. Verl.-Ges. Athenaion, 1969), 104–5.
137 RAO, RA-2174/Ef-L0034, Letter Klein to Todt, 22.11.40.
138 Ibid.
139 In July 1941, the GB Bau pointed out that due to the large construction projects in the *Altreich*, "[. . .] the use of Reich companies as well as the use of construction machinery from the Reich has to remain limited to the least extent": RAO, RA-2174/Ef-L0003, file 10, Letter GB Bau to the Technical and Transport Department, 3.7.41.
140 RAO, RA-2174/Ef-L0034, Letter Klein to Todt, 20.3.41.

In the Reich Commissariat, on the other hand, Klein clashed with his superior Carlo Otte, head of the Main Department for Economy (*Hauptabteilung Volkswirtschaft*), who tried to contain the expensive German construction activities. Circumventing Otte, Klein managed to interest Terboven in a motorway from the Swedish-Norwegian border in Halden to Trondheim.[141] As Hitler had decreed that a future Neu-Drontheim would become the northernmost metropolis in the German sphere of power and a crucial naval base, the area was to become the endpoint of a European north-south axis. The Technical and Transport Department conducted survey works, but the project was abandoned due to a lack of labour and funding before the first metre of road had been asphalted.[142]

Since Klein was not an NSDAP member, he did not have a power base within the party either. Last but not least, Klein had strained relations with the *Wehrmacht* leaders in Norway. In 1940–41, Klein constantly tried to gain superordinate control over all construction programmes in the country and over the allocation of construction materials. It is no surprise that *Wehrmacht's* reaction to this would-be GB Bau was rather frosty.[143]

Thus, when Terboven promised the "*Führer*" his railway in Northern Norway in late 1941, Klein pulled out all the stops. The first vague signs of the railway constructions project date to early December.[144] On 7 December 1941, a railway bureau was established within the Department for Technology.[145] The bureau was staffed predominantly with railway experts from Austria and Southern Germany who had experience of constructions in the Alps.[146] The entire surveying works to Kirkenes were carried out by the 40-year-old engineer Hans Ferkow. Ferkow, born in Graz, was an experienced railway engineer who from the late 1920s on had worked for the consortium of Philipp Holzmann, Julius Berger and Siemens-Bauunion, which had built parts of the Trans-Iranian Railway.[147] Only two months after Fritz Todt had declined the projects in the High North with a categorical "I don't build

141 *Bundesarchiv Koblenz* (hereafter: BAK), N 1550/1, Hans Clausen Korff, 'Norwegens Wirtschaft im Mahlstrom der Okkupation' (Unpublished manuscript, 1949), pt. I, 114.

142 Bohn, *Reichskommissariat Norwegen*, 356.

143 For more details on this topic, see chapter IV.4.

144 RAO, RA-2174/Ef-L0003, file 12, Note of the Department for Technology on two meetings in the Reich Commissariat on 1 and 3 December, 3.12.41; RAO, RA-2174/Ef-L0003, file 15, Note of Department for Technology, 5.12.41.

145 RAO, RA-2174/Ef-L0002, file 9, Organisational chart of the Department for Technology for the time before 1 April 1942, 11.4.42.

146 RAO, RA-2188/2/Hh-L0001, file 89-III, Overview over the staff of the railway bureau. See also: Bilfinger Archive, A 2797–1, Report on the line inspection for the railway project: Fauske – Narvik – Kirkenes, 15.6.42, fol. 30.

147 *Der Bauingenieur* 10, no. 33 (1929): 594. The project is mentioned in: Pohl, *Philipp Holzmann*, 189–91.

mountain railways",[148] his former co-worker Heinz Klein had the audacity to establish a bureau manned with – exactly – experts for mountain railways. There could hardly have been a bigger affront to Todt than this. With the backing of Terboven and especially Hitler himself, who had officially ordered the construction of the Polar Line on 14 January 1942, Klein was about to perform his biggest career move.[149]

In the following weeks, Klein was occupied with preparations. While the railway bureau's members began to plan the surveying works,[150] Klein negotiated on the deployment of Soviet POWs and concentration camp inmates from Serbia. Moreover, the erection of camps and the custody of the prisoners had to be organised.[151] On 4 January, Klein expected that 20,000 Soviet POWs could be set to work starting in May.[152] In February, the department started to negotiate with German construction firms on the mission in Northern Norway.[153] It was then that Heinz Klein and his co-workers began to use the title Main Department for Technology (*Hauptabteilung Technik*) instead of Department for Technology (*Abteilung Technik*), although the department did not officially become a main department until 1 April.[154]

In February 1942, events came thick and fast. By chance, Albert Speer was present in Rastenburg when Fritz Todt's airplane went up in flames. On the same day, Hitler summoned Speer and appointed him Todt's successor in all his positions. At the end of the meeting, Hitler sent the OT's new head on his way back to Berlin with an unambiguous order: "One more thing, Mr Speer, I want you to build me that railway in Norway that Terboven has proposed."[155] Hitler repeated his order to the OT in a meeting with Speer, Henne and Dorsch on 19 February.[156] Against this background, it is quite remarkable how reluctant the OT leadership were in the next months when it came to the railway project. Obviously not overly impressed by Hitler's order to their new boss Speer, Todt's old comrades were unwilling to support it, especially since it still was a RK project led by Heinz Klein. Klein's project depended completely on support from Germany, and in mid-February, requests with the GB Bau, the General Inspector and the RWM for construction materials,

148 NHM, FO.II/div. 24, Report and interrogation of Colonel General von Falkenhorst, Oslo, Sept./Oct. 1945, 43.
149 Andersen, 'Æresoppdrag', 212.
150 RAO, RAFA-2174/Ef-L0013, file 81.
151 RAO, RAFA-2174/Ef-L0003, file 15.
152 RAO, RAFA-2174/Ef-L0003, file 15, Letter Klein to the Higher SS and Police Leader, Oslo, 4.1.42.
153 RAO, RAFA-2188/1/F3a-L0001, Letter Philipp Holzmann AG to Thierbach, Oslo liaison office of the German building trades, concerning railway construction in Northern Norway, 21.2.42.
154 See for example RAFA-2174/Ef-L0018, Note by Main Department for Technology on the fortification of the western coast of Norway, 24.2.42.
155 NHM, FO.II/div. 24, Report and interrogation of Colonel General von Falkenhorst, Oslo, Sept./Oct. 1945, 44.
156 Boelcke, *Hitlers Konferenzen*, 67–68.

fuel, equipment, personnel, tonnage and firms were pending.[157] This gave Todt's loyal co-workers in Berlin the opportunity to throw obstacles in the way of Klein's and Terboven's prestigious undertaking. On 13 February, the GB Bau office prohibited the delivery of construction steel and non-ferrous metals to Norway, officially because the RK was not a claimant and had no competence to request materials from Germany.[158] However, as one official of the GB Bau in Berlin confided to two members of the RK, the real reason behind the blockade was that both the late Todt and the *Wehrmacht* "had been categorically against" the undertaking in the High North.[159]

It seems that there was a considerable amount of uncertainty about the railway project among the lower ranks of the OT and industry circles in late February and March 1942. The question was whether the whole undertaking was on the brink of being cancelled. While – as we have seen above – the establishment of the EW as a fortifications and road construction mission was in full swing, Wigru Bau had informed the construction firm Eugen Gärtner on 4 March 1942 that "the planned railway construction is newly called into question." Therefore, Eugen Gärtner enquired whether their tunnelling unit could be deployed for the construction of fortifications instead.[160]

On a meeting between Willi Henne and the Supply Staff on 17 March, details of the Wiking mission were discussed, such as transports, supply with fuel and building materials, and contracts. In the several pages of the document, the railway is only mentioned in one short paragraph on the mission's funding: "Particularly doubtful is [the funding of, S.G.] road and railway construction. [. . .] Clarity on road and railway construction will only exist when a Führer order exists." However, an official of OBR Frankfurt am Main underlined "road and railway construction" on 27 March and added a hand-written comment on the page's margin: *entfällt!* (cancelled).[161]

At this point in time, Willi Henne was already on his way to Norway, where he would take over the command of EW and the RK's Main Department for Technology. Henne's arrival in Oslo put an abrupt end to Heinz Klein's career. There was clearly no place for the overambitious man in the ranks of EW.[162]

---

157 RAO, RAFA 2188/1/E2b-L0006, file B.I.a. Band 2, telegram Neyer, Main Department for Technology, to Klein, 12.2.42.

158 RAO, RAFA 2188/1/E2b-L0006, file B.I.a. Band 2, Note by Neyer, Main Department for Technology, regarding road and railway construction in Northern Norway, 13.2.42.

159 Ibid.

160 RAO, RAFA 2188/1/F3a-L0001, Letter Eugen Gärtner, Würzburg, to Wigru Bau, Berlin, 9.3.42.

161 HHStAW, Abt. 485, 480, fol. 33–39, Minutes of a meeting concerning the Wiking mission on 17.3.42 in Berlin, 18.3.42.

162 On 9 March 1942, Klein still acted on the assumption that Henne's OT mission would be subordinated to his Department for Technology and believed himself to have achieved the central position in the Norwegian construction sector he had fought for during the last years: RAO, RAFA-2174,

On 1 April 1942, the newly established EW inherited a railway project from the former Department for Technology that had hardly got beyond the planning phase. No construction companies had been mobilised yet.[163] On 7 April, Hugo Koester in the Supply Staff informed the GB Bau in Berlin and the OBR Frankfurt am Main, Essen and Cologne that "for the railway construction between Narvik and Kirkenes, two bases in Skoganvarre and Nyborg must be erected and planning works pursued."[164] But Koester did not issue a command to mobilise firms or even start with the actual construction works.

The Polar Line was still more or less put on hold when Willi Henne and Xaver Dorsch met Hitler on 19 April. The OT leadership's reluctance is noticeable in the minutes of this meeting. Confronted with Hitler's demand that the line should be finished within two years, the rather arbitrary figure initially given by Terboven, Henne and Dorsch relayed that in Minister Speer's opinion it was then Reich Commissioner Terboven who should carry out the project and take responsibility for the deadline.[165]

After the elimination of Klein, however, the EW had no longer a "scapegoat" on whom it could dump the responsibility for the project. After months of hesitating, Henne had to follow Hitler's order and engage in what was an – at least during wartime – absurd project. The definite decision that EW was to take over the construction of the railway in Northern Norway must have been taken in the days after the meeting of 19 April. On 27 April, Koester informed the heads of the OBRs of the *Reichsautobahnen* that "on the order of the Führer, the tasks of the Reichsautobahnen on behalf of the OT in Norway have been expanded significantly *in recent days*. In addition to the previous tasks in the construction of roads and fortifications, a railway line of presumably more than 1,000 kilometres has to be built in the worst terrain and with numerous tunnels and bridges."[166] That the railway project became EW's responsibility only in the second half of April, and hence was not the initial reason for EW's engagement in Norway, is further confirmed by a summary report of Supply Staff Wiking from July 1942, stating that

---

Ef-L0001, file 7, Copy of a letter from Klein to the Main Department for Economy, 9.3.42. There are no leads towards the whereabouts and activities of Klein after his return to Germany. In the first half of the 1950s, he seems to have returned to Munich and was no longer a civil servant. The Munich telephone directory of 1953 lists him as *Ministerialdirektor a.D.* (meaning retired or off duty). In 1956, he participated in the fifth congress of the International Association for Bridge and Structural Engineering in Lisbon.

163 Cf. the notes on a meeting between officials of the Department for Technology and representatives of three railway construction firms on 25 March 1942: RAO, RAFA-2174/Ef-L0002, file 8.

164 HHStAW, Abt. 485, 480, fol. 70, Letter Koester, Supply Staff Mitte-Nord-Wiking, to Steffens, GB Bau, 7.4.42.

165 Boelcke, *Hitlers Konferenzen*, 104–5.

166 HHStAW, Abt. 485, 480, fol. 96, Koester, *Reichsautobahnen-Direktion*, to all OBRs, 27.4.42 [my own emphasis, S.G].

"the supply tasks for the Wiking mission were greatly expanded when, at the end of April, the Führer ordered the construction of a single-track line from Fauske via Narvik to Kirkenes with an overall length of ca. 1,200 km in addition to the OT's previous works in Norway."[167]

EW erected several new OBLs for the supervision of the railway construction. All OBRs in the Reich were supposed to provide the necessary personnel from their administrations. A squad of 169 topographers was dispatched in early May to support the line surveying works that Hans Ferkow had started around the turn of the year. Three professors from the *Technische Hochschulen* in Vienna, Danzig (Gdansk) and Brünn (Brno) accompanied the unit.[168]

23 German railway construction companies were mobilised in two waves. The first group of 14 firms was contacted in early May and took over works on the stretches that were deemed most crucial, between Fauske and Narvik, and between Nyborg and Kirkenes. Most of the firms of the second wave, mobilised in early June, were to work between Nordreisa, Kautokeino and Kirkenes.[169] On 12 May and 12 June, the Supply Staff requested that the RAM allocate additional construction workers on behalf of the firms. These express letters also clearly state that the railway construction was an additional task, which the "*Führer*" had only recently given to EW.[170]

EW invited several German firms on an inspection trip to Northern Norway during the first two weeks of June to discuss the project on site. If any of the participants had not been clear about the scope of the undertaking, they certainly were now. The line between Fauske and Kirkenes was to be 1,200 km long, including 90 km of tunnels. The tunnelling works alone would require moving 3 million m^3 of rock.[171] 200,000 m^3 of timber had to be transported to Northern Norway for the 1.7 million railway sleepers. The entire line would run north of the Arctic Circle, which limited the time of year when construction works could be executed. OT officials estimated that one would have to add 1,500 metres to the actual altitude in order to find similar conditions in the Alps. Grün & Bilfinger's representative harboured no illusions when he reported on the project:

---

167 HHStAW, Abt. 485, 480, fol. 114–22, Report on the activities of Supply Staff Wiking for the time from 1 March 1942 to 30 June 1942, 15.7.42.

168 Ibid. Professor Ladislaus Rabcewicz from Vienna had worked on the Trans-Iranian Railway, like Hans Ferkow.

169 Cf. the preliminary lists from 12 May 1942 and 12 June 1942 in R 3901/20211, fol. 9–10 and 12–14 with the list attached to the Supply Staff report of 15.7.42 in: HHStAW, Abt. 485, 480, fol. 128.

170 R 3901/20211, fol. 9–10 and 12–14.

171 As a comparison, between 1942 and the end of 1944, EW had excavated 952,711 m^3 of rock in occupied Norway: RAO, RAFA-2188/2/Hh-L0002, file 91-IV, Report on the construction of fortifications by the OT in Norway and Denmark 1942–1944, 1.5.45.

> The whole project of the railway from Fauske to Kirkenes is a huge task, which would not have been easy to fulfil in this northern latitude in peacetime, with sufficient labour and orderly logistics. [. . .] Under the current conditions of war and knowing the supply situation [. . .], it seems almost questionable that it will be possible to bring this project to a successful end.[172]

Even Grün & Bilfinger, which had been lucky enough to be assigned the line's southernmost section starting in Fauske, where the supply situation was best, only minor differences in altitude had to be overcome and only few kilometres of tunnel were involved, saw itself confronted with enormous challenges.[173] It is almost grotesque that the first ten kilometres of the 1,200 km railway already led through terrain so swampy that Bilfinger's engineer had a hard time to imagining a practicable solution to the problem. The companies' conclusion was rather fatalistic:

> Since the order for the railway's construction has been issued, the work will start for the time being, so that the firms that have participated in the trip will hardly have the opportunity to escape the task, unless they want to get into a very difficult relationship with the awarding authority.[174]

During the next months, tonnes of building materials and equipment were transported to the north, where EW had erected four OBLs in Fauske, Narvik, Nordreisa and Skipagurra, in addition to the existing OBL in Kirkenes. In February 1943, however, after the *Wehrmacht's* defeat at Stalingrad, the Polar Line's fate was sealed. After the war, WBN Nikolaus von Falkenhorst claimed that it had been him who convinced Hitler to drop the project. Before they had laid a single kilometre of tracks, EW's engineers packed up and moved the materials and machines back south.[175] Willi Henne ordered the liquidation of OBLs Nordreisa and Skipagurra. The plan was now to focus on the *Nordlandsbane* that was to be extended to Narvik. In April 1943, the *Wehrmacht's* railway engineers, who had worked on the *Nordlandsbane* together with the Norwegian State Railways (*Norges Statsbaner*, NSB) since 1940, backed out of the project and handed the responsibility over to EW. Until the end of the war, under the supervision of EW, German firms worked on the line's northern part between Fauske and Narvik, while on the southern

172 Bilfinger Archive, A 2797–1, Report on the line inspection for the railway project: Fauske – Narvik – Kirkenes, 15.6.42, fol. 31.

173 As a map from September 1943 show, the OT had subsequently assigned parts of Grün & Bilfinger's stretch to the firms Fanghänel, Leipzig, and Carl Rose, Berlin, which had been relocated from the abandoned northernmost construction sites of the Polar Line. The construction of the Trengselsund bridge had been given to the Ed. Züblin AG, Stuttgart, in 1944: RAO, RAFA-2188/2/Hfa-L0027, map of the railway Mo i Rana – Korsnes, 16.9.43; RAO, RAFA-2188/1/E9a-L0029, map of the railway Mo – Fauske – Korsnes, 3.3.44.

174 Bilfinger Archive, A 2797–1, Report on the line inspection for the railway project: Fauske – Narvik – Kirkenes, 15.6.42, fol. 31.

175 Bohn, *Reichskommissariat Norwegen*, 367.

stretch between Mo i Rana and Fauske, the NSB, Norwegian and German firms worked side by side.[176]

Throughout the war, Hitler could never be persuaded to abandon the railway in Norway. Time and again, he insisted on the construction works' continuation in his meetings with Speer, Dorsch and the *Wehrmacht* generals.[177] In March 1944, Hitler even issued a directive that ordered the deployment of another 4,000 German skilled workers and 15,000 Soviet POWs to ensure the project's timely completion.[178] In his eyes, extending the *Nordlandsbane* to Narvik would secure Germany's access to Swedish iron ore and save shipping space, which was indeed one of the biggest bottlenecks in the occupied country's economy.[179] Given the German military situation, however, the *Nordlandsbane* was a waste of resources and strategically useless. The *Wehrmacht's* railway engineers had estimated in 1943 that it would take nine years just to reach Narvik.[180] In November 1943, Hitler was informed that Narvik could be reached in 1947 at the earliest.[181] Hence, both the *Wehrmacht* and the OT leaders would have preferred to see the project cancelled, as for example Speer himself admitted in a letter to Henne in November 1944. Moreover, the *Wehrmacht* was unwilling to allocate more shipping space for railway construction at the expense of other military construction projects in Norway.[182] However, works went on until the last weeks of the war. Speer and Henne did not dare oppose Hitler's explicit will to continue construction. When interrogated after the war, Henne quite rightly concluded that the railway had been a long-term peacetime project that had been started in the middle of a brutal war.[183] The railway in Northern Norway – *as a wartime project* – remains EW's only project for which it is hard to identify a rationale beyond the will of Adolf Hitler. Not a single kilometre of the 1,200 kilometre Polar Line had been finished, and of the *Nordlandsbane*, little more than 90 track kilometres were open to traffic by 1 May 1945.[184] As has been shown, the OT's attitude towards the railway construction project was remarkably reluctant both in spring 1942 and during the following years. This might help to explain why to date, EW is the only known OT mission Adolf Hitler felt the need to launch by means of a "*Führer*" order.

176 Andersen, 'Æresoppdrag'.
177 Boelcke, *Hitlers Konferenzen*, 118–19, 257, 311–12, 381–82, 429.
178 Moll, *Führer-Erlasse*, 405; Hatlehol, 'Norwegeneinsatz', 109–10.
179 Boelcke, *Hitlers Konferenzen*, 118–19, 311–12.
180 Petersen, *Polareisenbahn*, 128.
181 Boelcke, *Hitlers Konferenzen*, 311.
182 BArch, R 3/1593, fol. 127–30, Letter Henne to Speer, 30.5.44.
183 Andersen, 'Æresoppdrag', 220–22.
184 Hjeltnes, *Hverdagsliv*, 69.

c Organisational Structure and Administrative Personnel

A lean organisation and a reliance on industrial self-responsibility were the most important characteristics of *Einsatzgruppe Wiking's* administrative body between 1942 and 1945. Like Fritz Todt before him, Albert Speer advocated industrial self-responsibility.[185] Both men regarded this principle as the most effective tool for mobilising the German industry for the war effort, as Speer never tired of proclaiming publicly.[186] And as the war went on and public administrations, too, were combed for potential soldiers, this principle became a necessity, not least in the occupied territories. In 1944, Albert Speer explicitly referred to industrial self-responsibility as a means to free up personnel in the building administration.[187] In order to tackle its vast tasks, EW therefore relied heavily on private industry, not only by contracting firms, but by integrating them into its administrative apparatus.

EW's headquarters were located in the heart of Oslo, at Kirkegata 15. Its organisational structure changed constantly over the course of the occupation in order to adapt to its changing tasks. Generally speaking, however, three departments were concerned with the construction of fortifications, roads and piers, and railways, respectively, while other departments had cross-sectional responsibilities. Among others, there were departments for supplies, contracts, transport and medical service.[188]

EW's staff was simultaneously manning the Main Department for Technology in the RK. EW's and the RK's regional offices, which monitored the works around the country, the so-called *Oberbauleitungen*, merged in September 1942.[189] The same men were also representing the GB Bau in Norway.[190] EW also supervised the construction projects in Denmark, which were executed by local Danish businesses to a much higher degree, though. According to its mission, the *Einsatzgruppe* oversaw the army's construction projects and, like in all other parts of occupied Europe, took over the administration of navy and air force construction during the first half of 1944.[191] The OT's previously independent unit *Einsatz Finnland* also became part of EW in the summer of 1944.[192] This means that virtually every German construction firm

185 Tooze, *Wages*, 349–53, 570–76, 634–36.

186 Albert Speer, 'Selbstverantwortung in der Rüstungsindustrie', *Der Vierjahresplan*, no. 7 (1943): 242–46; Albert Speer, 'Die Grundlagen der deutschen Rüstung. Reichsminister Speer über Bewährung und Aufgaben der Selbstverantwortung', *Berliner Börsen-Zeitung*, 18 June 1944.

187 BAK, N 1318/5, Speer Chronicle, 1944, fol. 59–60.

188 Kåre Olsen, *Innledning til katalog for arkivet etter Organisation Todt Einsatzgruppe Wiking 1940 (–42) – 1945* (Oslo: Riksarkivet Oslo, 2012).

189 RAO, RAFA-2188/1/E5a-L0035, Letter Terboven, 20.9.42.

190 See chapter IV.4.

191 Frøland and Hatlehol, 'Forced Labour', 56.

192 The incorporation of *Einsatz Finnland* was a process of several weeks in July 1944. On *Einsatz Finnland*, see the upcoming doctoral thesis to be submitted by Mari Olafsson Lundemo at the

in Northern Europe was contracted and supervised by this administrative body by August 1944.

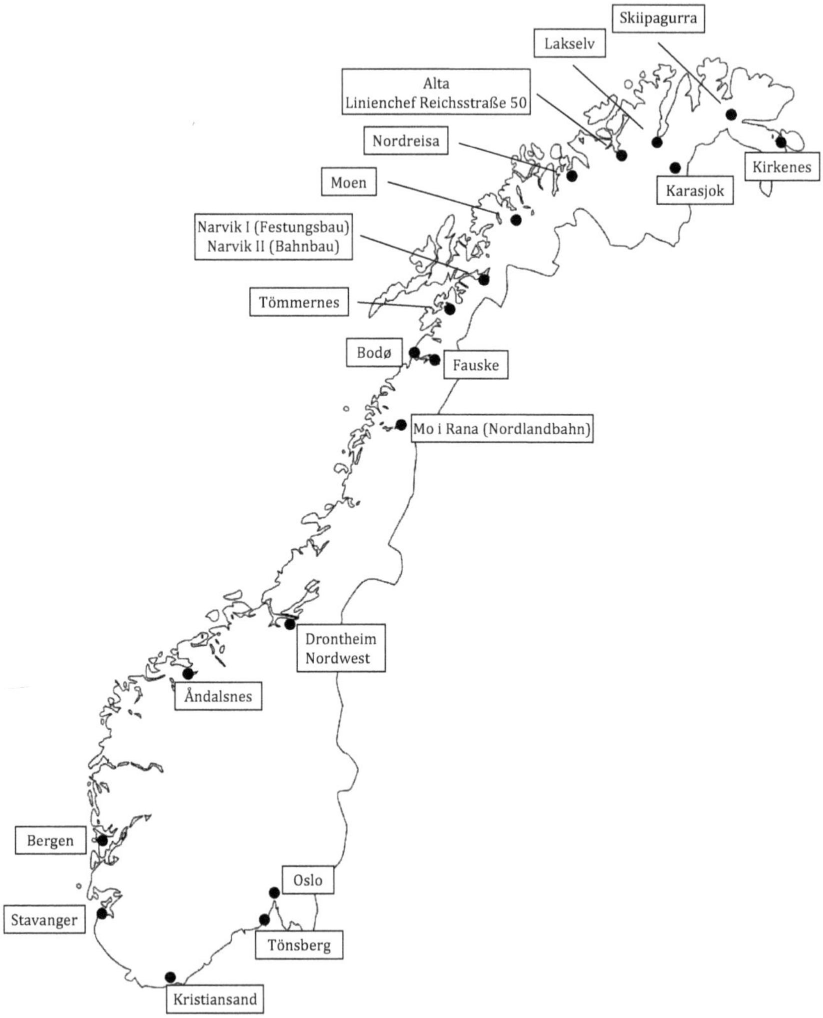

Figure 14: Einsatzgruppe Wiking's Oberbauleitungen in occupied Norway. Source: Own illustration. Not all OBLs existed simultaneously or throughout the entire occupation.

EW's administration thus reached its maximum size in late August 1944, before construction activities were reduced significantly and firms relocated to Germany

European University Institute, Florence (preliminary working title: *An Alliance of Snow, Ice and Fire: Organisation Todt in Finland, 1941–1944*).

during the autumn. On 25 August 1944, 6,096 men and women were working in the administration of *Einsatzgruppe Wiking* in Norway, Denmark and Finland (Table 16).

Table 16: Staff of *Einsatzgruppe Wiking*, 25 August 1944.*

	Technical staff	Administrative and clerical staff	Others	Women	Sum
Norway	1,812	1,357	481	866	4,516
Denmark	332	338	31	189	890
Finland	350	115	99	126	690
EW	2,494	1,810	611	1,181	6,096

*RAO, RAFA-2188/2/Hh-L0002, file 90-III.

30 percent (1,854) of the staff were non-Germans, most of them Norwegians, Danes and Finns. This shows the important role that locals within the German administration played for the maintenance of the German rule in occupied Europe, as researchers of the German occupation have repeatedly argued in recent decades. The statistics do not provide exact information on what the female staff members were occupied with, but it is known that most of them worked as typists and secretaries, operated telephones and telegraphs (so-called *Nachrichtenhelferinnen*) or worked in facilities like camp kitchens.[193] However, as more detailed staffing plans from EW's railway bureau show, women also worked in the Organisation as technical drawers.[194] An overview of 25 January 1945 registered 779 female salaried employees under EW in Norway. 458 of them were German and 308 Norwegian.[195]

At first glance, the number of more than 6,000 staff members suggests a massive apparatus, which should have been able to monitor its contractors closely. It is somewhat misleading, though. First, the majority of men and women working for EW were either technical staff, concerned with planning and designing the construction projects, conducting surveying works and making geological investigations, or they managed the provisioning of the workforce and the supply with building materials, machines and spare parts. Second, the figure contains a significant number of staff that were employees of the construction firms. Their exact share is not given, but a comparable statistical overview from January 1944 shows that 36 percent of EW's staff were employees of the firms.[196] What is more, the remaining 64 percent, identified as OT members, still contained an undisclosed number of persons who came

193 *Handbook of the Organisation Todt*, 123.

194 RAO, RAFA-2188/2/Hh-L0001, files 89-III and 89-VI, Staffing plans of EW's railway construction bureau.

195 Ibid., file 90-I, Chart A1d on the number of EW's staff members and workers, 25.1.45.

196 This means before the incorporation of the Finnish OT units and the construction apparatus of the navy and the air force.

from private construction firms and were on the OT's payroll only temporarily.[197] It is not unreasonable to assume that they expected to return to their companies after the war and hardly would have pursued any policy that would have harmed their industry.[198]

In EW's headquarters in Oslo, two small departments were responsible for all aspects of contracting, pricing, and monitoring the firms' book-keeping. Department XIa (later VA1), also referred to as Department for Contracting and Pricing, under architect Josef Fröhler, led the negotiations with the construction firms. It consisted of 10 to 11 persons, including the typists, in 1943–44.[199] The number of personnel in Department XIb (later VA2), also referred to as the Department for Accounting, varied from 6 to 9, including its head Fritz Motsch.[200] In the autumn of 1944, the two departments had to release 30 percent of their staff to the *Wehrmacht* and armaments industry.[201] The two departments were supported by a handful of officials placed with EW's *Oberbauleitungen* around the country. Most of the work, however, seems to have rested on the shoulders of the staff in Oslo. Negotiations on contracts and prices, for example, were often carried out personally by Josef Fröhler.[202]

In occupied Norway, as in other occupied territories before, the German Court of Audit continued its policy to monitor awarding state authorities while their projects were still under execution. A department of its own at the OTZ in Berlin was commissioned with monitoring the expenses of the *Einsatzgruppen*. At EW in Oslo, it was Fritz Motsch and his men in Department VA2 who were in charge of the so-called pre-auditing office (*Vorprüfdienst*), a position Motsch had already held at the *Westwall*.[203] This somewhat problematic arrangement, in which the Court of Audit left the monitoring of EW to officers of the Organisation itself, was justified with the fact that it would save staff and avoid travel between the Reich and Norway.[204] The additional task was an enormous burden for the small department, as the men were

197 RAO, RAFA-2188/2/Hh-L0001, file 90-II, EW employment statistics, 25.1.44.

198 For this argument in the case of price auditors monitoring German construction firms, cf. Streb, 'Preisregulierung', 46–47.

199 Josef Michael Fröhler was born on 24 March 1896 in Schwandorf. Fröhler came from a family of builders. Like one of his brothers, he was a trained architect. Fröhler died on 13 January 1956 in Munich. I would like to thank Josef Fischer at Schwandorf town archives for these details.

200 See the various staffing plans from 1943 and 1944 in: RAO, RAFA-2188/1/E5a-L0003, file Personal.

201 RAO, RAFA-2188/1/E5a-L0004, file Personal A-K 1942–45, Letter Fröhler to Ludwig Alter, OT *Einsatzgruppe VII*, 4.10.44. This was presumably a result of the "*Führer*" decree concerning the war deployment of the construction administrations of 24 August 1944: RGBl. 1944, I, 207.

202 *Einsatzgruppe West* in France reported a lack of qualified accountants, too: Streb, 'Negotiating Contract Types', 375.

203 RAO, RAFA-2188/1/E5a-L0003, file Personal.

204 Ibid., Note by Köbele regarding accounting, 23.11.42.

travelling for weeks and months to visit EW's remote outposts all around the country. Repeated requests for support from Berlin, that is, for additional staff, seem to have gone unanswered.[205]

For an increasingly overtaxed administration, German industry's self-governing bodies became important partners during the years of occupation. Construction was no exception in this respect. When both the construction tasks and the problems in Norway were growing in the autumn of 1941, Heinz Klein had suggested establishing a liaison office of Wigru Bau in Norway, which among other things was to gain an overview over the German firms' construction machinery in Norway.[206] In late October, Fritz Todt put Friedrich-Joachim Thierbach, head of Wigru Bau's office in Kiel, in charge of representing the association in occupied Norway. In line with Todt's ideas of industrial self-responsibility, Thierbach's duties as the liaison office secretary were wide-ranging. He was not only to gather information on the number of workers and machines on German construction sites in Norway, but also to support the RK in the fields of price and wage policy and labour deployment.[207] The liaison office was affiliated with the RK as *Verbindungsstelle der Wirtschaftsgruppe Bauindustrie, Oslo*, although it was somewhat unclear during the first weeks whether Thierbach was supposed to represent the entire German construction sector, including the handwork firms, or merely the industry firms organised in Wigru Bau.[208]

The liaison office's importance grew after the start of EW's mission, which implicated a massively extended deployment of German firms. It was Willi Henne who explained in October 1942 that he desired a greater involvement of the association in Norway. Given the OT's ever-growing tasks, construction firms were to work as self-responsibly as possible. The model that Henne had in mind was Wigru Bau's self-regulating committee at the *Westwall*. The liaison office's leader was, on the one hand, to act as a counsel to Henne and forward requests and complaints by the firms. On the other hand, he was to be authorised to independently implement measures concerning the construction industry.[209]

Gustav Klumpp from the Stuttgart-based construction firm Ed. Züblin AG became the office's new leader. Following the consent of the Reich Association of the

205 RAO, RAFA-2188/1/E5a-L0007, file Prüfdienst OT. Zentrale, Letter by Nolte, head of the pre-auditing office at OTZ, to Berger, department P5 at OTZ, 24.3.44.
206 RAO, RAFA-2188/2/He-L0004, file 18 (Dr. Klein Mappe 2), Monthly report of Klein to Todt, 18.10.41.
207 BArch, R 13-VIII/264, First report of Thierbach, Oslo liaison office of Wigru Bau, 27.12.41; Ibid.: Second report of Thierbach, Oslo liaison office of Wigru Bau, 18.1.42.
208 Ibid.
209 RAO, RAFA-2188/1/E1l-L0044, Copy of a note on a meeting between representatives of EW, Wigru Bau, the Reich Association of the Building Handwork Guilds and of the Ed. Züblin AG on 20 October 1942 in Berlin, 13.11.42.

Building Handwork Guilds, the office now represented both industry and handwork firms, and accordingly was renamed *Verbindungsstelle der deutschen Bauwirtschaft beim Reichskommissar für die besetzten norwegischen Gebiete*. Thierbach remained the office's executive (*Geschäftsführer*). Joachim Schlimper, who had already been Thierbach's co-worker during the last months, became deputy executive.[210] EW promised to ensure Klumpp's and Schlimper's exemption from military service.[211] Thierbach was recalled from Norway in May 1944, apparently after some personal conflicts with Klumpp. Joachim Schlimper replaced him in his post, while Thierbach returned to his former position in Kiel.[212]

3 Cost of EW's Projects and their Impact on the Norwegian Economy

This chapter describes EW's funding mechanism and assesses the German construction activities' impact on the Norwegian economy. Based on novel source material, I show that German construction activities in Norway – including EW's projects – were not exclusively financed from Norwegian sources, as historians have believed for decades, but were to a substantial degree funded through the German war budget. In doing so, I substantiate the findings of Norwegian historians, who in recent years have argued that German investments in occupied Norway were far higher than hitherto assumed and thus contributed to Norway's rapid post-war economic recovery. In this respect, Norway was probably an exception among the occupied territories, due to the exceptional imbalance between the country's own economic capacity and the high volume of German demand. Another major finding concerns the Norwegian construction industry: I argue that the German demand was primarily an *additional* demand and did not stifle "purely Norwegian" construction activities. The post-war defence put forward by many Norwegian entrepreneurs, who claimed that not cooperating with the occupiers would have meant committing economic suicide, is thus stripped of much of its basis.

210 Dr. Joachim Schlimper, born in 1914, was delegated by Wigru Bau's branch office in Stuttgart. It is unclear, however, whether he was in any way associated in with the Ed. Züblin AG, like Gustav Klumpp. Schlimper was a pricing expert, as he had earned a doctorate in 1940 with a thesis on contracts and pricing in *Hochbau*: Joachim Schlimper, 'Die Preisermittlung im Hochbaugewerbe' (Handels-Hochschule Leipzig, 1940).
211 RAO, RAFA-2188/1/E1l-L0044, Copy of a note on a meeting between representatives of EW, Wigru Bau, the Reich Association of the Building Handwork Guilds and of the Ed. Züblin AG on 20 October 1942 in Berlin, 13.11.42.
212 BArch, R 13-VIII/87, fol. 175, 194, 198.

a The Cost of the German Occupation of Norway

In the summer of 1945, dozens of former officials of the Reich Commissariat were interned at the Bogstad camp, located between the famous *Holmenkollbakken* and Lake Bogstad on the northern outskirts of Oslo. Among them was the former head of the Department for Finance in the Reich Commissariat, Hans Clausen Korff, who had been the leading German financial expert in occupied Norway.[213] Korff himself approached the Norwegian Reparations Office (*Erstatningsdirektorat*) and offered his help in organising and evaluating thousands of invoices and other documents that could provide information on the OT's, the *Wehrmacht's* and the RK's expenses during the occupation. He argued that there were still 20 to 40 Germans at Bogstad who had dealt with financial questions during the occupation and would be able to provide valuable information. After a meeting at the Norwegian Ministry of Finance on 25 July 1945 and having obtained statements from various institutions like the Central Bank (*Norges Bank*), the Norwegian Court of Audit and several other departmental offices, the Reparation Office's Jannik Lindbæk issued a mandate to Korff on 14 August to commence his work.[214] Although somewhat sceptical about letting the former occupiers access the captured documents, the Norwegians were interested in Korff's work both because they were trying to gather information on Norwegian legal subjects that had collaborated economically with the Germans and because an overview over the occupier's finances could be used as the basis for subsequent compensation negotiations with Germany.

The "Korff group" consisted of up to 12 men and women who over the next two years were mostly occupied with setting up a card file registering all German payments in Norway. Furthermore, on Norwegian request, they compiled reports on Norwegian firms that were being investigated for economic treason. The group initially worked in the rooms of the Allied "document pool" at Klingenberggata 7, before the material was handed over to the Norwegians and transferred to Hovedøya, a small island in the Oslofjord, in May 1946.[215]

On 29 September 1947, Hans Clausen Korff submitted a 31-page report on his group's work to the Reparations Office. The report starts with a short, almost parenthetical statement on the overall cost of Nazi Germany's occupation of Norway. According to Korff, the occupation of Norway cost between 25 and 30 billion kroner

---

213 Hans Clausen Korff was born in 1905 in Stranderott, today's Stranderød (Denmark). After having studied law, he worked in the public financial administration during the 1930s. In autumn 1940, he became head of the RK's Department for Finance. After the war, Korff continued to be a high-ranking official in the West German Federal Ministry of Finance. He died in Bonn in 2000.

214 RAO, S-1543/Da-L0006, file 3, Mandate by Jannik Lindbæk, Reparations Office, to Hans Clausen Korff, 14.8.45.

215 RAO, S-1543/Da-L0006, file 3, Final report by Hans Clausen Korff regarding the German occupation authorities' business relations in Norway, 29.9.47.

(that is approximately 14–17 billion RM). 11 billion kroner were *internal* occupation costs, borne by Norway. An estimated 14 to 19 billion kroner were *external* occupation costs, borne by the German Reich.[216] This report, which does not seem to have been used by earlier research, contributes to understanding EW's financial *modus operandi*, how EW's projects were funded, how much they cost and who footed the bill. But perhaps more importantly, it fundamentally challenges our established knowledge on the financial aspects and costs of the German occupation of Norway.

The foundation of this established knowledge was already laid in 1945, when the economists Odd Aukrust and Petter Jakob Bjerve published their seminal book *Hva krigen kostet Norge* ("What the war cost Norway").[217] The study can be regarded as the quasi-official history of the economic consequences of Nazi Germany's occupation of Norway and, in the words of the Norwegian economic historian Harald Espeli, is "the most politically influential book written by Norwegian economists in the 20th century."[218] It painted a gloomy picture: Norway had been forced to provide the funds both for supplying and accommodating several hundred thousand German soldiers on its territory and for the German construction activities, which included EW's projects. All of these investments were foisted on the Norwegian Central Bank and, ultimately, the Norwegian tax payer. As a result, nowhere in occupied Western Europe was the per capita burden of occupation costs higher than in Norway.[219] Moreover, Germany had seized goods without paying for them and had disadvantaged Norway by manipulating import and export prices to its own advantage. The authors also estimated war damages, including among other things the damage that resulted from Norway being cut off from many of its trade partners, particularly Great Britain.[220]

The central means of exploiting occupied Norway were Germany's withdrawals from the so-called occupation account. On 24 April 1940, the Norwegian Central Bank had given the *Wehrmacht carte blanche* to withdraw money from this account to finance the German garrison and – most importantly – the massive construction activities all over the country.[221] In doing so, the Norwegian Central Bank was following the general strategy of the self-appointed Norwegian Administrative Council

216 Ibid., 1.

217 Odd Aukrust and Petter Jakob Bjerve, *Hva krigen kostet Norge* (Oslo: Dreyers Forlag, 1945). On Aukrust and Bjerve and the history behind their book, see Olav Bjerkholt, *Kunnskapens krav: om opprettelsen av Forskningsavdelingen i Statistisk sentralbyrå* (Oslo, Kongsvinger: Statistisk Sentralbyrå, 2000), 93–103.

218 Espeli, 'Economic Consequences', 503.

219 Cf. Hein Klemann and Sergei Kudryashov, *Occupied Economies. An Economic History of Nazi-Occupied Europe, 1939–1945* (London, New York: Berg, 2012), 202.

220 Aukrust and Bjerve, *Hva krigen kostet Norge*, 21–49.

221 Harald Espeli, '"Det gavner ingenting å gjøre store vanskeligheter i små saker. Dette er ikke store saker". Norges Bank, Administrasjonsrådet og etableringen av okkupasjonskontoen i 1940', *Historisk Tidsskrift* 90, no. 4 (2011): 559–84; Harald Espeli, '"Cooperation on a Purely Matter-of-Fact

(*Administrasjonsrådet*) that had been established after the Norwegian government and King had fled the capital on 9 April 1940. The Council instructed the Norwegian public to keep calm, refrain from sabotage and maintain normal economic activity as far as possible. The wheels had to keep turning, not least in order to spare the population unnecessary hardship. Thus the Norwegian leaders neither discouraged businesses from accepting orders from German authorities, nor forbade citizens to take on jobs on German construction sites.[222]

In principle, the Norwegian Central Bank's accommodating stance towards the occupiers was in accordance with international agreements at the time. According to the *Convention respecting the Laws and Customs of War on Land*, the fourth treaty of the Hague Convention of 1907, occupied countries had to pay for the administration of the country and the sustenance of the occupying troops. Nazi Germany, however, stretched and violated the rules of the Convention in the territories it occupied.[223] For example, forcing the occupied countries to pay for the enormous war-related construction projects and especially forcing POWs to work on these war-related construction sites was a clear breach of the Hague Convention. In occupied Norway, these construction activities in particular gained a dominant position in the economy. In the first months after the invasion, half of the construction volume was accounted for by the *Wehrmacht*, and in 1941, the share of projects on behalf of the German armaments industry and the military had risen to 90 percent.[224]

In *Hva krigen kostet Norge*, Aukrust and Bjerve documented that the German occupiers had withdrawn 11,676 million kroner from the occupation account by 7 May 1945. From this sum, Aukrust and Bjerve subtracted German deposits in other accounts worth 335 million kroner and Norway's excess imports from Germany (760 million kroner) while adding 1,785 million kroner for German requisitions and confiscations. Thus they assessed direct occupation costs at approximately 12,300 million kroner.[225] Hans Clausen Korff also arrived at very similar figures in an unpublished manuscript on the

Basis": The Norwegian Central Bank and Its Relationship to the German Supervisory Authority during the Occupation, 1940–1945', *Scandinavian Economic History Review* 62, no. 2 (2014): 188–212.

222 Erstatningsdirektoratet, *Undersøkelser vedrørende norske myndigheters stilling til tyske arbeidsoppdrag den først tid etter 8. april 1940* (Oslo: Erstatningsdirektoratet, 1946); Espeli, 'Det økonomiske forholdet mellom Tyskland og Norge 1940–45', 151–55; Espeli, 'Economic Consequences', 505–6.

223 Jonas Scherner, 'The Institutional Architecture of Financing German Exploitation: Principles, Conflicts, and Results', in *Paying for Hitler's War: The Consequences of Nazi Hegemony for Europe*, ed. Jonas Scherner and Eugene N. White (New York: Cambridge University Press, 2015), 43–66; Jürgen Kilian, *Krieg auf Kosten anderer: Das Reichsministerium der Finanzen und die wirtschaftliche Mobilisierung Europas für Hitlers Krieg* (Berlin, Boston: De Gruyter Oldenbourg, 2017), 69–73.

224 RAO, RAFA-2174, Ef-L0002, file 9, Draft of a report by Wick, construction industry group in the Main Department for Technology, for the overall report on the activities of the RK, 14.4.42.

225 Aukrust and Bjerve, *Hva krigen kostet Norge*, 28. Cf. *NOS X. 102. Nasjonalinntekten i Norge 1935–1943. Realkapitalen 1939 og kapitalreduksjonen under krigen. Okkupasjonskostnadene.* (Oslo: Statistisk Sentralbyrå, 1946), 164.

Norwegian economy under the occupation. This study, which was more than 500 pages long, was finalised between 1947 and 1949, when he had finished his work for the Reparations Office. In it, Korff subtracted from the 11,676 million kroner the aforementioned 335 million kroner, and 287 million kroner that had been repaid through the German-Norwegian clearing account. Hence, he arrived at a slightly lower figure for the German occupiers' net withdrawals from the occupation account, 11,054 million kroner.[226] According to Korff, almost half of the sum was invested in construction projects.[227] The monthly German withdrawals were highest in 1941 before stabilising at around 200 million kroner per month from 1942 onwards (Figure 15).

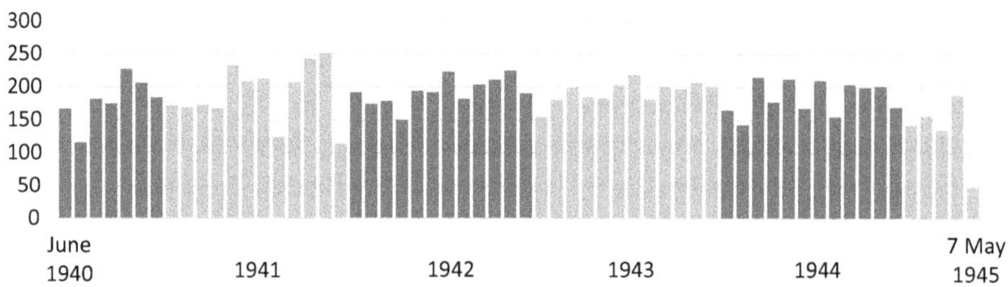

Figure 15: Monthly German withdrawals from the occupation account in Norges Bank, June 1940–7 May 1945, in million kr. Source: Norges Bank, *Norges Bank under okkupasjonen*, 125–26.

Aukrust and Bjerve were already working on their book during the second half of the occupation. They were in close contact with the men and women of the Norwegian resistance movement, such as the head of the Norwegian Statistical Office, Gunnar Jahn, who became Norway's first post-war Minister of Finance and wrote the preface for Aukrust and Bjerve's book. Moreover, the calculations must be seen in the light of the subsequent negotiations on German reparation payments. The study provided the background before which the narrative of a nation rising from the ashes could unfold in post-war Norway, and it remained the authoritative account on the topic for decades.

In a 1950 book that attracted little attention in the Norwegian post-war society, economist Johan Vogt already criticised some parts of the calculation. He claimed that the country's capital loss under the occupation had been smaller than Aukrust and Bjerve assumed and that they had undervalued the facilities

226 BAK, N 1550/2, Korff, 'Norwegens Wirtschaft', 1949, pt. II, 243. This is also the figure given by the Norwegian Central Bank in 1945: Norges Bank, *Norges Bank under okkupasjonen* (Oslo: Norges Bank, 1945), 126.

227 Korff, 'Norwegens Wirtschaft', 1949, pt. II, 251.

built by the Germans.[228] But it was only from the 1980s onwards that more historians began to challenge the dominant narrative, which had focused almost exclusively on the occupation's destructive economic effects. They pointed out that the country had got back on its feet again surprisingly quickly and that many of the enforced German investments had contributed to the Norwegian post-war upswing after all.[229] Investments in the aluminium industry provided the foundation for the sector's crucial position in the Norwegian post-war economy; so did the erection of deep-freeze plants for the fishing industry. Capacities for the production of electricity were expanded by one third, the length of telegraph lines increased by 200 percent and the number of telephones by 65 percent. The railway network had been expanded by 450 km, and for the first time, a road now connected the northern parts of the country with the south. The number of airports that could be used for civilian air traffic had risen from two (Sola, Stavanger and Fornebu, Oslo) to more than twenty.[230] Einar Lie showed that Aukrust and Bjerve were aware of the fact that the level of production and consumption during the first post-war years was considerably higher than what official statistics showed, meaning that they had underestimated the Norwegian economy's capability at the time of the liberation.[231]

There was a basic assumption, however, that remained unchallenged also by the literature critical of Aukrust and Bjerve's calculations: the assumption that all German investments in Norway were paid for by the Norwegians. International historiography, too, has more or less implicitly equated the Norwegian financial contributions, that is, the *internal* occupation costs, with the occupants' total expenses: Boelcke writes that the "total costs of occupation exceeded 12 billion Norwegian kroner", leaving the reader in the dark about whether he is referring to *total costs*

---

228 Johan Vogt, *Vår økonomiske stilling: En kritisk vurdering av gjenreisningsperioden* (Oslo: Aschehoug, 1950), 11–27.

229 Wicken, 'Industrial Change'; Hanisch and Lange, *Veien til velstand*, 44–46; Hjeltnes, *Hverdagsliv*, 67–95; Tjelmeland, Hallvard, 'Kva gjorde krigen med Norge?'; Lange, *Samling*, 107–11. Aukrust and Bjerve were conscious to the problem that they had omitted the value of German investments in their calculations, as a short remark in *Hva krigen kostet Norge* shows: "From this figure [what the war cost Norway, S.G.] one has to subtract the value of fortifications, airfields, roads, Nordag's facilities, etc., which were built during the war on behalf of Germany and which can be now taken over by the state, and moreover the value of confiscated enemy [that is German, S.G.] property and Norway's share in the Allied war booty. However, we have not considered these items here, partly because we lack an overview of them, partly because it is a matter of opinion how we want to judge, for example, the German fortifications that are often more harmful than good": Aukrust and Bjerve, *Hva krigen kostet Norge*, 49.

230 Lange, *Samling*, 107–11.

231 Einar Lie, *Ambisjon og tradisjon: Finansdepartementet 1945–1965* (Oslo: Universitetsforlaget, 1995), 115–16. Another critique of Aukrust and Bjerve's calculations is provided by Espeli, 'Det økonomiske forholdet mellom Tyskland og Norge 1940–45', 160–63.

or to *total costs for Norway* only.[232] Klemann and Kudryashov, on the other hand, are very clear when they write: "Norway was even worse off, as it had to pay for the enormous defence works built by the Organisation Todt [. . .]."[233] Eichholtz likewise states that the *Wehrmacht* and the OT built the fortifications in the occupied territories with materials, money and labour provided by the countries themselves.[234]

Obviously, Aukrust and Bjerve, Korff in his 1949 manuscript, and all subsequent research only considered Norway's *internal* occupation costs. The question is of course why Korff omitted his 1947 estimate on the *external* occupation costs from his unpublished 1949 manuscript. There are indications that Korff's manuscript was censored by Norwegian authorities, which must have been aware of the information in the 1947 report, but did not want so see it published at a time when Norwegian reparation claims against Germany were still on the table.[235]

Only very recently have Norwegian historians become aware of the fact that the math hardly adds up.[236] In his 1949 manuscript, Korff broke down what the Germans had spent the 11,054 million kr. from the occupation account on. The *Wehrmacht* itself spent 96 percent of the total sum, including the expenses of the three *Wehrmacht* branches, the OT, the RAD, the SS, the National Socialist Motor Corps (*Nationalsozialistisches Kraftfahrkorps*, NSKK) and the police forces. According to Korff's manuscript, EW spent only 9 percent of the *Wehrmacht* expenses, that is 953 million kr., between 1940 and 1945.[237] But could this figure really have been plausible, given the fact that EW organised the construction of coastal fortifications, roads, and railways with the help of hundreds of Norwegian and German companies and using thousands of Norwegian, German and foreign labourers and prisoners of war? Can we realistically assume that the organisation whose activity had a huge impact on the Norwegian economy for more than three years would have received less than 1 billion kr. in funding? And how can it be possible that the *Wehrmacht's* monthly withdrawals from the occupation account, as shown in Figure 15, remained stable also after spring 1942, when German construction activity reached an unprecedented scale under EW – and while at the same time the number of German soldiers in the country doubled between late 1941 and summer 1943?[238]

---

232 Willi A. Boelcke, *Die Kosten von Hitlers Krieg. Kriegsfinanzierung und finanzielles Kriegserbe in Deutschland 1933–1948* (Paderborn: Ferdinand Schöningh, 1985), 112.
233 Klemann and Kudryashov, *Occupied Economies*, 204. Similar: Kilian, *Krieg auf Kosten anderer*, 194.
234 Eichholtz, *Kriegswirtschaft II/2*, 379.
235 RAO, S-1543/Da-L0006, file 3, P.M. from the Reparations Office regarding the publication of Korff's manuscript, 24.9.47.
236 Espeli, 'Economic Consequences', 507; Hans Otto Frøland, 'Organisation Todt som byggherre i Norge', *Historisk Tidsskrift* 97, no. 3 (2018): 172–76.
237 Korff, 'Norwegens Wirtschaft', 1949, pt. II, 244.
238 Frøland, 'Organisation Todt i Norge', 175–76.

b Einsatzgruppe Wiking's Funding Mechanism

Together with Hans Clausen Korff's report of September 1947, an analysis of *Einsatzgruppe Wiking's* funding mechanisms helps to answer these questions. The various financial aspects of the OT's activities in occupied Europe are among the biggest desiderata in the historiography on the Organisation. Moreover, little is known about the impact of the OT's construction projects on the "occupied economies". Existing literature has only seldom dealt with these questions, mostly contenting itself with a short note that the *Wehrmacht* paid for the OT's projects in most cases.[239] However, the deeper one digs into the piles of invoices, remittance orders, statistics and account books that EW has left behind, the clearer it becomes that the issue was much more complicated.

In its funding guidelines, the OT specified that all of its bodies paying invoices or buying materials (that is the OBLs in most cases) were to receive only those amounts in local currency that were absolutely necessary.[240] That means that EW's OBLs received Norwegian kroner to pay Norwegian entrepreneurs and their workers, as well as to cover any expenses within Norway such as construction materials, renting machinery, transport costs and taxes. The German contractors were allowed to request only as much Norwegian currency from the OBLs as necessary to pay their German workers' allowances, OT pay, travel costs and certain premiums.[241] The OBLs also refunded the 10 percent turnover tax that German companies had to pay to the Norwegian state in kroner.[242] The OT disclosed these funding guidelines

239 Two short pages in Seidler's book on the OT are the most detailed contribution to the topic: Seidler, *Die Organisation Todt*, 216–17.

240 In fact, also the Wiking order had specified that EW's mission was to be financed through the German war budget.

241 BArch, R 2301/7125, fol. 188–90, Funding guidelines for OT orders, December 1943. Cf. RAO, RAFA-2188/1/E5a-L0046, Terms and conditions by the General Inspector for German Roads/ Organisation Todt for the execution of construction works at fixed prices (*OT-Leistungsvertrag*) as of January 1942, 19. On the topic of allowances and the OT pay, see chapter V.3.b.

242 The turnover tax was the only tax German construction companies had to pay in occupied Norway. The tax was also paid on invoices that were settled by the OT in RM in the Reich, but not on the supplies that came to Norway as *Wehrmacht* goods: RAO, RAFA-2188/2/Hfa-L0007, file 25, Interne meddelelser fra OT Einsatzgruppe Wiking, Letter by Einsatzgruppe Wiking regarding the turnover tax in Norway, 27.7.42. As the firms' mission was regarded as "temporary", all other corporate taxes were paid in Germany: BArch, R 2/358, fol. 8–9. For the financial authorities in Germany, this arrangement led to considerable control problems. It seems that systematic controls of German construction sites in Norway by German tax auditors never materialised. Cf. ibid., fol. 39–41. As the Norwegian state collected the turnover tax, Norwegian tax auditors, too, were interested in checking the German firms' book-keeping. However, they were often denied both access to German construction sites and insight into the firms' books for reasons of military secrecy. Financial experts in the RK, such as Hans Clausen Korff, were unhappy with this situation. Korff argued that Norway would only be able to contribute financially to the occupation if it had sufficient tax revenue: Ibid., fol. 33.

also in the contract documents and made clear that the firms were not allowed to hoard larger sums of local currency.[243]

As a result of this differentiation, EW had both a kroner account in Oslo to handle its payments in kroner and a completely independent reichsmark account in Berlin for all transactions in German currency.[244] Both accounts were refilled by EW's customer, the *Wehrmacht*. In addition, the *Wehrmacht* often provided EW with construction materials without demanding payment for it.

Hence, EW was economically reliant on three sources. The first source, for all payments in Norwegian currency, was the occupation account at the Norwegian Central Bank. From this account, the Chief Quartermaster at the WBN transferred kroner to EW's Norwegian account in Oslo. Either on a monthly basis or after an OBL request, EW's headquarters in Oslo then transferred the operating funds to the various OBLs around the country, which had accounts with local Norwegian banks.[245] When money was paid out to contractors, this was normally done by the OBLs.

The second funding source was EW's reichsmark account at the OTZ in Berlin. There was only one reichsmark account, since the OBLs had no reichsmark accounts of their own. This account was debited whenever EW – on behalf of the various OBLs – paid German firms and workers. EW in Oslo then forwarded the transfer orders to the OTZ in Berlin, which carried out the payments and charged them to EW's reichsmark account.

Other OT *Einsatzgruppen* and *Einsätze* also had their accounts in Berlin, which made it possible to settle financial transactions between *Einsatzgruppen* by simple transfers in the books within the OTZ in Berlin. It seems that these transactions were not a huge phenomenon, but the OT was bypassing the official clearing system for the trade between the countries in the German economic sphere of control. In 1942, for example, OBL Copenhagen bought medical equipment in Denmark on behalf of *Einsatzgruppe Wiking* and sent it to Norway. The payment was made in Berlin through a money transfer from EW's reichsmark account to OBL Copenhagen's reichsmark account.[246] This method could not be used for purchases in Sweden, since the OT was not present in the neutral country. Therefore, the OTZ transferred money via the official clearing channel to a special account

---

243 RAO, RAFA-2188/1/E5e-L0001, Supplement to the efficiency-output contract of 11 and 14 December 1943 between EW and construction firm Diehl, Essen, 15.12.43.

244 For a short explanation of how EW and its OBLs handled payments in kr. and RM, see RAO, RAFA-2188/1/E8a-L0040, file OTW Schriftwechsel, Report by OTZ auditors Werner and Schalla regarding their audit of EW's revenue and spending between 30 March 1942 and 31 March 1943, 9.4.44.

245 For examples of these so-called *Betriebsmittelverstärkungen*, see RAO, RAFA-2188/1/E8a-L0002, file Rj. 1945 Kronen-Konto Ausgabe.

246 RAO, RAFA-2188/1/E8a-L0039, Letter EW to OTZ, 7.9.42.

(the so-called *Sonderkonto Schwedenkronen*) of EW at the German legation in Stockholm. The OTZ charged EW's reichsmark account with the expenses.[247]

The third of EW's funding sources were construction materials that were bought by the *Wehrmacht* or the OTZ in Berlin and provided to EW for free. They were probably significant, but it has been impossible to determine their value because no detailed German war budget has survived. Furthermore, neither the Norwegian nor German customs officials were authorised to open and register trains and ships bringing *Wehrmacht* goods from the Reich to Norway. These hidden imports (and vice versa Norway's hidden exports) were thus not brought to account in the official clearing statistics.[248] The Statistical Office of Norway also confirmed after the war that "[. . .] there have been imported a lot of goods at the occupying power's expense, and for these imports, customs has normally not been able to provide information."[249] After 1 July 1942, deliveries for A/S Nordag, too, were no longer registered by the Norwegian customs authorities, as the enterprise became one of EW's responsibilities.[250]

In Table 17, EW's payments in RM and kr. have been reconstructed from account books and statements. They amounted to approximately 1.6 billion kroner. The table only shows the payments carried out by the *Einsatzgruppe* itself, either from the kroner account in Oslo or the reichsmark account in Berlin. These

Table 17: Expenses of Einsatzgruppe Wiking (without Denmark), accounting years 1942–1944 (1 April 1942 to 31 March 1945) and 1 April to 8 May 1945.*

	1942	**1943**	**1944**	**1945 (1 April to 8 May)**	**Total**
RM account, Berlin	27.6	92.5	164.3	0.3	**284.7**
Converted to kr.	*48.3*	*162.3*	*288.3*	*0.5*	***499.4***
Kr. account in Oslo	242.6	358.6	472.5	21.6	**1,095.3**
Converted to RM	*138.3*	*204.4*	269.3	*12.3*	***624.3***
Total in RM	165.8	296.9	433.7	12.6	**909**
Total in kr.	291	520.9	760.8	22.1	**1,594.7**

*Source: appendix, Table A8.1.

247 RAO, RAFA-2188/1/E8a-L0032, file Sonderkonto Schwedenkronen. These funds were used to make minor purchases like machines, spare parts, lamps and pots. Thus, by 12 December 1944, the expenses in Swedish kroner had only amounted to the equivalent of a moderate 80,000 RM.

248 Jonas Scherner, 'Der deutsche Importboom während des Zweiten Weltkriegs. Neue Ergebnisse zur Struktur der Ausbeutung des besetzten Europas auf der Grundlage einer Neuschätzung der deutschen Handelsbilanz', *Historische Zeitschrift* 294, no. 1 (2012): 86–90.

249 *Statistisk-økonomisk utsyn over krigsårene* (Oslo: Statistisk Sentralbyrå, 1945), 48.

250 *Statistisk-økonomisk utsyn*, 48.

payments alone are two thirds higher than the 953 million kr. mentioned in Korff's manuscript.

The figures in Table 17 are confirmed by a report of September 1945, compiled by *Einsatzgruppe Wiking's* settlement centre (*Abwicklungsstelle*).[251] Staffed with former EW members who were now in Norwegian captivity, the settlement centre arrived at an estimate of 1.8 billion kroner for EW's expenses. Since the sum contains an estimated 260 million kr. for construction materials and machinery that the OTZ had bought in Germany and sent to Norway without charging one of EW's accounts,[252] the figure confirms the information in Table 17 quite well.

What remains unknown is the value of the *Wehrmacht's* "free" deliveries of construction materials to EW. We must assume that their value was considerable. The small Norwegian economy was simply not capable of providing the resources required for the German mammoth projects. Even before 1942, the OT already had to supply its bunker construction site in Trondheim, for example, with materials and machines from abroad. Most construction materials, equipment and even the locomotives needed to transport the materials from a pier outside Trondheim to the *Dora* construction site had to be delivered from Germany.[253]

Cement was also delivered from abroad in significant amounts, mostly from Germany, but also from Denmark and Sweden. In 1942, consumption amounted to 531,000 tonnes of cement, while the country's three cement plants in Slemmestad, Kjøpsvik and Brevik produced 363,000 tonnes of cement. 192,000 tonnes of cement were imported from Germany and Denmark that year.[254] In 1944, the Norwegian production amounted to approximately 310–320,000 tonnes.[255] The consumption, however, was still at 470,000 tonnes.[256] Therefore *Wehrmacht* transports continued to play an important role. Almost 170,000 tonnes of cement were shipped from Germany to Norway between January 1944 and the end of September 1944. Minor deliveries also came from Denmark and Sweden (6,000 tonnes each) during

251 NHM, FO.II/11.4, box 27, file 56b, RK – EW settlement centre, report on the construction expenses by EW between 1942 and 1945 with 11 supplements, 3.9.45.

252 Ibid., supplement 11.

253 Neitzel, *Die deutschen Ubootbunker*.

254 RAO, RAFA-2188/1/E1l-L0044, Draft of a letter by Wick, Department for Labour and Welfare, 3.4.43. A slightly different value for Norwegian cement production in 1942, 374,000 tonnes, is given in: *Statistisk-økonomisk utsyn*, 193.

255 RAO, RAFA-2188/2/Hfa-L0025, file 111, Letter with attached statistics by Seidensticker, GB Bau, Oslo, to *Amt Bau-OT*, Berlin, 13.12.44. To produce the cement, the Norwegian plants depended on coal. Before the transport situation worsened in autumn 1944, between 100,000 and 200,000 tonnes of coal were shipped from the Reich to Norway every month: Bohn, *Reichskommissariat Norwegen*, 442. Coal, however, which was also used for heating and the railways, was registered as a regular import.

256 RAO, RAFA-2188/2/Haa-L0021, Construction volume Norway 1945, 12.2.45.

this period, which do not seem to have been registered in the countries' clearing accounts.[257]

While it seems to have been possible to cover the demand for construction timber through Norwegian production, construction iron and steel seems to have come to the country through *Wehrmacht* transports. In early 1943, for example, the RK estimated that the approximately 90,000 tonnes of construction iron that had been used in 1942 came entirely from German production.[258]

The noticeable gap between what Norway was able to provide and what EW and the *Wehrmacht* were consuming indicates that unregistered deliveries from Germany played an important role in the Norwegian wartime economy. In the official clearing statistics, Norway was already the only occupied country that had a surplus in its balance with the German Reich.[259] The hidden *Wehrmacht* transports added to this – admittedly unwanted – Norwegian import surplus. While the *Wehrmacht* transports usually disguised the degree of German exploitation in occupied territories, in Norway's case they brought a considerable amount of resources to an occupied territory. In this respect, Norway was probably an exception within occupied Europe.

The conclusion is that German construction investments in occupied Norway were much higher than hitherto assumed. However, neither the total cost of EW's activities nor the overall value of German construction investments in occupied Norway will be possible to calculate unless documents on the hidden *Wehrmacht* transports are found. As shown above, the literature on the Norwegian economy between 1940 and 1945 has hitherto only considered one funding source, the occupation account in the Norwegian Central Bank, and has thus massively underestimated German construction investments in the country. From an economic perspective, Norway was in fact an expensive country to occupy.

The way EW was funded not only made it possible to finance its large-scale projects, but provided also a mechanism to do so without further increasing the burden on the Norwegian budget. The Germans were well aware of the fact that their plans were straining Norway's economic capacity to the limit, with far-reaching effects on the wage and price structure particularly in construction. When Friedrich Thierbach, the head of Wigru Bau's recently established liaison office, arrived in Oslo in late 1941, he found the construction sector "in shambles". As he was told by Heinz Klein, price policy had slipped almost entirely from the German

257 Ibid., Undated chart on the cement management in Norway (Norwegian production and imports). After the war, the Norwegian Statistical Office indicated that only 380 tonnes of cement had been imported in 1944. The discrepancy between these official figures and the German wartime figures is striking: *Statistisk-økonomisk utsyn*, 68.

258 RAO, RAFA-2188/1/E11-L0044, Draft of a letter by Wick, Department for Labour and Welfare, 3.4.43.

259 Klemann and Kudryashov, *Occupied Economies*, 210.

authorities' control. Due to the high time pressure, the *Wehrmacht* was drawing on the same firms again and again and thus eliminating the cost-reducing effects of competition. "There are several sites in Norway", Thierbach learned, "that cost a multiple of comparable sites in Germany." Even unskilled Norwegian construction workers could earn up to 2,500 kr. (c. 1,400 RM) per month.[260] Hans Clausen Korff recalled after the war that in Northern Norway, weekly wages of 250 kroner were "normal", and 500–650 kroner not uncommon.[261] A member of the RK put it drastically around the turn of the year 1941–42: "The entire economy of Norway is on the verge of being shattered by the unsolved questions in the construction sector."[262]

At this time, the conflict between the financial experts of the Reich Commissariat and the *Wehrmacht* had been smouldering for almost two years. For the WBN von Falkenhorst, it was beyond question that all budgetary concerns had to be subordinated to the needs of the military. Reich Commissioner Josef Terboven, the Reich Ministry of Finance and even the OKW in Berlin, however, feared that a collapse of the Norwegian economy might bring about the downfall of the entire German regime in the country. After tedious negotiations, it was agreed that Norway's financial contribution to the German construction projects had to be limited and that more material and financial resources had to be provided by the Reich. As Hans Otto Frøland has recently argued, this financial dimension is essential in explaining why it was the OT's *Einsatzgruppe Wiking* that was ordered to carry out the tasks laid down in the Wiking order. The OT provided a solution to the conflict about *Wehrmacht* spending in occupied Norway.[263]

EW followed the principle of limiting the burden that the German construction activities meant for the Norwegian economy throughout the war. When EW took over the air force construction apparatus in the second half of 1944, it once more outlined the OT's guideline that German firms in Norway were only to receive as much local currency as absolutely necessary.[264] In the last months of the occupation, the distinction between expenses that had to be covered in reichsmark and those paid for from the kroner account sometimes became blurred. It seems, however, that this usually meant that firms were paid in reichsmark in Germany for expenses that actually would have required reimbursement in Norwegian currency. In March 1945, the construction firm H.C. Hagemann from Hamburg asked for reimbursement of several

---

260 BArch, R 13-VIII/264, First report of Thierbach, Oslo liaison office of Wigru Bau, 27.12.41.

261 Korff, 'Norwegens Wirtschaft', 1949, pt. II, 259. As a comparison, the gross weekly wage of a skilled construction worker was 88 kr. (based on a 48 hours week) according to the national Norwegian wage agreement of 1941: RAO, RAFA-2188/2/Hca-L0001.

262 BArch, R 13-VIII/264, Second report of Thierbach, Oslo liaison office of Wigru Bau, 18.1.42, 6.

263 Frøland, 'Organisation Todt i Norge', 174–75.

264 RAO, RAFA-2188/1/E5a-L0008, file Reichskommissar für die Preisbildung, Letter without sender and addressee on the occassion of the OT's take-over of the air force's construction apparatus, 3.8.44.

kroner payments it had made to Norwegian suppliers. The firm asked EW to transfer the sum in reichsmark to the firm's German account.[265] Before returning to Germany, the firms also paid in their cash and bank deposits. The OT then paid out the equivalent value in reichsmark to the firms' headquarters in Germany.[266] Obviously, the companies did not want to accumulate larger sums of Norwegian currency, which they would probably lose in the event of the imminent German defeat.

c Norwegian Construction Activities during the Occupation

The German construction activities of course had a major impact on the Norwegian construction sector, too. During the first months of the occupation, the *Wehrmacht* already accounted for half of the Norwegian construction volume, and in 1941, the share of projects on behalf of the German armaments industry and the military had risen to 90 percent.[267] Registered unemployment, which had been around 6 percent in the last pre-war years, had more or less vanished in the first autumn of the occupation. The Norwegian Statistical Office estimated that in the summer of 1941, up to 165,000 Norwegians were working on German construction sites. After that, the numbers decreased constantly, as more and more forced labourers were brought to Norway. But in June 1944, there were still approximately 145,000 workers, employees and auxiliary workers such as drivers and cleaners in the sector.[268] In terms of hours worked by Norwegian construction workers, the years of the occupation outmatched anything the country experienced both before and after the war (Figure 16).

The Germans paid generous wages in order to attract the scarce labour to their construction sites. Reports and complaints about Norwegian work gangs earning fabulous wages were legion during the occupation years. In regions with high construction activities, like Nordland, Rogaland or Sør-Trøndelag, for example, the development could have a great effect on entire communities.[269] In the Nordland region, where German authorities already forced the construction of the *Nordlandsbane* between Mosjøen and Mo i Rana during the first months of the occupation, "a certain Klondyke [sic] atmosphere" evolved and "the business community of Mo i Rana had never before

265 RAO, RAFA-2188/1/E8a-L0001, Payment authorisation for H.C. Hagemann's invoices no. 587 and 589, 21.3.45.
266 See for example: RAO, RAFA-2188/1/E8a-L0032, file Deutsches Konto Ausgabe Verrechnungen, Request by Arbeitsgemeinschaft Norden, 13.4.45.
267 RAO, RAFA-2174, Ef-L0002, file 9, Draft of a report by Wick, construction industry group in the Main Department for Technology, for the overall report on the activities of the RK, 14.4.42.
268 *Statistisk-økonomisk utsyn*, 218–23; Morten Tuveng, *Arbeidsløshet og beskjeftigelse i Norge før og under krigen. En statistisk analyse med henblikk på beskjeftigelsesspørsmålet etter krigen* (Bergen: John Griegs Forlag, 1946), 124–37.
269 Tuveng, *Arbeidsløshet*, 161–62.

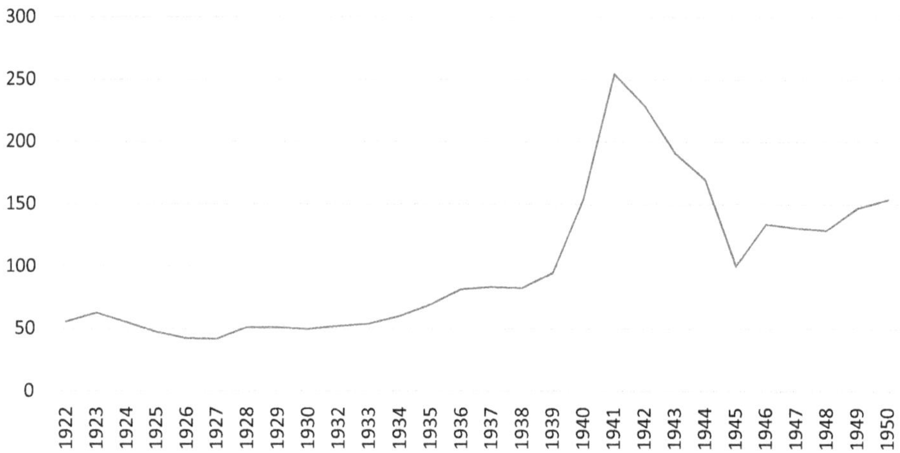

Figure 16: Hours worked by Norwegians in construction, in millions, 1922–50. Source: *NOS XII. 291. Historisk Statistikk 1978* (Oslo: Statistisk Sentralbyrå, 1978), 245. No value available for 1931.

experienced anything similar to the Christmas sales in 1940: the shelves were literally emptied of goods."[270] To be sure, the Norwegian workers' purchasing power decreased during these years and empty shelves were of course also the result of a genuine scarcity of consumer goods. At the end of the day, however, the German activities certainly brought real prosperity to these communities. In the 1930s, both the Norwegian municipalities and many private households had been suffering under high levels of debt. By 1945, most of them were debt-free.[271]

Even though most of the Norwegian contractors working for the Germans were not organised in the construction industry's trade association (*Entreprenørenes Landssammenslutning*), the member firms of the latter were still able to more than quadruple their number of workers between 1939 and 1941.[272] The so-called barrack barons (*brakkebaroner*), craftsmen who erected wooden barracks for the occupying troops, became the synonym for war profiteers in post-war Norway. In 1943, Norwegian and German financial experts estimated that on the construction sites of the *Sørlandsbanen*, prices granted by the *Wehrmacht* to Norwegian firms were four times higher than the self-costs of the NSB.[273]

270 Svein Fygle, Svein Lundestad, and Inge Strand, *Banken, folket og fylket. Nordlandsbanken og Nordlands næringsliv gjennom 100 år: 1893–1993* (Bodø: Nordlandsbanken, 1993), 209.
271 Espeli, 'Economic Consequences', 511; Tjelmeland, Hallvard, 'Kva gjorde krigen med Norge?', 30.
272 Espeli, 'Det økonomiske forholdet mellom Tyskland og Norge 1940–45', 142.
273 Bohn, *Reichskommissariat Norwegen*, 369.

Given how widespread a phenomenon working for the Germans was, the question of whether this happened under compulsion has occupied the Norwegian public and historians ever since. During the *landssvik* trials, the Norwegian post-war trials of the late 1940s and early 1950s that investigated citizens for treason and collaboration, many entrepreneurs claimed that German authorities had threatened them with closure of their business, confiscation of machinery and even imprisonment and death. Parts of the literature accepted this narrative.[274] More recent research, however, paints a much more nuanced picture. There are no documented cases of Norwegian businessmen who were punished with imprisonment, torture or even execution because they had turned down a German order in the crucial first months of the occupation, as Norwegian investigators already concluded in 1946.[275] In 1947, after having interviewed around 30 German procurement officers of the OT and the *Wehrmacht*, Hans Clausen Korff reported that the officers had consistently stated that "actual pressure through threats of force and imprisonment had not happened. This would have not been necessary, as enough firms were willing to work for the Wehrmacht."[276] Accordingly, recent literature rather speaks of indirect threats, a menacing atmosphere or the *possibility* of severe consequences under the general power imbalance between the occupiers and the occupied.[277] Many contributions to the discussion, though, to this day rest on one assumption: that there was more or less no market left for Norwegian construction firms beyond the German orders.[278] This underlying premise is ripe for revision.

It is indeed correct that German projects represented the by far largest share of construction. However, it is often forgotten that the German demand was primarily an additional demand and that the purely Norwegian construction market had not

274 Steen, *Selmer*, 54 ff; Per Helge Pedersen and Øyvind Steen, *Skaperevne og initiativ. Byggenæringen i 100 år, 1905–2005* (Oslo: Byggeindustrien / Bygg og Anlegg Media AS, 2005), 40–48.
275 Erstatningsdirektoratet, *Undersøkelser*, 3. Cf. also: Vidar Eng, 'Hvorfor arbeidet gode nordmenn på flyplassene i Oslo og Trondheim i april 1940?', *Historisk Tidsskrift* 89, no. 3 (2010): 438–39. Eng provides a multicausal explanation for why "good Norwegians" worked on German construction sites during the early phase of the occupation, sporadic threats and a generally menacing atmosphere being just one of them. However, he ignores the most obvious motive for economic cooperation: profit-seeking and the wish to be on good terms with the power that in 1940 appeared it would be the European political, economic and military hegemon for the foreseeable future.
276 RAO, S-1543/Da-L0006, file 3, Final report by Hans Clausen Korff regarding the German occupation authorities' business relations in Norway, 29.9.47.
277 Johs. Andenæs, *Det vanskelige oppgjøret: rettsoppgjøret etter okkupasjonen* (1979; repr., Oslo: Tano Aschehoug, 1998), 72; Eng, 'Gode nordmenn', 440; Sæveraas, 'Beton macht Geschichte', 277; Sæveraas, 'Festung Norwegen', 203. Anders Kirkhusmo, on the other hand, stresses the seriousness of German threats in his reply to Vidar Eng: Anders Kirkhusmo, 'Kommentar til Vidar Engs artikkel: Hvorfor arbeidet gode nordmenn på flyplassene i Oslo og Trondheim i april 1940?', *Historisk Tidsskrift* 90, no. 4 (2011): 587–91.
278 Andenæs, *Oppgjøret*, 72; Sæveraas, 'Beton macht Geschichte', 213.

contracted as dramatically as often suggested.[279] As shown above, it is not possible to assess the overall volume of construction activities in Norway during the occupation. However, the Norwegian Statistical Office assessed the gross production of the Norwegian construction sector up to and including 1943. From 1940 to 1943, the gross production amounted to 3.64 billion kroner, measured in constant 1939 prices, with 1941 being the peak year.[280] Assuming that the production decreased between 1943 and 1944 by the same factor as it had between 1942 and 1943, and assuming for the sake of simplicity that there was no significant Norwegian production at all in the first months of 1945, one arrives at a very conservative estimate of 4.26 billion kroner for the occupation years. In order to exclude the share that Norwegian firms and workers earned on German construction sites from this sum, we can compare this figure to the sums flowing from the occupation account in *Norges Bank* to German construction projects (predominantly to those of the *Wehrmacht* and OT). They amounted to approximately 2.50 billion kroner, again in constant 1939 prices. Hence, one arrives at a sum of 1.76 billion kr. that Norwegian construction firms were able to earn on "Norwegian" construction sites during the occupation years. This again is a very conservative estimate: an unknown share of the 2.50 billion kr. spent by German agencies actually went into the pockets of German firms and their German workers in Norway. Thus the German-financed share of the 4.84 billion kr. was probably smaller and the share of Norwegian projects accordingly larger. This means that purely Norwegian activities were responsible for around 41 percent of the turnover of the Norwegian construction industry during the occupation.

In a final step, we can compare statistics on hours worked by Norwegians in construction during the occupation (1940–44) with an equally long period before the war (1935–39) to get an idea of the Norwegian construction sector's expansion. Between 1935 and 1939, Norwegians in construction worked 411.7 million hours, against 996.9 million hours during the occupation.[281] The construction sector's Norwegian part thus shrank from 100 percent of 411.7 million hours between 1935 and 1939 to approximately 41 percent of 996.9 million hours during the occupation. In summary, we can roughly estimate that during the five years of occupation, the Norwegian construction sector had a price-adjusted volume that equalled that of

279 In the following, the purely Norwegian construction market is understood as those projects that were not ordered and financed by the German occupiers or – if they were – were conducted by Norwegian authorities like the NSB or *Statens Vegvesen*. German projects that had positive economic effects in post-war Norway, like the construction of airports and the projects of Norsk Hydro, A/S Nordag or A/S Nordisk Lettmetall, are considered part of the German construction market in Norway. Cf. *Statistisk-økonomisk utsyn*, 221.

280 For the following calculations and the indices used for deflation in this paragraph, see appendix A7.

281 *NOS XII. 291*, 245.

the 5-year period before the war. Norwegian construction firms carried out projects for the state and municipalities, repaired houses that had been destroyed during the German campaign in 1940, or were engaged in railway and road construction projects on behalf of Norwegian agencies. The Norwegian industry also continued to award contracts to local construction firms. The German construction activities did not suppress Norwegian activities, but generated an additional demand.[282]

In addition, it is reasonable to assume that the new construction businesses that mushroomed during the occupation in order to profit from the building boom primarily capitalised on the very lucrative German projects. Hence, a purely Norwegian market of considerable size remained between 1940 and 1945, on which an established Norwegian construction company could focus if it wanted to keep its business relations with the German occupiers to a minimum. Turning down German orders involved economic risks and certainly decreased earnings, but it did not mean committing economic suicide.[283]

A good illustration of the argument presented in this chapter is the wartime history of the firms A/S Betonmast, Oslo, and A/S Impregnertbygg, Molde. The two firms were more or less the only companies in Norway in the crucial field of power line construction. This made it very difficult to simply refer the Germans to other companies, as A/S Impregnertbygg's manager Alf Holst Høstmark reported after the war. Hence, the firms were in a particularly exposed position vis-à-vis the occupying power. Still, they managed to keep their business relations with the Germans at a minimum level. A/S Impregnertbygg's turnover on German construction sites only amounted to approximately 15 percent of total turnover, that of A/S Betonmast to approximately 30 percent. Without much competition when it came to the construction of power lines, Høstmark was convinced that his company could have earned much more money, if it only had chosen a slightly more accommodating stance towards the *Wehrmacht* and the OT during the occupation.[284]

---

282 This is confirmed by German estimates on the consumption of construction iron and steel in 1942. Of the approximately 90,000 tonnes of iron and steel that had been used on construction sites in Norway, c. 13,000 tonnes were allocated to civilian Norwegian projects. RAO, RAFA-2188/1/E1l-L0044, Draft of a letter by Wick, Department for Labour and Welfare, 3.4.43; RAO, RAFA-2188/1/E1b-L0006, file 2230, Letter Feuchtinger to the GB Bau, Berlin, 20.2.43.

283 It is important to keep in mind that the total construction volume in Norway was significantly higher due to payments made in RM to German firms and their German workers. They are not considered here. Neither are the thousands of forced labourers and POWs, as they normally only worked under German firms. But since Norwegian firms working for account of the German occupiers were paid in kroner from the occupation account in *Norges Bank*, both RM payments and forced labour could remain unconsidered in this context.

284 RAO, S-3138/0001/Dg-L0450/0003, Hnr. 5317, trial against Carl Josef Christensen, general manager of A/S Betonmast, Oslo; RAO, S-3138/0036/Da-L0039/0019, Anr. 459, trial against A/S Impregnertbygg, Molde. See especially: ibid., document 35, report on the interrogation of Alf Holst Høstmark, 29.11.46.

It must be noted that my estimates made in this part of the chapter are again based on rough estimates of the annual construction volume and hours worked. The Norwegian Statistical Office's ability to obtain detailed figures was of course limited during the occupation. Also German authorities were not always able to gather absolutely reliable data. I have tried to address this problem by always choosing the most conservative estimate at the various stages of my estimation.

4 GB Bau in Occupied Norway

The *Wehrmacht's* construction activities, and from spring 1942 those of EW, were a shock to the Norwegian economy. In many ways, the situation resembled that of the German construction sector in 1938. Labour and construction materials were scarce, and wages and prices were rising accordingly. Without some kind of regulation, Germany's military and economic ambitions in Norway would be seriously jeopardised. Since the occupied country depended increasingly on German supplies and the deployment of German construction firms, it seemed natural that any regulatory body in Norway would be closely associated with the GB Bau's office in Berlin. The role of GB Bau's representative in occupied Norway will thus be the topic of this chapter.

From 1942 onwards, Willi Henne held the office in Norway. Henne thus was in a triple position as head of the OT, head of the RK's Main Department for Technology and GB Bau's representative. Literature on the GB Bau in occupied Norway is scarce, but those historians who have investigated its role have consistently highlighted the power lying in this triple position. At the same time, they all have pointed to the limits of this power.[285] This chapter will take the existing research as a point of departure and ask about how far-reaching the GB Bau's powers in occupied Norway really were and to what extent it steered the German companies and limited their room for manoeuvre.

Part a will for the first time describe the process behind the establishment of a GB Bau office in occupied Norway. Once again, it was the ambitious Heinz Klein in the Reich Commissariat who had tried to attain the position of some kind of superordinate Norwegian construction commissioner since 1940. Parts b and c will then describe the tasks of the GB Bau's representative in Norway and analyse how successful the office was in monitoring and steering the German construction activities in the occupied country.

---

285 Botzet, 'Generalbevollmächtigter'; Hatlehol, 'Norwegeneinsatz', 111–12; Sæveraas, 'Beton macht Geschichte', 75–77, 137, 141; Mats Ingulstad, 'I nyordningens tegn? Organisation Todt og reguleringen av den norske byggebransjen', *Historisk Tidsskrift* 97, no. 3 (2018): 256–76.

a Establishing a Central Construction Authority in Occupied Norway

Given its tasks in Germany, there were several fields a construction commissioner in Norway could *potentially* take control over: permits or bans of new construction in the Norwegian construction sector, the allocation and distribution of construction materials, labour and machinery, the deployment of firms, and measures in the field of wages and pricing. Before any of these measures could be achieved, however, the GB Bau needed to gain an overview over the rampant construction activity in order to reconcile what was planned with what was viable.

First attempts to coordinate and control German and Norwegian construction programmes – most importantly the *Wehrmacht* projects – dated back to autumn 1940. At the time, the men within the Reich Commissariat feared that the *Wehrmacht's* reckless use of Norwegian resources was about to kill the cow that Terboven still wanted to milk. While the RK's financial experts tried to contain the *Wehrmacht's* excessive spending, it was Heinz Klein's Technical and Transport Department that looked for a way to gain control over the actual construction projects. At the time, Klein was the GB Bau's regional representative (*Gebietsbeauftragter*) in Norway, although his exact scope of responsibility at this point in time remains somewhat unclear.[286] However, as we have already seen, Klein was a man with grander ambitions. In October 1940, he envisaged that his department should become a superordinate office, similar to the GB Bau office in Berlin. All awarding authorities in Norway should report their planned projects, stock of machinery and required building materials to the RK. The procurement of materials, the import of construction machinery from Germany and the deployment of German companies should depend on his approval. As Klein emphasised, effective regulation would only be possible if the *Wehrmacht* reported to him.[287] There is no sign that Klein's proposal, which would have made him the central figure on the construction sector in Norway, resulted in a formal transfer of responsibilities to his department in late 1940. At the time, his department suffered from staff shortages, which rendered the implementation of his plans impossible.[288]

However, in February 1941, Klein was able to claim new responsibilities at last. In the Reich, the GB Bau not only was responsible for the fixing of construction iron and steel quotas, the office was also in charge of controlling the economical use of the materials. However, the OKW for *Wehrmacht* projects and the Reich Centre for Economic Development for the light metals programme increasingly transported

286 *Die Bauindustrie* 9, no. 8 (1941): 295. The source lists Klein as "*Landstraßenbevollmächtigter beim Reichskommissar für die besetzten norwegischen Gebiete*", indicating that Todt had sent him to Norway with a mandate limited to the supervision of road construction. Cf. the conflict between Todt and Klein described in chapter IV.2.b, which suggests that Todt was reluctant to support Klein's grander ambitions.
287 RAO, RAFA-2188/2/He-L0004, file 17 (Dr. Klein Mappe I), Copy of a note by Klein, 21.10.40.
288 Ibid.

materials to Norway that came from their German quotas. The Berlin GB Bau had no chance of controlling whether these materials were being used thriftily in Norway. Thus Günther Schulze-Fielitz from the GB Bau in Berlin contacted the RK and asked whether Klein's department could assume this task.[289] Klein accepted and requested that the construction plans be forwarded to him.[290]

It seems, however, that the department did not manage to gain any significant control over the *Wehrmacht* projects during the following months. In Berlin, the GB Bau had in the meantime obtained a decree by its formal superior Hermann Göring, which obligated all claimants to report their construction projects to the GB Bau.[291] The Göring decree of 20 June 1941, however, only referred to the Reich territory.[292] The development in Berlin nevertheless seems to have encouraged Klein to launch a new attempt at becoming the strong man of the Norwegian construction sector. First, he assured himself of Terboven's backing. On 26 June, Klein conferred with the Reich Commissioner and obtained approval for his department (now *Technik*, no longer *Technik und Verkehr*) to assume responsibility for building permits, the compilation of priority lists and the allocation of construction materials. Moreover, the Department for Technology was to gain control over labour deployment.[293] On 1 August 1941, a construction industry group (*Gruppe Bauwirtschaft und Landesplanung*) was established in the Department for Technology.

Klein now reported to Fritz Todt on a monthly base. In his first report, Klein insisted on being assigned the same responsibilities and powers as the GB Bau in Germany. In particular, this meant that all construction projects in Norway, particularly those of the *Wehrmacht*, were to be ranked in priority lists and that Klein alone would control the allocation of iron and cement to the *Wehrmacht*. Earlier attempts in this direction, Klein added, had been repeatedly blocked by the *Wehrmacht*. He was particularly upset by the fact that the military organised its transports of building materials from Germany to Norway independently, and he thus had no overview over the amount of materials imported into and consumed in Norway.[294]

---

289 RAO, RAFA-2188/2/He-L0005, file 20 (Dr. Klein Mappe IV), Letter Schulze-Fielitz to Reich Commissioner Terboven, 24.2.41.

290 Ibid., Letter Klein (draft) to the GB Bau, 19.3.41.

291 BArch R 3/1449, fol. 2–3, Decree by the Head of the Four Year Plan, Hermann Göring, regarding the curtailment of construction activity, 20.6.41.

292 RAO, RAFA-2174/Ef-L0001, file 7, Note by Klein on a meeting with Steffens on 16 September 1941, 19.9.41.

293 RAO, RAFA-2174/Ef-L0001, file 7, Note by Klein, 26.6.41, and letter Klein to the Department for Labour and Welfare, 26.6.41.

294 RAO, RAFA-2174/Ef-L0001, file 7, Monthly report by Klein, Department for Technology, to Todt, 7.8.41.

Klein achieved a partial success in August 1941. In a circular decree of 21 August, the GB Bau ordered that construction projects in occupied Norway could only receive construction iron after the Department for Technology's approval.[295] On 8 September, Reich Commissioner Terboven officially made Heinz Klein GB Bau's representative in Norway. In a letter to Todt, Terboven also demanded that the GB Bau should make it obligatory for the three *Wehrmacht* branches to report their Norwegian construction projects to Oslo and obey all orders of the RK's Department for Technology. The decree of 21 August had only stated that the Department for Technology had to authorise all *Wehrmacht* projects in Norway, but the allocation of building materials was still a *Wehrmacht* task. In several talks with the GB Bau's Joachim Steffens, Klein tried to push through that *Wehrmacht* projects in Norway would not only depend on his approval, but that the Department for Technology would officially become a claimant responsible for the allocation of all construction materials.[296]

Although Klein continued to push for a decision in Berlin,[297] the results of his advance in the summer of 1941 were meagre. This has to be seen in the light of the ongoing power struggles between Fritz Todt and the *Wehrmacht* branches in Germany,[298] but even more so between the RK and the *Wehrmacht* in Norway. In the winter of 1941–42, Klein tried to gain an overview over all projected construction works in Norway and attempted to determine the construction volume of 1942, but the *Wehrmacht* branches refused to provide full particulars on their projects. A WBN officer explained this blockade with the fact "that the construction commissioner in Norway is an office of the RK. Under these circumstances, the Wehrmacht branches fear that the RK will gain influence on the execution of the ordered military constructions."[299] It is worth noting that the *Wehrmacht* was not generally against a restriction of construction activities. In the autumn of 1941, the military, too, had realised that Norway's economic and financial capacity had reached its limits. Accordingly, the *Wehrmacht* cut back some of its construction plans.[300] However, the WBN officer proposed that the Norwegian construction commissioner should be a *Wehrmacht* office or be directly subordinate to the GB Bau in Berlin, a solution that of course met with scant approval among the men of the RK's

295 BArch, R 3/1445, fol. 21–24, Circular Decree by the GB Bau regarding the administration of construction iron and timber quotas, 21.8.41.
296 RAO, RAFA-2174/Ef-L0001, file 7, Note by Klein on a meeting with Steffens on 16 September 1941, 19.9.41.
297 RAO, RAFA-2174/Ef-L0001, file 7, Letter Klein to GB Bau, Berlin, 23.10.41.
298 Botzet, 'Generalbevollmächtigter', 123–24.
299 RAO, RAFA-2174/Ef-L0001, file 7, Note on a meeting between Klein and *Wehrmacht* representatives in Oslo on 18 February 1942, 19.2.42.
300 Frøland, 'Organisation Todt i Norge', 175.

Department for Technology.[301] Only in late February did the *Wehrmacht* branches agree to provide lists of their planned projects for 1942.[302]

When Willi Henne assumed office as head of EW and head of the RK's Main Department for Technology on 1 April 1942, the road construction engineer Max-Erich Feuchtinger took over the department's construction industry group. Initially, the office's tasks and responsibilities did not change. Todt's successor Albert Speer was only able to consolidate his influence in occupied Europe after August 1942, when Hitler decreed that the GB Bau's competencies also applied to the occupied territories. Speer was able to appoint construction commissioners who were subordinate only to him and authorised to issue directives to all other German civilian and military bodies.[303]

Officially, Speer only appointed Willi Henne his construction commissioner in Norway on 30 September 1942.[304] Day-to-day, it was Max-Erich Feuchtinger who managed the GB Bau's affairs in Norway.[305] The office was now inseparably linked to the OT, not only on the highest levels in Berlin (Speer) and Oslo (Henne), but also in the Norwegian regions, where the OBLs of the RK's Main Department for Technology became the OBLs of *Einsatzgruppe Wiking*. The men under Henne now presided over EW's OBLs, were the heads of the Main Department for Technology's regional offices (*Dienststellen*) and at the same time were the local representatives of the GB Bau. Terboven confirmed this mode of operation in a decree on 24 February 1943.[306] Both contemporaries and historians have referred to the enormous potential that lay in this triple position.[307] The

---

301 RAO, RAFA-2174/Ef-L0001, file 7, Note on a meeting between Klein and *Wehrmacht* representatives in Oslo on 18 February 1942, 19.2.42.

302 RAO, RAFA-2174/Ef-L0001, file 7, Copy of a letter from Klein to the Main Department for Economy, 9.3.42.

303 Moll, *Führer-Erlasse*, 274–75.

304 His official and rather cumbersome title was "*Der Generalbevollmächtigte für die Regelung der Bauwirtschaft – Der Bevollmächtigte beim Reichskommissar für die besetzten norwegischen Gebiete*".

305 Max-Erich Feuchtinger, born in Kiel in 1909, had studied civil engineering in Munich and Berlin. After receiving his doctoral degree, he worked for OBR Essen of the *Reichsautobahnen* Corporation from 1939 onwards. He was 31 years old when he came to Norway. After the war, he founded an engineering consulting firm in Ulm and from 1957 on was professor for road construction and transportation at the *Technische Hochschule Stuttgart*. Feuchtinger died in Ulm in 1960. See Bernd Ottnad, ed., *Baden-Württembergische Biographien 1* (Stuttgart: Kohlhammer, 1994), 82–84.

306 *Verordnungsblatt für die besetzten norwegischen Gebiete*, no. 3 (1943). Cf. RAO, RAFA-2188/1/E1m-L0046, Letter by the GB Bau, Oslo, 15.3.43, as well as: RAO, RAFA-2188/1/E5a-L0035, Letter Terboven, 20.9.42.

307 RAO, Oslo Politikammer, B-sak 3061, box 2, file 3, Carlo Otte, 'Die Hauptabteilung Volkswirtschaft der Behörde des Reichskommissars für die besetzten norwegischen Gebiete' (1945), 31–32; Korff, 'Norwegens Wirtschaft', 1949, pt. I, 115–16; Bohn, *Reichskommissariat Norwegen*, 179–80; Hatlehol, 'Norwegeneinsatz', 89.

question is, however, how powerful an institution the GB Bau actually was in occupied Norway.

b Monitoring and Steering Norwegian and German Construction Activities

By autumn 1942, GB Bau's representative in Norway had finally assumed the powers that Heinz Klein had fought for during the past two years. At least formally, his powers were far-reaching. All construction projects, either civilian or military, Norwegian or German, had to be approved by Feuchtinger's office. In particular, awarding authorities were asked to disclose building costs, required labour, as well as the amount of construction iron, cement, bricks and timber necessary for the projects' execution. Based on this information, the GB Bau set the annual construction volume of the *Wehrmacht*, the OT, the RK for civilian German projects and the Norwegian Economic Ministry (*Næringsdepartement*) for civilian Norwegian projects.[308]

One major field of policy was to secure capacity and resources in construction for the German occupiers by regulating the Norwegian part of the sector. In the following years, the construction commissioner tried to tackle the problem from multiple angles. Possible strategies included limiting Norwegian construction activities and closing down construction sites not vital to the war effort, or relocating labour from other sectors to construction. Furthermore, the construction commissioner tried to take action against the unauthorised Norwegian businesses that had mushroomed since the Germans had occupied the country. Those firms in particular were blamed for the exploding prices and for employing workers who had fled conscription.

From the German perspective, it was self-evident that Norwegian civilian construction activities had to be reduced to a minimum. The RK left this duty to Norwegian legislation. Two ordinances issued by the Ministry of Provisioning (*Forsyningsdepartement*) in February and March 1941 restricted the sale and use of building materials and regulated Norwegian public orders.[309] Projects above a certain construction value had to be approved by the Germans, minor projects by the Norwegian Economic Ministry. However, the German authorities constantly

---

308 RAO, RAFA-2188/2/Hfa-L0007, file 27 Regelung des Bauschaffens in Norwegen, Undated document "Tasks of the General Commissioner for the Regulation of the Construction Industry and their execution by the Commissioner for the occupied Norwegian territories". In April 1943, the Ministry of Provisioning (*Forsyningsdepartement*) and the Ministry of Commerce and Industry (*Handels- og industridepartement*) were merged to the Norwegian Economic Ministry under Eivind Blehr.

309 RAO, RAFA-2188/1/E1a-L0002, file 2100, Ordinance by the Ministry of Provisioning on the ban of sale of construction materials, 7.2.41, and ordinance by the Ministry of Provisioning on the control of public orders, 14.3.41.

complained about ongoing Norwegian construction projects that were not vital to the war effort and had not been authorised. For years, the RK's construction industry group and later the construction commissioner tried to force the Ministry of Provisioning into issuing a complete ban of new construction. In 1942, Feuchtinger drafted several ordinances on a construction ban, but none of them got as far as being implemented.[310] The conflict intensified during the following months,[311] reaching its peak in the summer of 1943, when Henne once again wrote a threatening letter to the Economic Minister, Eivind Blehr, because the ministry was still ignoring the construction commissioner's regulations. Henne even prompted Reich Commissioner Terboven to personally write to Vidkun Quisling.[312] The Norwegians, however, successfully blocked a complete construction ban for the duration of the occupation.

The RK also struggled to control the sector. Its regional offices were too few in number and too poorly manned to keep track of all Norwegian activities. For instance, Feuchtinger suspected that cement allocated through the Norwegian Economic Ministry for the construction of potato cellars was being used to construct entire farm buildings.[313] In fact, a Norwegian firm claimed after the war that it had systematically diverted construction materials and workers allocated for German projects to its illegal Norwegian construction sites. At one point, a German inspection found that 7–8,000 bags of cement were missing in one of the firm's German construction sites, but was not able to prove the fraud.[314] A secret Norwegian report stated in 1944 that most Norwegian businesses had tried to prepare themselves for the time after the occupation. Between 1940 and 1942, they had often managed to modernise their businesses by tapping the Germans for machinery and "even in 1943, many industries succeeded in carrying out construction and modernisation works by means of false statements."[315] Therefore, from July 1942 onwards, construction commissioner Henne used EW's network and tasked the organisation with controlling and closing down unauthorised Norwegian construction sites.[316] EW's

310 RAO, RAFA-2188/1/E1a-L0002, file 2100.

311 See the documents in RAO, RAFA-2188/1/E1a-L0002, file 2100, especially the letter from the Ministry of Provisioning to the RK's Department for Technology, 21.4.42.

312 Bohn, *Reichskommissariat Norwegen*, 180–81.

313 RAO, RAFA-2188/1/E1a-L0002, file 2100, Letter Feuchtinger, GB Bau, Oslo, to the Ministry of Provisioning with copy for the Food and Agriculture Department, 6.10.42.

314 RAO, S-3138/0026/Db-L0005, dommer 75/45, trial against A/S Betongbygg, Kristiansand, document 13, Statement by engineer A. Sørvig, 26.7.45. It should be noted that this statement was given after the war when the Norwegian firm was tried for economic treason. Hence, it is of course possible that the firm's management exaggerated the extent of its sabotage of German construction activities.

315 RAO, S-2079/Da-L0009, Secret "report from home" to the Norwegian Industry Committees in London and New York, August 1944.

316 RAO, RAFA-2188/1/E1m-L0046, Letter by Henne to all RK offices and to all OBLs of *Einsatzgruppe Wiking*, 16.7.42.

units also checked all construction plans of projects with a value over 5,000 kroner for their significance for the German war effort and whether all materials were being used thriftily.[317] When closing down sites, EW was able to transfer the workforce to German construction sites and – albeit against payment – confiscate construction materials. This procedure serves as an example of the effects of office accumulation in the German administration in the occupied territories. Ultimately, the GB Bau's representative in Norway thrust powers onto the head of *Einsatzgruppe Wiking* – two offices that were both held by Willi Henne.

The construction commissioner was able to close down illegal Norwegian construction sites. However, the Germans had no authority to punish Norwegians caught for unauthorised construction activities, since they – strictly speaking – were only violating Norwegian regulations. In a letter to Blehr of March 1944, Feuchtinger complained that during the last two years, he had repeatedly informed the Ministry of Economic about unauthorised construction activities and requested penalties: "To this day, I have not heard about a single effective, exemplary punishment of a violator, which should be made public in order to act as a deterrent against further infringements." Of about ten cases that Feuchtinger had reported to Blehr in the last months, only one had resulted in a "practically non-effective" fine of 50 kroner.[318] Obviously, Henne and Terboven's interventions had failed to impress Blehr.

Despite this recalcitrance, the Germans seem to have had little interest in a full escalation. A tougher stance on illegal Norwegian construction activity might have provoked resistance among the population. The same was true for a complete construction ban. As Feuchtinger had already written to Terboven in August 1942, a ban of new constructions was desirable both as a matter of principle and to highlight the seriousness of the situation. But 90 percent of the construction volume in Norway consisted of German projects, and most of the remaining Norwegian works were also to a certain degree in the German interest.[319] The Germans were well aware of the fact that a construction ban would only have had a small effect, anyway.

A second strategy was to relocate the workforce from other industries to the construction sector and to gain control over the deployment of conscripted workers in particular. From 9 June 1943 onwards, German labour authorities all over Europe were only able to displace conscripted workers after the construction commissioner's approval.[320] In Norway, this meant that the Department for Labour and

---

317 Ibid., Letter Feuchtinger, GB Bau, Oslo, to all RK offices and to all OBLs of *Einsatzgruppe Wiking*, 15.9.43.

318 RAO, RAFA-2188/1/E1n-L0049, file 5360, Letter Feuchtinger, GB Bau, Oslo, to Blehr, 4.3.44.

319 RAO, RAFA-2188/1/E1a-L0002, file 2100, Letter Feuchtinger, GB Bau, Oslo, to Terboven, 19.8.42.

320 RAO, RAFA-2188/1/E3g-L0027, Circular by GBA Sauckel and GB Bau Speer, 9.6.43.

Welfare (*Abteilung Arbeit und Sozialwesen*) in the Main Department for Economy became increasingly dependent on GB Bau's representative when it came to the deployment of Norwegian conscripted workers. Max-Erich Feuchtinger was eager to point out to the WBN that this included also relocations of workers between different *Wehrmacht* construction sites.[321] Once again, Willi Henne was able to use his authority as construction commissioner to serve his interests as EW's head, in this case to provide his *Einsatzgruppe* with labour. Henne's strategy clashed particularly strongly with the intentions of Carlo Otte, head of the RK's Main Department for Economy. Henne had cancelled industrial construction projects without conferring with Otte, and had relocated the workers to military construction sites.[322] In autumn 1942, Feuchtinger had already requested that the Department for Road Construction (*Abteilung Straßenbau*) in the Main Department for Economy cancel road works in order to free labour for the OT's construction projects.[323] It is therefore hardly surprising that the group around Otte feared that the Norwegian economy would become oriented completely towards construction, which resulted in the Department for Labour and Welfare's (unsuccessful) December 1943 attempt to suspend the Speer/Sauckel decree in Norway.[324]

Similar rivalries and infightings within the German administration also hampered measures against illegal Norwegian construction businesses. The construction boom of the occupation years created the breeding ground for many dubious entrepreneurs. In a secret report from October 1942, the Norwegian resistance movement remarked that "there was a saying that if you just had a carpenter's pencil and a saw, you already were an entrepreneur."[325] Other "firms" with no machinery and experience in construction only existed to recruit workers, which they then placed on German construction sites – naturally pocketing considerable agency fees. In coordination with the construction commissioner, the Norwegian government issued a law on 15 April 1943, stating that only authorised Norwegian and foreign (non-German) firms were allowed to conduct construction works in the country. Every firm with more than five workers and/or working on a construction site with a project value of more than 5,000 kroner had to be tested to assess their technical and commercial qualifications. The county governors (*fylkesmenn*) were able to license firms in their counties, while the Norwegian Economic Ministry was

321 RAO, RAFA-2188/1/E1m-L0046, Letter Feuchtinger, GB Bau, Oslo, to the WBN, 20.9.43.
322 Bohn, *Reichskommissariat Norwegen*, 181.
323 RAO, RAFA-2188/1/E1m-L0046, Letter Feuchtinger, GB Bau, Oslo, to the Department for Road Construction, 28.9.42.
324 Hatlehol, 'Norwegeneinsatz', 112–13.
325 Arbeiderbevegelsens arkiv og bibliotek (ARBARK), LO/Ja 20, Anonymous report to London, 12.10.42, as quoted (in Norwegian) in Ingulstad, 'Regulering', 272.

able to issue nation-wide licenses.[326] However, in July 1944, the construction commissioner discovered that German awarding authorities like the OT or the *Wehrmacht* were ignoring the law and still contracting unauthorised firms. Moreover, the premiums paid out to the contractors were often higher than price regulations permitted and it was reported to the construction commissioner that some of the unauthorised entrepreneurs were not even paying taxes or social contributions.[327] Finally, there were also renowned and authorised firms that did not always live up to expectations, but still profited from the enormous German demand. The firm Bakke & Dybvik, which had been operating a quarry at Gardermoen, lost its contract with the OT in autumn 1944 because it had worked inefficiently, completely run down the site and paid excess wages to its Norwegian workforce. However, when the German firm that replaced Bakke & Dybvik returned to the Reich already a few weeks later, the OT apparently saw no option other than contracting Bakke & Dybvik once again.[328]

c The Allocation of Building Materials

Neither the restriction or even ban of Norwegian construction activities nor infightings over the distribution of Norwegian labour alone would ease the situation on the construction sector. If the construction commissioner wanted to fulfil his task, which was to regain control over wages and prices, to cut back on inflated construction projects and to monitor that valuable construction materials were used economically, he would have to address the construction activities of EW and the *Wehrmacht*. As we have seen in the previous chapter, while Terboven's men had managed to contain the financial burden that the Norwegian economy had to bear, they had no influence on EW's and the *Wehrmacht's* spending in the Reich. Hence, the regulation of other input factors gained importance, in particular the rationing of construction iron and steel, cement, and timber.

As we have seen in chapter III.3, the GB Bau's ideal had been to register all planned construction projects, set an annual construction volume, cancel those projects that exceeded the volume, allocate the construction materials to the claimants and monitor the economical use of the materials. Measured against this ideal, the construction commissioner's powers were very limited. The main obstacle lay in a short paragraph of the "*Führer*" order of 18 August 1942. Hitler had given Albert

326 RAO, RAFA-2188/1/E1m-L0046, Letter by Feuchtinger, GB Bau, Oslo, to all German awarding authorities in Norway, 6.7.44. For details on this topic, see Ingulstad, 272–75.

327 RAO, RAFA-2188/1/E1m-L0046, Letter by Parzich, GB Bau, Oslo, to all German awarding authorities in Norway, 22.7.44.

328 RAO, RAFA-2188/1/H1b-L0014, file tysk korrespondanse, Letter construction firm Hermann Wegener, *Einsatz Wiking*, to its head quarter in Adelebsen, 13.9.44. Ibid., Letter construction firm Hermann Wegener to OBL Oslo, 16.11.44.

Speer the right to appoint construction commissioners and steer the construction sectors of the occupied territories. However, as the order continues, "this does not apply to constructions that serve the troops' direct operative mission."[329] The bearing of this reservation was soon to become apparent in the case of construction iron and steel.

Almost all construction iron and steel used on German construction sites in occupied Norway came from Germany and was not registered in official import and export statistics. All German customers in Norway reported their demand for construction iron and steel to the construction commissioner in Oslo every quarter. Willi Henne then travelled to Berlin on a regular basis to participate in the negotiations in the Central Planning Office. As described in chapter III.3.c, the GB Bau received a share of the German iron and steel production in order to distribute it between the claimants. Some of the allocations were earmarked for occupied territories like Norway. Five claimants received construction iron and steel for use in Norway.[330] The OT headquarters in Berlin received iron and steel for the construction sites of EW and the IdLN. The Plenipotentiary of Special Issues in Chemical Production (*Generalbevollmächtigter für Sonderfragen der chemischen Erzeugung, GB Chem*) received a quota for the Norwegian light metals programme, that is, the construction sites of A/S Nordag and A/S Nordisk Lettmetall. The RWM was responsible for the allocation of German materials to industrial construction sites in the occupied and annexed territories.[331] It thus received iron and steel for a civilian Norwegian contingent, which was – via the RK – made available to the Norwegian Ministry of Provisioning. The Norwegian private industry was provided with German iron and steel through this channel, although the overall volume of these supplies is yet unknown. The construction commissioner in Oslo was not directly involved in the management of construction iron and steel when it came to these three claimants. This was a minor problem in the case of the OTZ, since Henne and Feuchtinger were OT men, anyway. But in the case of the RWM and GB Chem quotas, they were unable to monitor whether iron and steel were really allocated to the designated construction sites and whether the materials were used thriftily in accordance with GB Bau regulations.

329 Moll, *Führer-Erlasse*, 274–75.

330 The information on the rationing system of construction iron and steel in occupied Norway described in this and the following paragraphs has been reconstructed as far as possible from three sources: RAO, RAFA-2188/1/E1b-L0005, file 2223–2228, Letter Feuchtinger, GB Bau, Oslo, to GB Bau, Berlin, 22.3.43; RAO, RAFA-2188/2/Hfa-L0007, file 27, Regelung des Bauschaffens in Norwegen, Undated document "Tasks of the General Commissioner for the Regulation of the Construction Industry and their execution by the Commissioner for the occupied Norwegian territories"; *Ministerialblatt des Generalbevollmächtigten für die Bauwirtschaft, des Generalinspektors für das deutsche Straßenwesen, des Generalinspektors für Wasser und Energie und für den Rüstungsausbau, herausgegeben vom Reichsministerium Speer* 1, no. 11 (15.10.43).

331 Generalbevollmächtigter für die Regelung der Bauwirtschaft, *Regelung*, 62.

The remaining two claimants, the *Wehrmacht* and the GB Bau itself, transferred their construction iron and steel quota for Norway to a so-called relocation contingent (*Verlagerungskontingent*) of the Reich Commissariat. This contingent was administered by the construction commissioner in Oslo.[332] The GB Bau in Berlin was the claimant for civilian German construction projects in Norway. By transferring its quota to the relocation contingent, the GB Bau in Berlin tasked its representative in Oslo with monitoring the allocation and consumption of iron and steel from its quota in Norway. The *Wehrmacht* likewise transferred its huge construction iron and steel quota to the relocation contingent. However, the abovementioned "*Führer*" order of 18 August 1942 effectively prohibited the construction commissioner gaining any control over this quota. The representatives of the GB Bau in occupied Europe were not responsible for materials used on construction sites "serving the direct operative mission" of the *Wehrmacht*. WBN von Falkenhorst made it clear that he regarded more or less all construction projects of the *Wehrmacht* in occupied Norway as "serving the direct operative mission".[333] Max-Erich Feuchtinger complained in March 1943 that he had no chance to check the requisition notes, the allocation or the use of construction iron and steel by the *Wehrmacht* in Norway. He saw his office degraded to not much more than an exchange office for iron and steel cheques.[334]

Feuchtinger pointed out another shortcoming that limited the powers of his office. Unlike in the Reich, there was no differentiation between iron and steel products for construction purposes on the one hand and for the purchase and repair of machinery on the other hand. The construction commissioner in Oslo was therefore unable to control whether iron and steel from German production, which had been allocated to the construction sector in tedious negotiations in Berlin, were actually used on construction sites in Norway. Both German and Norwegian customers could use their iron and steel cheques to order machinery for the companies working on their construction sites. As Feuchtinger wrote in an urgent letter to his superiors in Berlin in March 1943,

> [t]he lacking differentiation between construction iron and machinery iron renders it impossible not only for me, but also for the department for internal economy of the Reich Commissioner for the occupied Norwegian territories, to conduct any control of the consumption of construction iron and thus a proper steering of the construction industry in Norway. Thus, one of the most essential means of controlling construction activities in Norway is missing.[335]

332 RAO, RAFA-2188/1/E1b-L0006, file 2230, Circular decree of the GB Bau regarding the rationing of construction iron, timber and wood fibreboards, 25.9.43.

333 RAO, RAFA-2188/1/E1b-L0005, file 2223–2228, Letter Feuchtinger, GB Bau, Oslo, to GB Bau, Berlin, 22.3.43.

334 Ibid.

335 Ibid.

So far, there is no direct evidence that companies working in Norway diverted iron from the construction quota in order to modernise their machinery. However, similar cases have been documented in other sectors of the German war economy and in occupied territories, where materials were diverted into clandestine production.[336] Moreover, the fact that the office of the construction commissioner itself identified this as a major problem suggests that it was indeed a problem. Max-Erich Feuchtinger concluded that "the current conditions are unacceptable." The fact that his office was chronically understaffed certainly did not strengthen the position of the GB Bau's representative in Norway.

The complicated system of regulations only applied to construction iron and steel from German production. The Norwegian iron and steel production, in contrast, was not under the construction commissioner's control. In April 1943, Willi Henne registered with some surprise that the Holmenkollen railway in Oslo had modernised their tracks and possessed large stocks of railway tracks. His GB Bau office informed him that it did not monitor the Norwegian production and did not even know how much iron and steel the Norwegians produced. Norwegian customers were able to buy iron and steel from Norwegian producers at those producers' discretion.[337] For the *Wehrmacht* and the OT, the Norwegian production certainly was a welcome source of additional, non-rationed iron and steel goods. The share of German purchases and the prices paid are unknown, however.

It seems that the construction commissioner was able to gain a better overview and more control in the case of cement. The *Wehrmacht*, EW and the RK reported their construction projects and the required amount of cement to the construction commissioner before the start of the construction season. The RK's Main Department for Economy controlled the production in the three Norwegian cement plants and reported the expected annual production to the GB Bau.[338] These numbers became the basis for Willi Henne's negotiations in Berlin on the amount of construction materials Norway would receive from German production. Based on the expected Norwegian production and transports from Germany and Denmark, the GB Bau distributed the available quantities between the *Wehrmacht*, EW, the RK and the civilian Norwegian sector. The latter received cement exclusively from Norwegian production according to a fixed ratio. In February 1941, 40 percent of the Norwegian production was used by the Norwegians themselves. However, after the commencement of EW's enormous fortification programme, which consumed large quantities of cement, the quota was

336 Scherner, 'Keine primäre moralische Verantwortung?', 166–69; Klemann and Kudryashov, *Occupied Economies*, 310–11.

337 RAO, RAFA-2188/1/E1b-L0005, file 2223–2228, Note by the office of the GB Bau in Oslo, 29.4.43.

338 RAO, RAFA-2188/1/E1m-L0046, Minutes by Feuchtinger of a meeting on the rationing of cement in Norway on 1 February 1943 in Oslo, 15.2.43.

reduced to 20 percent in June 1942.[339] It was the Norwegian Ministry of Provisioning that distributed the cement, but the GB Bau still made sure that it had a say in the process. 5 percent of the Norwegian production was reserved for civilian German purposes, mostly for the RK and the SS. The remaining 75 percent of Norwegian production and all supplies from Germany and Denmark went to the military construction sites of the *Wehrmacht* and EW. The distribution of the military quota between the three *Wehrmacht* branches and the OT was discussed at monthly meetings. Most of the transports were organised and operated by the quartermaster (*Wehrmacht-Quartiermeister*) under the WBN.[340]

Hence, at least on paper, after 1942 the GB Bau gained a central position when it came to the rationing of cement. Similar regulations existed wherever a GB Bau representative had been installed in the occupied territories. The small office in Berlin was thus the centre of a European network of cement distribution, encompassing France and Belgium, the *Reichkommissariate* of Norway, Ostland and the Netherlands, the Protectorate of Bohemia and Moravia as well as the General Government.[341] However, there were still many factors beyond the influence of the GB Bau. In the case of Norway, the lack of tonnage in particular became a major problem.[342] Ships had to transport from Germany both cement and especially the coal needed for the production of cement in Norway. Providing the Norwegian industry with German coal was a major problem during the occupation years after imports from Great Britain had stopped in April 1940. Before the war, the British had satisfied 70 percent of the Norwegian demand of coal. Now, it was the Germans who had to provide the coal, but more and more often the supplies did not even leave the Reich because Allied air strikes on the German railway network and harbours hampered shipment.[343] Due to the lack of coal, the Norwegian cement plants had to cease their production completely in January 1945. By the end of the month, their combined stocks had fallen to the equivalent of approximately one month's production.[344]

---

339 *Statistisk-økonomisk utsyn*, 193. Allocations to the Norwegian civilian sector decreased further in the following years. In 1944, around 13 percent of the annual Norwegian cement production was allocated to the Norwegian civilian sector: RAO, RAFA-2188/2/Hfa-0111, Undated scheme on the Norwegian cement production 1944.

340 RAO, RAFA-2188/1/E1m-L0046, Minutes by Feuchtinger of a meeting on the rationing of cement in Norway on 1 February 1943 in Oslo, 15.2.43. However, the OT seems to have carried out some cement transports to Norway on its own: RAO, RAFA-2188/2/Hfa-0111, Feuchtinger to transport group regarding cement transports, 19.2.43.

341 RAO, RAFA-2188/1/E1m-L0046, Letter by the OKW to the *Wehrmacht* branches regarding the cement demand of the *Wehrmacht* outside the Reich, 15.2.43.

342 BA-MA, RH 11-III/219, fol. 153, General of the Pioneers and Fortifications at the OKW to the Navy Fortifications Pioneers Inspectorate at the OKM, 2.12.44.

343 Bohn, *Reichskommissariat Norwegen*, 434–43.

344 RAO, RAFA-2188/2/Haa-L0021, Undated scheme on the Norwegian cement production 1945.

The rationing of timber was organised similarly to that of cement. The *Wehrmacht*, OT and RK reported their annual demand to the GB Bau in Oslo who tried to balance it with the availability of other materials and transport capacities.[345] The Norwegian civilian sector received a quota, too, which was managed by the Norwegian Economic Ministry.[346] The GB Bau in Oslo only decided on the distribution, not the production of construction timber, which was supervised by the RK's Department for Forestry and Timber Industry (*Abteilung Forst- und Holzwirtschaft*). The department reported the annual production available for construction to the GB Bau. The wood harvest itself was organised by a forestry office within the Norwegian Ministry of Agriculture.[347]

As the GB Bau emphasised, the quotas allocated before every building season could only be preliminary values. The *Wehrmacht* and OT often imported timber or organised additional felling in the course of the year. They then had to report the quantities gathered from these sources to the GB Bau, which deducted them from their quotas. However, there was one important exception. Timber that was felled by foreign workers and POWs was not deducted from the quota.[348] Thus the deployment of forced labourers became a way for EW and the *Wehrmacht* to increase the timber supply of their construction sites beyond the restrictions of the rationing system. Sources on EW's workforce indeed reveal that the organisation deployed forced labourers not only as construction workers, but also as tree fellers.[349]

All in all, the power of the construction commissioner in Oslo appears to have been rather limited. Albert Speer's representative managed to curtail Norwegian construction activities in particular. However, the office was far from exercising total control in this field. Wherever it depended on the Norwegian administration to control the sector or implement its policy, the latter seems to have been able to retain a certain room for manoeuvre and to act as a buffer between German economic interests and Norwegian businesses.[350] With regard to the German part of the sector, the *Wehrmacht* in particular refused to let any other institution gain significant power over its projects. Both Max-Erich Feuchtinger in his abovementioned report of 22 March 1943 and contemporary observers like Hans Clausen Korff recognised that the *Wehrmacht* could effectively veto any attempt at regulation that ran

345 RAO, RAFA-2188/1/E1m-L0046, Guidelines by the GB Bau, Oslo, on the timber rationing for German customers in Norway, 21.7.43.

346 Ibid., Guidelines by the GB Bau, Oslo, on the timber rationing for civilian Norwegian customers, 13.10.43.

347 Bohn, *Reichskommissariat Norwegen*, 177–78.

348 RAO, RAFA-2188/1/E1m-L0046, Guidelines by the GB Bau, Oslo, on the timber rationing for German customers in Norway, 21.7.43.

349 For the example of Polish forced labourers, see Hatlehol, 'Norwegeneinsatz', 345–46.

350 Kjetil Braut Simonsen, 'Nazifisering, kollaborasjon, motstand. En analyse av Politidepartementet og Forsyningsdepartementet (Næringsdepartementet), 25. september 1940–8. mai 1945' (Universitetet i Oslo, 2016), 230–36.

counter to its aims.[351] The construction volume set by GB Bau's representative could only be a vague guideline. Table 18 shows the discrepancy between the initial demands for construction materials by awarding authorities in occupied Norway compared to the allocations agreed on after negotiations with the construction commissioner and, finally, the actually delivered materials. So far, comprehensive figures have only been found for 1944, and therefore conclusions should be drawn with some caution. Nevertheless, some trends can be identified. The mismatch between plans and results was biggest for construction iron. The situation was better for timber and cement, which could mainly or partly be produced within the country. This underscores the crucial role that the transport situation played in construction activities during the occupation. In the summer of 1944, the *Wehrmacht* identified the lack of fuel as the biggest obstacle for the completion of its construction projects in Norway.[352] Furthermore, it is striking that for all materials, the mismatch between initial demands, allocations and actual deliveries is smallest in those fields controlled personally by Willi Henne, that is, the OT and the German civilian projects. Based on the available sources, it is not possible to say whether this was a result of Henne using his triple position as construction commissioner, head of OT and head of Department for Technology to his own advantage.[353] An alternative explanation could be that Henne, by virtue of his central position, was able to gather information that allowed him to draft more realistic plans from the outset. In the case of the *Wehrmacht*, however, the mismatch was probably just an expression of the three *Wehrmacht* branches trying to be on the safe side and

Table 18: Intended allocation of construction materials by the construction commissioner and actual deliveries, in % of initial demands, occupied Norway, 1944.*

	Construction iron		Timber		Cement	
	Allocated	**Delivered**	**Allocated**	**Delivered**	**Allocated**	**Delivered**
Wehrmacht	45	26	68	61	65	61
OT	101	85	99	98	104	74
German civilian projects	105	8	92	102	111	49
Norwegian civilian projects	37	29	68	63	55	54
Sum	**62**	**42**	**71**	**63**	**78**	**64**

*RAO, RAFA-2188/2/Haa-L0021, Construction volume Norway 1945, 12.2.45.

351 Korff, 'Norwegens Wirtschaft', 1949, pt. II, 338.
352 BA-MA, RH 11-III/219, fol. 6–7, Letter by the IdLN to the General of the Pioneers and Fortifications at the OKW, 28.7.44.
353 Christiane Botzet has pointed out the potential that lay in this accumulation of offices: Botzet, 'Generalbevollmächtigter', 135–36.

claiming as much material as possible. This was a common problem in all rationing systems of the German war economy.

Ultimately, the amount of construction materials delivered to construction sites in Norway was nowhere near the construction commissioner's schemes. Too many factors were beyond the office's influence, above all the sea transport of construction materials to Norway, which depended not least on the military situation and decisions taken in the high commands in Berlin. Consequently, it was the *Wehrmacht* itself, not the GB Bau, that stopped the supply of construction materials to occupied Norway and thus put an end to most of *Wehrmacht's* construction activities in the summer and autumn of 1944.[354]

The construction commissioner also met resistance within the Reich Commissariat. Henne's relation to the head of the Main Department for Economy, Carlo Otte, was conflictual. Henne tried to push Otte's department aside when it came to controlling the Norwegian production of building materials (Henne only controlled the distribution of the materials). However, the attack failed and Henne failed to get this area under his direct influence as long as the occupation lasted.[355]

Although Willi Henne can be called the single most important man in the Norwegian construction sector after his arrival in Norway,[356] this chapter has shown that there were limits to his powers. The impression is that the construction commissioner in Oslo was less powerful in his domain than his colleagues in the German Reich. There were some peculiarities, like the lacking differentiation between construction iron and machine iron, that hindered full control. The existence of a domestic part of the construction sector – both in terms of building activities and production of construction materials – supervised by Norwegian institutions added a source of potential resistance and friction. Moreover, Norway's dependence on supplies from abroad, mostly by ship, constantly threatened to mess up schedules and plans.[357]

Nevertheless, *Einsatzgruppe Wiking* as an organisation certainly profited in certain regards from its close institutional and personal integration with the GB Bau. This becomes particularly clear where the mobilisation and steering of German construction firms by EW is concerned, which will form one of the topics of the following chapter.

---

354 Sæveraas, 'Beton macht Geschichte', 114–15.

355 Bohn, *Reichskommissariat Norwegen*, 181.

356 Bohn, 180.

357 There are no indications, on the other hand, that the *Wehrmacht's* importance in the construction sector – which was greater in Norway, comparatively speaking, than it was in Germany – led to major conflicts with Henne.

V German Companies under Einsatzgruppe Wiking

1 Entrepreneurial Motives and Scope of Action

In the winter of 1941–42, the construction industry began to feel the drain of Germany's war in the east. During the past two years, at least the larger firms had been able to compensate the reduction of construction activities. Their order books had still been full, while the first waves of draft calls to the *Wehrmacht* had already hit smaller companies much more severely. By this point at the latest, however, German construction firms were no longer able to rely on domestic projects alone. They had to look for tasks in the occupied territories. The following chapters will focus on the German construction firms working under *Einsatzgruppe Wiking*. Based on source material left behind by construction firms, most of which has not been used in earlier research, I identify several motives for accepting an OT commission during the Second World War.

Part a reconstructs EW's mobilisation process in early 1942. I show that German construction firms were not forced to follow the *Einsatzgruppe* to occupied Norway. Freedom of contract was preserved. In part b, I compare my findings to research on German companies in other parts of occupied Europe as well as non-German firms under the OT.

Part c then investigates three entrepreneurial motives for taking an OT order in occupied Europe. I focus firstly on the firms' aspiration to gain privileged access to construction machinery, and secondly on their aim to obtain valuable know-how when working for the OT. Thirdly, I argue that construction companies sometimes accepted less attractive OT orders in order to maintain good business relations with what was perhaps Germany's most important construction client.[1]

Finally, part d focuses on the so-called transport ban for construction machinery between 1942 and 1944, which prevented firms from returning to Germany without the permission of the GB Bau. The chapter argues that the transport ban was a clear interference with German construction firms' entrepreneurial freedom of action during their engagement in Norway.

a The Mobilisation of Einsatzgruppe Wiking's contractors

In early 1942, frantic activity set in at the branch offices of the road construction company Straßenbau AG in Weimar. Local manager Alex Korffmacher had just returned

1 However, the sources give the clear impression that the most important motivations were to earn money with favourable contracts and to safeguard the workforce against being called up by the *Wehrmacht*. These motives will therefore be discussed seperately in chapters V.2 and V.3.

https://doi.org/10.1515/9783110694291-005

from a meeting at Wigru Bau in Berlin, where the association had informed the convened business leaders about Fritz Todt's intention to cut back construction activities in the Reich significantly in the upcoming building season. Korffmacher feared that the Weimar branch would run out of work in the next months and lose its workforce to the *Wehrmacht*.[2] Accordingly, in the following weeks and months, Strabag Weimar left nothing undone to reel in orders in occupied Europe.

The Straßenbau Aktiengesellschaft (Strabag) was founded in Niederlahnstein in 1923, when the triumph of the car had created a rapidly growing market for companies specialised in the construction of tar and asphalt roads after the First World War. The company jumped on the bandwagon and by the 1930s, it had become the leading road construction company in Germany.[3] The first branch offices were founded in Berlin and Weimar in 1932, followed by many more all over Germany in the mid-1930s. Subsidiaries operating under the name Vianova existed in Vienna, Munich, Breslau and Stuttgart.[4] Between 1938 and 1941, the Weimar branch was awarded the status of a "National Socialist Model Company" four times in a row.[5]

Correspondence from early 1942 has survived in Strabag Weimar's archive and there is probably no other construction firm under EW whose mobilisation can be reconstructed in such great detail.[6] In January and February 1942, however, Eastern Europe still seems to have held more promising opportunities in the eyes of Alex Korffmacher in Weimar. Accordingly, he looked for ways to supply his firm with lucrative orders in this region.

Strabag's case shows the importance of personal networks for the mobilisation process. In peacetime, Korffmacher had become acquainted with officials of building authorities from Weimar and Erfurt who had now been assigned to the civilian administration in the Reich Commissariat Ukraine. The official from Erfurt had worked together with Strabag for 13 years. Both had indicated that they would need construction firms for their projects in Kiev and Zhytomyr, respectively. When they were on furlough in Weimar around Christmas 1941, the men informally (in one case during a night at the theatre) discussed opportunities for the firm to take up a mission in Ukraine. Later, Korffmacher personally contacted the wife of one of the officials in order to obtain his current address in Ukraine. Moreover, he used their subsequent correspondence to inform the officials about

2 ThHStAW, Straßenbau AG Weimar Nr. 105, fol. 503 f., Letter from Korffmacher, Strabag Weimar, to Schmidt, Vianova Straßenbau m.b.H. Breslau, 30.1.42.

3 On the history of the Strabag, see Pohl, *Strabag*.

4 Pohl, 190–201.

5 Straßenbau-Aktiengesellschaft, business reports 1937, 1939, 1940.

6 The material at the ThHStAW has been used by Botzet and Pohl, however, not for a comprehensive study of Strabag's mission in occupied Norway: Botzet, 'Generalbevollmächtigter'; Pohl, *Strabag*.

Strabag's branch office in Dresden and two other companies that were interested in working in Ukraine, too.[7]

The contact between the firm and the awarding authorities' representatives seems to have been trusting and quite frictionless, and there are no indications that the latter tried to force Strabag Weimar into a contract. In fact, the company successfully refrained from taking up a project in Ukraine when a more attractive option emerged in March: works for EW's fortification programme in southwestern Norway.

Strabag Weimar had already become active in occupied Norway shortly after the occupation in 1940. The company built airfields, bunkers, barracks, gun batteries and piers for the *Wehrmacht* and Reich Commissariat in a working combine with Sager & Woerner (Munich) and Norwegian A/S Høyer-Ellefsen (Oslo).[8] The working combine's leader was Strabag Weimar's director himself, Ingo von Ingersleben.

Von Ingersleben must have heard of the Wiking programme around the second half of March. On 27 March, he contacted the RK's Department for Technology and asked whether Strabag Weimar could be deployed in the Stavanger region. The firm knew the area and still had machinery in Stavanger. In the light of increasing transport problems, keeping the machines in use where they were was by far the best alternative for the company.[9]

As a matter of fact, EW's first reaction to Strabag Weimar's approach was rather reluctant. As von Ingersleben was told in May, the OT had already mobilised 127 companies, almost twice as many as initially intended: "Above all, an exceptionally large number of mid-sized companies, which no longer have works in Germany, seem to have applied for the mission in Norway."[10] However, with the resolute backing of Wigru Bau's representative in Oslo, Friedrich Thierbach, von Ingersleben travelled to Kristiansand and Stavanger to negotiate with the local OT units and visit the designated building sites for bunkers. Again, Strabag was able to rely on contacts from earlier construction projects, as the head of the OT's BL Stavanger, Haag, knew the company from motorway construction works in the Essen region. By the end of May, Strabag Weimar and the OT had come to an agreement.[11] The firm received the official "order" (*Einsatzbefehl*) to send workers from Weimar to Stavanger and take up

7 ThHStAW, Straßenbau AG Weimar Nr. 105, fol. 500–7.
8 Ellingsen, *Krigsprofitørene og rettsoppgjøret*, 37–49.
9 ThHStAW, Straßenbau AG Weimar Nr. 107, fol. 358 f., Letter v. Ingersleben to Department for Technology, 27.3.42.
10 ThHStAW, Straßenbau AG Weimar Nr. 108, fol. 276, Notes by v. Ingersleben about meetings on 11 to 13 May 1942 with *Reg.Baumeister* Köbele and *Baurat* Schneider (OT) and Thierbach (Wigru Bau) in Oslo, 30.5.42.
11 ThHStAW, Straßenbau AG Weimar Nr. 108, fol. 298 ff., Summary of the mission's progress, 19.7.42.

work immediately on 22 June 1942.[12] At this point in time, machines and some employees were already on their way to Stavanger.

The example of Strabag Weimar is representative of the situation of German construction firms in spring 1942. In general, they proactively sought work under the OT. Numerous application letters written by firms have survived in *Einsatzgruppe Wiking's* archives. Usually, firms submitted lists with their available machines and specified their know-how so that the awarding authorities could assess whether the firm could be of use for them.[13] Although it soon became apparent that there was more than enough work for an even larger number of firms, EW certainly did not have to resort to threats and coercion in order to cover its demand for German firms.[14]

Wigru Bau became an important intermediary between the larger construction industry firms and the OT. Firms often contacted the organisation when they were looking for orders in the occupied countries and, as the example from Strabag has shown, Wigru Bau represented the firms' interests towards the OT. At the same time, state agencies like the OT used Wigru Bau's network in order to inform firms about new missions and to mobilise them. The association was asked to name and contact firms that possessed sufficient numbers of workers and machines to take over construction sites in the occupied territories. This arrangement was an efficient way for the state to mobilise firms, while the latter avoided being disrupted by conscriptions of their workforce. Wigru Bau's competence to select firms for projects in the occupied territories was eventually formalised in a joint decree of the GB Bau and the General Commissioner for Labour Deployment (*Generalbevollmächtigter für den Arbeitseinsatz*, GBA) in June 1942.[15]

Smaller firms often joined forces in a working combine before applying for an OT mission, but these were not necessarily just *ad hoc* co-operations. Several of the working combines in occupied Europe had already been established at the *Westwall* and then decided to continue their partnership during wartime. Rights and duties were laid down in a contract and profits shared. This allowed the firms to take over bigger and more complex tasks by pooling their machines and their technical and administrative staff. Authorities had an interest in the establishment of Arge, too, as the combines promised a more rational use of personnel and scarce

12 ThHStAW, Straßenbau AG Weimar Nr. 107, fol. 299. On the chronology, cf. also: ibid., fol. 305–7, Strabag Weimar to *Einsatz Wiking*, *Nachschubgruppe Wiking-Mitte*, 3.6.42.

13 RAO, RAFA-2188/1/F3a-L0001. Strabag Weimar also attached such a list to its application of 27 March: ThHStAW, Straßenbau AG Weimar Nr. 107, fol. 358 f., Letter v. Ingersleben to Department for Technology, 27.3.42.

14 Fabian Lemmes and Steen Andersen have come to analogous results in the case of French construction firms and Danish Christiani & Nielsen, respectively: Lemmes, 'Arbeiten für das Reich', 383–411; Andersen, 'Between Imperative and Risk'.

15 Botzet, 'Generalbevollmächtigter', 130–31.

resources. Similarly to Wigru Bau's role in the construction industry, the handwork associations tried to act as agents between their member firms and German authorities in the occupied territories. In December 1941, for example, the Reich Guild Association of Carpentry Craft (*Reichsinnungsverband des Zimmerhandwerks*) asked their local branches to report handwork firms that were interested in working on air force construction projects in Norway.[16] Another option for smaller companies was to get directly in touch with a bigger German building company and offer themselves as subcontractors. An example was Jesendorfer Kieswerke, who contacted Strabag Weimar in summer 1942, hoping to use their equipment on the building sites around Stavanger.[17]

The proactive approach of many construction firms in spring 1942 was certainly motivated by the increasingly problematic situation within the Reich. Equally, however, the relation to the officials of the OT seems to have been characterised by mutual trust. Beginning in 1933 on the *Autobahn* construction sites and especially at the *Westwall*, Todt's men had gained such a reputation among German entrepreneurs that the latter obviously did not hesitate to follow the OT to occupied Norway, convinced that almost any deal would guarantee them profitable business. As Allied intelligence concluded in 1945:

> [. . .] Enough large building firms offered their services so as to put the entire construction [of the *Westwall*, S.G.] on a voluntary basis. [. . .] Nor is there any basis later for assuming that firms in any large numbers became so reluctant to work for the OT as to make mass conscriptions of such concerns necessary. This willingness is due to the attractive profits obtainable from OT contracts.[18]

The construction firm Steffen Sohst from Kiel, working in and around Bergen, had no reason to doubt the profitability of its mission under the OT: "There is plenty of work", Steffen Sohst's site manager was able to report in July 1942, "And as far as I could hear, everything is paid for very well."[19] As the following years would show, also the payment moral of the *Einsatzgruppe* was good, even in the last weeks of the war. Still in April 1945, EW transferred money to German firms that had already returned to Germany.[20] In September 1945, EW's settlement centre calculated (partly based on estimates) that EW owed its German contractors circa 41.8 million RM – a significant, but not vast sum, considering its overall expenditures (cf. above, Table 17).[21] The firms'

16 BArch, R 13-VIII/264, Ancillary report to the first report of Thierbach, Oslo liaison office of Wigru Bau, 27.12.41.
17 ThHStAW, Straßenbau AG Weimar Nr. 107, fol. 119.
18 *Handbook of the Organisation Todt*, 53.
19 RAO, RAFA-2188/1/H1a-L0009, Letter from the Bergen office of Steffen Sohst to the Kiel headquarters, 16.7.42.
20 RAO, RAFA-2188/1/E8a-L0032, file Rj. 1945 Deutsches Konto Ausgabe Betriebsmittel.
21 NHM, FO.II/11.4, box 27, file 56b, RK – EW settlement centre, report on the construction expenses by EW between 1942 and 1945 with 11 supplements, 3.9.45.

chance to collect outstanding OT invoices after the war, however, seems to have been rather slender.[22]

Given this set of incentives and opportunities, it does not come as a surprise that the Organisation had no problem mobilising a substantial number of the construction firms in a rather short period of time in spring 1942. According to a post-war manuscript on EW's history by Willi Henne, EW contracted more than 200 German (industry) firms and an almost equally high number of handwork firms.[23] Mid-sized firms were mobilised from Western Germany for the construction of fortifications, for harbour construction specialised firms from Northern Germany, and large *Tiefbau* companies from Berlin, Southern Germany and Austria for railway construction.[24] A list compiled by Wigru Bau already comprised more than 200 German firms in Norway in mid-July.[25] The number was to increase in the following months and years, especially because of the firms that were integrated into the OT apparatus in the summer of 1944 after working for the air force and the navy in the previous years. Hans Clausen Korff stated after the war that around 350 German firms were working under EW at the time of the capitulation.[26] Since many firms had already left Norway at this point in time, Korff's number certainly is a minimum estimate. It should also be noted that the OT usually counted working combines that consisted of several firms as one company. Based on several lists that can be found in the OT archives at the National Archives in Oslo,[27] and particularly a card file on EW's firms,[28] one arrives at a qualified estimate of around 500 German construction firms that worked under *Einsatzgruppe Wiking* at some point between 1942 and 1945, either as stand-alone firms or members of a working combine.

By the second half of 1942, however, EW's mission in occupied Norway was struggling with massive start-up difficulties. Bitter complaints reached Wigru Bau in Oslo and Berlin, especially from Northern Norway. In June, the Westmark branch of the Reich Association of the Building Handwork Guilds already received news from the Arge Kaiserslautern, which had been mobilised with 125 men to work in Northern Norway by OBR Frankfurt am Main. Upon arrival, the men learned that the local camp leader had not been informed they were coming. The workforce was

22 Pohl, *Philipp Holzmann*, 293–300.

23 Henne, 'Organisation Todt in Norwegen', pt. III, 15.

24 Henne, pt. II, 3.

25 BArch, R 13-VIII/264, Letter by Thierbach, Oslo liaison office of Wigru Bau, to Wigru Bau, Berlin, with attached list on German construction firms in Norway (as of 10 July 1942), 11.7.42.

26 Korff, 'Norwegens Wirtschaft', 1949, pt. II, 336.

27 RAO, RAFA-2188/2/Gbb-L0005, Undated list on German firms working under the air force; RAFA-2188/1/E9a-L0048 and L0049; RAFA-2188/2/Haa-L0029, List on German firms divided by *fylker*; for an early list of July 1942, see BArch, R 13-VIII/264, Letter by Thierbach, Oslo liaison office of Wigru Bau, to Wigru Bau, Berlin, with attached list on German construction firms in Norway (as of 10 July 1942), 11.7.42.

28 RAO, RAFA-2188/2/Haa-L0055.

split up and assigned to two German road construction firms in Finland.[29] The Reich Association of the Building Handwork Guilds made an identical complaint in a meeting with representatives of EW on 15 August 1942. Several of their working combines had been torn apart and the workers allocated to different German construction industry firms. For example, the OT had placed four work gangs of carpenters, who had been guaranteed their own construction project, on Sager & Woerner's construction sites in Trondheim. Sager & Woerner had assumed that workers provided by the OT would be at its free disposal. The OT proposed establishing a working combine between Sager & Woerner and the carpenters and sharing the profits. Sager & Woerner, however, refused and EW saw itself compelled to "pick up the bill for the operating expenses and shares of the profit that the firms were entitled to. The OT", it was agreed, "will be answerable to the pricing authorities for the resulting rise in construction costs."[30]

Much of the situation seems to have resulted from logistical problems, and hence was worst in Northern Norway. In November, firms deployed in the Narvik region were still complaining of improper accommodation for their German workforce, a lack of washing facilities and lavatories, a lack of winter clothes and shoes, as well as poor food.[31] Several of them also reported that machinery had been lost during the transport.[32] To be sure, EW was bound by contract to compensate the firms for the loss and, as far as possible, to provide them with the reference numbers necessary to buy rationed construction machinery.[33] However, given the general shortage of iron and steel products and constant delays regarding their delivery, it is understandable that the firms were not satisfied with this solution. They would have preferred a compensation in kind.[34]

b Freedom of Contract under the OT – Findings from Occupied Europe

The findings on EW's mobilisation process conform with examples both from missions of German firms in other parts of Europe and from local firms working for the

29 RAO, RAFA-2188/1/F3a-L0012, file Korrespondenz mit der OT, Reich Association of the Building Handwork Guilds, Westmark branch, to Reich Association of the Building Handwork Guilds, Nordmark branch, 22.6.42.
30 BArch, R 13-VIII/265, Note on a meeting regarding the deployment of construction workers to external firms on 15 August 1942 in Oslo, 18.8.42.
31 RAO, RAFA-2188/1/F3a-L0002, Minutes of the monthly meeting between German construction firms and the Oslo liaison office of Wigru Bau, on 7 November 1942 in Oslo, 8.11.42.
32 Ibid.
33 Ibid., Talk of Fröhler, head of EW's Department for Contracting and Pricing, during the meeting on 7 November 1942 in Oslo, 8.11.42.
34 Ibid., Minutes of the monthly meeting between German construction firms and the Oslo liaison office of Wigru Bau, on 7 November 1942 in Oslo, 8.11.42.

German occupiers. In Eastern Europe, transport and communications may have been even more complicated than in Norway and the risk of losing machinery was allegedly higher, but this does not seem to have stopped firms from applying for projects in this area, too. Examples of the proactive role of German (and Austrian) construction companies in the mobilisation process include the Viennese Allgemeine Baugesellschaft – A. Porr AG, which sent one of its board members to Eastern Silesia and Galicia to secure deals for the company,[35] or the firm Franz Winter Tiefbau from Thuringia, which applied for deployment in Eastern Europe twice in 1942.[36] According to Christian Gerlach, the activities of German construction firms under the OT in Belorussia were also based on voluntary agreements.[37] Other firms seem to have been rather indifferent when it came to the destination of their foreign assignment, as attested by an application for deployment in "France or the east".[38]

Shortly before the Wiking programme opened up even more opportunities, the run on projects in occupied Eastern Europe was immense. In early 1942, Fritz Todt intervened with what was to become his last directive as GB Bau, issued one day before his death. Todt noted that such a high number of companies were volunteering for projects in the east that construction projects within the Reich were jeopardised. The firms' motivation was that they "[. . .] expected more steady orders there than in the Reich and hoped for a better protection of their workforce. Those applications are often combined with offers of labour, which are only dictated by the attempt to keep the workforce together, and do not consider the essential requirements of construction projects in the armaments sector in the homeland."[39] Todt ordered that construction firms had to obtain the approval of the GB Bau's regional representatives before they could take orders abroad. The firms had to prove that their construction works in Germany were not being set back due to the relocation of workers and machinery to the east.[40]

While the wish to participate in the OT and other state agencies' construction projects was one side of the coin, the option to decline a request of the OT was the other. This negative freedom of contract has been identified as a crucial touchstone of entrepreneurial freedom under the Nazi regime and it has been proven for several sectors that firms were indeed able to turn down state orders between 1933 and

---

35 Matis and Stiefel, *Porr*, 422.

36 BArch, R 50-I/323, fol. 51–54.

37 Christian Gerlach, *Kalkulierte Morde. Die deutsche Wirtschafts- und Vernichtungspolitik in Weißrußland 1941 bis 1944* (Hamburg: Hamburger Edition, 1999), 431–42.

38 RAO, RAFA-2188/1/F3a-L0001, Letter by construction firm Oltsch & Co., Zweibrücken, 11.3.42.

39 RAO, RAFA-2188/1/F3a-L0012, Copy of a letter by Todt to all heads of *Prüfungskommissionen* and all of GB Bau's regional representatives regarding the deployment of German construction companies in the occupied eastern territories, 7.2.42.

40 24th decree of the GB Bau regarding the deployment of German construction companies in the occupied eastern territories, 7.2.42, in: *Deutscher Reichsanzeiger und Preußischer Staatsanzeiger*, no. 56 (7.3.42). Cf. Botzet, 'Generalbevollmächtigter', 125–26.

1945.[41] Examples of this negative freedom of contract can be found in the case of Norway, too. Both Karl Plinke Beton- und Tiefbau from Hanover and the Philipp Holzmann AG, Frankfurt am Main, turned down requests to work in occupied Norway without any negative consequences.[42] At Grün & Bilfinger, the company's branch offices repeatedly declined requests by public authorities, as Munich and Hamburg declared in their reports for 1940. Cologne reported in 1941 that they had to refuse construction works on the French Atlantic coast for the OT's *Einsatzgruppe West* because of a lack of machinery.[43]

Significant entrepreneurial room for manoeuvre also existed for local construction businesses in the occupied territories of Northern and Western Europe. Numerous examples show that freedom of contract was normally respected, even though we have no reason to assume that the Nazis treated local construction firms more leniently than German firms. Fabian Lemmes has shown that French companies actively sought and could choose between orders of German agencies like the OT.[44] They could turn down requests by the occupiers, too. The price was of course reduced economic activity, but Lemmes lists several companies that managed to secure their economic survival without working for the German occupiers.[45] Steen Andersen has shown the Danish construction industry's ample scope of action and proactive attitude.[46] Many Danish firms like Christiani & Nielsen, Monberg & Thorsen or Højgaard & Schultz declined the *Wehrmacht's* wishes to build airfields in occupied Denmark in 1940.[47] When the OT projected the construction of the submarine bunkers in Trondheim in spring 1941, Xaver Dorsch initially tried to compel several Danish firms to conduct the works. This time, too, all of the firms managed to turn down the request without any negative consequences.[48] Finally, with their international post-war businesses in mind, the Danish multinationals backed out of German projects in autumn 1944, when it was clear that Germany's defeat had become a question of when rather than if.[49]

---

41 Buchheim and Scherner, 'Anmerkungen zum Wirtschaftssystem des "Dritten Reichs"', 86–87; Scherner, 'Zwangs oder Kooperation?'

42 For Karl Plinke see HHStAW, 485/480, fol. 55. For Philipp Holzmann see RAO, RAFA-2188/1/F3a-L0001, Philipp Holzmann AG to the RK's Technical and Transport Department, 21.2.42.

43 Bilfinger Archive, A 13, Report by the Cologne branch office for 1941; ibid., A 12, Reports by the Hamburg and Munich branch offices for 1940.

44 Lemmes, 'Arbeiten für das Reich', 383–411.

45 Lemmes, 406.

46 Andersen, *De gjorde Danmark større*. A particulary striking example given by Andersen is Wright, Thomsen & Kier's entry into the construction of fortifications on Bornholm: Andersen, 196–205. See also Andersen, 'Between Imperative and Risk'; Andersen, 'French Opportunity'.

47 Andersen, *De gjorde Danmark større*, 149, 157–63, 216, 441.

48 Andersen, 242–49.

49 Andersen, 392.

In Norway, there were hardly any local firms that had not worked for EW, the *Wehrmacht* and other German authorities at some point during the occupation. Thus, Norwegian construction firms came under investigation for economic treason (*økonomisk landssvik*) after the war. Threatened with severe punishment, entrepreneurs stressed that they had been forced into working for the Germans and that almost all construction activities in the country were ordered by the Germans. Hence, there would have been basically two options: either to go along with the German demands, or to see their workers and machines confiscated by the Germans.[50] However, as it has already been shown in chapter IV.3, this argument does not stand up to closer examination.

The OT, the *Wehrmacht* and other German authorities clearly circumvented regulations and used the price mechanism to attract firms and labour. Rising wages, prices and profits were the result. This behaviour by German awarding authorities would have been incomprehensible and pointless in a command economy primarily based on threats and coercion. Instead, local construction businesses in occupied Northern and Western Europe had significant economic room for manoeuvre, which was responsive to contractual incentives. The same holds true for the German firms. The available archival material gives the clear impression that the OT normally respected entrepreneurial autonomy and freedom of contract during the mobilisation process. German firms actively sought to get involved in the OT's projects in Norway and elsewhere and they entered contract negotiations with the OT as independent economic entities. There is nothing in the archival material to suggest that construction firms gave up their identity under the OT, not least since the OT was usually only one of a firm's several customers. Nor does anything suggest that firms replaced economic rationality with a kind of military logic. Erhard Knechtel's claim that the OT's orders were in fact commands and German firms were conscripted and absorbed by the Organisation is indefensible.[51] Xaver Dorsch decidedly opposed such views in March 1943:

> A misconception of the relation between the OT and the entrepreneur prevails in the public. Here and there, you hear the opinion that the OT is a large government-operated system, sometimes the OT units are even compared to the construction battalions. Nothing could be more wrong than this. We know from the history of technology that for example the majority of all inventions were carried by what is called entrepreneurial initiative. In recognition of this fact, the OT has deliberately deployed entrepreneurs from the outset. Even the smallest formation is somehow carried by the entrepreneur.[52]

Although it is possible that Dorsch's statement also was an attempt to dispel the industry's concerns, it is still clear that the firms' motives mattered. They have to

50 Cf. the post-war reasoning of one of Norway's leading industrialists, Lorentz Vogt: Lorentz Vogt, *Det organiserte, offisielle, økonomiske samarbeide med okkupantene* (Oslo, 1947), 1–2.
51 Knechtel, *Arbeitseinsatz im Baugewerbe*, 23.
52 Franz Xaver Dorsch, 'Das Wesen der OT', *Der Deutsche Baumeister* 5, no. 3 (1943): 2–3.

take centre stage in any analysis of the relation between EW and the German construction industry. Among these motives, two in particular dominated the firms' reasoning and negotiations with the state, the first being profitability, the second being the protection of their German workforce and access to additional (conscripted or forced) labour. Hence, they will be discussed in greater detail in two separate chapters (V.2 and V.3).

c Entrepreneurial Motives for an OT engagement

Beyond these two major reasons for accepting OT commissions, there were of course more aspects that mattered for the firms, especially in an economy as regulated as the National Socialist one and under the conditions of war and occupation. Three effects of cooperating with the OT stand out. First, the possibility of maintaining good relations with a crucial state authority. Second, the hope that the OT would grant the firms privileged access to machinery and protect said machinery from the grasp of other state authorities. And third, the fact that the OT's wartime construction sites offered an opportunity to test new construction methods and gain access to valuable know-how. All this of course still served the underlying long-term aim of economic success, that is, entrepreneurial profit seeking and the securing of the firm's survival.

Construction firms certainly had to weigh risk and benefits when refusing to enter a contract with the OT. The OT was almost omnipresent in wartime Europe and there are indications that firms took orders only because they wanted to maintain good relations with the Organisation. Grün & Bilfinger, for example, was rather sceptical about working on the Polar Line. As we have already seen in chapter IV.2. b, the company knew full well that the railway was an ostensibly unfeasible project and stated that the firms "will hardly have the opportunity to escape the task, unless they want to get into a very difficult relationship with the awarding authority."[53] At first glance, this could be interpreted as evidence for a company being forced into a contract with the OT. However, it is not clear what Grün & Bilfinger meant with "a very difficult relationship". There are no documented examples of non-commercial sanctions (for example imprisonment) by National Socialist state authorities against German construction firms that turned down an offer. It is more likely that Grün & Bilfinger expected that the OT would continue to be the most important client for the rest of the war and perhaps even in post-war Germany. From this perspective, following the call to work on the Polar Line most likely was part of a long-term strategy to secure future orders. Accepting less attractive orders just to

53 Bilfinger Archive, A 2797–1, Report on the line inspection for the railway project: Fauske – Narvik – Kirkenes, 15.6.42, fol. 31.

stay on the ball, however, was a strategy that construction firms had also pursued during the crisis of the early 1930s and with regard to business contacts abroad during the peace years of the National Socialist reign.[54] In the 1950s, too, firms engaged in projects abroad that held little attraction on their own, but helped to maintain business connections.[55] Finally, it should be noted that Grün & Bilfinger was earning money on its 17 km stretch of the railway. Whether the entire project of a line to Kirkenes was feasible or not, was, strictly speaking, not Grün & Bilfinger's problem.

The sheer mass of construction activities all over Europe made construction machinery scarce goods. What is more, the firms lost machinery during transport or due to enemy action. Thus the privileged access to construction machinery became another important incentive to work for the OT. As it has been mentioned above, executing projects that were vital to the war effort made firms eligible to receive iron and steel cheques that they could use to order new machines.[56] Moreover, the OT provided its contractors with machinery that the Organisation had bought from other German state authorities or in the occupied territories. In 1941, *Einsatzgruppe West*, for example, began to buy construction machinery in the unoccupied part of France for its OBLs and its contractors for more than 4.5 million RM. As the German Court of Audit criticised, *Einsatzgruppe West* was later unable to clarify the whereabouts of more than half of the machines.[57] There are in fact indications that raise the suspicion that German construction firms were able to stock up on construction machinery at favourable conditions during their OT missions. At the time when Philipp Holzmann, for example, was heavily involved in construction projects of the OT in France and Italy, the firm's stock of machinery grew considerably, even though investment sums were cut back significantly.[58]

In occupied Norway, EW already began to buy machinery from closed-down construction sites of A/S Nordag in 1942 in order to sell or rent it to its contractors.[59] In January 1944, for example, four firms working on the *Nordlandsbane* (Philipp Holzmann, Stuag, Mayreder, Kraus & Co and Universale Hoch- und Tiefbau) received machine tools from Nordag and IG Farben stocks via OBL Narvik II.[60] By

54 See for example: Allgemeine Baugesellschaft Lenz & Co., business report 1934; Julius Berger Tiefbau AG, business reports from the first half of the 1930s. Whether the OT really would have penalised and cold-shouldered one of the leading German construction firms, which possessed more expertise in large *Tiefbau* projects than most others in Europe, is another question.

55 Pohl, *Philipp Holzmann*, 336.

56 See above, chapter III.3.c.

57 BArch, R 2301/644, fol. 308.

58 Cf. above, chapter III.3.c, Figure 9, and chapter IV.1.a, Figure 13.

59 RAO, RAFA-2188/1/E1m-L0046, Circular to all OBLs regarding the acquisition of Nordag machinery, 7.12.42.

60 RAO, RAFA-2188/1/E7a-L0001, file perm uten merking, Note by EW on the allocation of machine tools for the firms of OBL Narvik II, 19.1.44.

January 1945, EW had acquired from A/S Nordag machinery, tools and building materials with a total value of more than 14 million kroner.[61] In February 1944, EW decided to sell OT-owned machines to the companies that were working for the organisation and hitherto had rented the equipment. Wigru Bau welcomed the opportunity and informed its member firms that only the most productive and efficient German firms were allowed to buy.[62] In this process, EW apparently rewarded reliable firms that had worked under the organisation for a longer time. Allgemeine Baugesellschaft Lorenz, which had previously worked for the air force and had only recently become an OT contractor, received a negative reply from OBL Kristiansand to its request to buy OT machinery.[63] Strabag Weimar, on the other hand, had worked for the OT for years, and was thus considered a legitimate buyer. Hence, Strabag gave its Stavanger office free hand to buy OT machines as long as it was economically justifiable and advantageous.[64] Site manager Korffmacher assured his superiors in Weimar that he would only buy equipment he had inspected personally "in order to be independent of the OT [. . .] and to make sure that after the war, machines can be handed over to Strabag that are 100 % operational."[65] As late as in November 1944, just a few weeks before Strabag Weimar returned to Germany, Korffmacher was still able to buy construction machines in Oslo on behalf of the Weimar headquarters. These machines were never destined for sites in Norway, but in all likelihood were shipped directly to Germany.[66]

Having a contract with EW could also help to protect equipment from the grasp of other state agencies. Between 1942 and 1945, Kiel-based Steffen Sohst was not only engaged in construction projects and the quarrying of gravel for EW. With its boats and barges, the firm also transported building materials on behalf of the OT in the vicinity of OBL Bergen. In doing so, the firm from the very first day came into conflict with *Transportflotte Speer*, which regarded the private firm a bothersome competitor.[67] Throughout the entire occupation, *Transportflotte Speer* tried to eliminate Steffen Sohst's transport business in Western Norway by trying to entice Sohst's crews away and, above all, to confiscate the firm's boats.[68] However, Steffen Sohst

---

61 RAO, RAFA-2188/1/E5f-L0005, file Durchschriften ot/v/254 1944–1945 [forts.], Note by EW, 16.1.45.

62 ThHStAW, Straßenbau AG Weimar Nr. 144, fol. 107, Copy from circular no. 6 of the Oslo liaison office of the German building trades, 13.7.44.

63 LAB, A Rep. 250-05-13, nr. 149, Letter by Allesch, Allgemeine Baugesellschaft Lorenz, Lunde construction site, to the firm's headquarters in Berlin, 23.8.44.

64 ThHStAW, Straßenbau AG Weimar Nr. 144, fol. 126.

65 ThHStAW, Straßenbau AG Weimar Nr. 144, fol. 106.

66 ThHStAW, Straßenbau AG Weimar Nr. 144, fol. 10 ff.

67 RAO, RAFA-2188/1/H1a-L0001, file Schriftwechsel Oberbauleitung 1942, Note on a meeting between representatives of *Transportflotte Speer* and Steffen Sohst on 6 August 1942, 7.8.42.

68 RAO, RAFA-2188/1/H1a-L0002, file Schriftwechsel Oberbauleitung 1944–1945, Letter by Steffen Sohst, Oslo construction site, to EW, 9.1.45

was able to parry these repeated attacks as the firm had the backing of the OT, which was obviously satisfied with Sohst's performance.[69] In the last weeks of the occupation, the seaborne transport of building materials in the Bergen area was still in the hands of Steffen Sohst. The conflict between the private business and *Transportflotte Speer* sheds light on two central aspects covered in this thesis. First, the firm was not only *not* forced to enter a contract with *Einsatzgruppe Wiking*, it even spent years fighting against being edged out of a contractual relationship with the OT by another state authority. Second, the *Transportflotte's* unsuccessful attempts to take over Sohst's boats shows that National Socialist state authorities could not simply commandeer the private property of a German firm without further ado – at least when the firm mobilised the support of another authority.

The OT's missions offered the opportunity to acquire know-how and develop new construction methods. In 1958, Wolf von Niebelschütz of Eduard Züblin AG described the company's wartime projects in occupied Europe primarily as "great challenges" or "adventures" that gave Züblin the chance to "prove its genius". He left no doubt that the projects were technologically interesting, like the complicated construction of a pier in Banak (Lakselv) or the railway bridge crossing the Trengselsund near Fauske, "[. . .] one of those tasks, for which alone it would have been worth to becoming an engineer [. . .]".[70]

While the economic value of EW's railway in Northern Norway was rather limited for the German and Austrian construction firms participating in its construction, especially taking into account the machinery lost in 1945, they did indeed acquire valuable know-how. As Hermann Häusler has shown in his research, the airport, bunker and railway construction sites in Norway became the source of important scientific developments in the field of geology, especially in post-war Austria. During the occupation, around 60 military geologists and engineers from Germany and Austria were in the service of the *Wehrmacht* and EW in Norway. They conducted ground surveys and advised the OT officials and firms that were building the tunnels on the railway line in Northern Norway. One of the geologists working for EW was the Austrian Leopold Müller, who after the war gathered many of his former colleagues in his "Salzburg Circle", where they established the discipline of geomechanics based on their experiences in Norway.[71] Closely connected to the Salzburg Circle was the Viennese professor Ladislaus Rabcewicz, the man who had accompanied EW's railway engineers in spring 1942.[72] Based on the

69 RAO, RAFA-2188/1/H1a-L0003, file Schriftwechsel Oberbauleitung 1944–1945 [forts.], Letter by Köbele, EW, Oslo, to OBL Bergen, 31.5.44.

70 Ed. Züblin AG, *75 Jahre Züblin-Bau, 1898–1973*, 37.

71 Hermann Häusler, 'Von der Wehrgeologie in Norwegen 1940–45 zum „Salzburger Kreis" der Geomechanik', in *14. Wissenschaftshistorisches Symposium 'Geologie und Medizin', 4. Dezember 2015*, Berichte der Geologischen Bundesanstalt 113 (Wien: Geologische Bundesanstalt, 2015), 56–84.

72 See above, chapter IV.2.b, fn. 168.

tunnelling works in Norway among other things, after the war Rabcewicz became the pioneer of the so-called New Austrian Tunnelling Method (NATM). The NATM became a trademark of the Austrian construction industry and is still widely used today.[73] Hence, the National Socialist mammoth project yielded some valuable technological long-term effects for the industry.

The technological gains from a cooperation with the OT can also be illustrated by an example from OT's *Einsatzgruppe Rußland-Nord*. In 1943, the OT had been sent to Estonia in order to build facilities for shale oil extraction for the Baltische Öl GmbH. As usual, a significant share of the men in the OT's construction bureaus seems to have come from private construction firms. While in Estonia, one of the engineers obtained the technology for the production of autoclaved aerated concrete from a Swedish-owned plant that had fallen into the hands of the *Einsatzgruppe*.[74] Already thinking "of the time after the war, [and] the coming reconstruction of the destroyed homeland", the engineer quickly realised the material's potential as this type of concrete had various advantages and could be used in residential building in particular. The OT first asked the building materials producer Rank in Fürstenfeldbruck to use some of its idle capacity to produce the new concrete. However, the firm refused the OT's request, once again demonstrating that negative freedom of contract still existed at the height of the war. The OT then offered a contract to the construction firm Josef Hebel from Memmingen.[75] As the following decades would show, the product became a huge success. Even today, "Hebel autoclaved aerated concrete" is used all over the world.

Research on businesses in the Nazi economy often identified yet another motive in the process of entrepreneurial decision-making. When confronted with investment wishes of the state, companies tried to avoid overcapacities whenever they doubted the long-term value of an investment.[76] There are some examples of construction firms, too, that were anxious not to build up overcapacities. In construction, this concerned particularly investments in *Tiefbau* machinery, as this business area increasingly superseded *Hochbau* activities during the war. Grün & Bilfinger feared overcapacities would build up at the *Westwall*[77] and Philipp Holzmann did in fact report overcapacities in this field in 1950.[78] By and large, however, the

73 Bertrand Perz, 'Die "Neue Österreichische Tunnelbaumethode" und ihre weniger bekannten Bezüge zum Nationalsozialismus', in *Wissenschaft, Technologie und industrielle Entwicklung in Zentraleuropa im Kalten Krieg*, ed. Wolfgang L. Reiter et al. (Berlin, Münster, Wien, Zürich, London: LIT Verlag, 2017), 245–64.

74 The engineer was Fritz Leonhardt (1909–1999), one of the leading structural engineers in Germany after the war. In 1954, he founded a construction consulting firm in Stuttgart and was a lecturer at Stuttgart University. From 1967 to 1969, he was the vice-chancellor of the university.

75 Fritz Leonhardt, *Baumeister in einer umwälzenden Zeit: Erinnerungen* (Stuttgart: Deutsche Verlags-Anstalt, 1984), 96–98.

76 Scherner, *Logik der Industriepolitik*.

77 Stier and Krauß, *Bilfinger Berger AG*, 135.

78 IfS, W1-2/265, Philipp Holzmann AG, business report 1950.

danger of overcapacities in construction was probably not overly great. The purchase of mobile construction machinery could hardly be compared to building a new factory. Moreover, the high level of construction activities during the war put an enormous strain on the machinery of German construction firms. In the daily round, the firms had to cope with a shortage of new machinery and a lack of spare parts rather than overcapacity. Most importantly, the deferred construction investments within the Reich and even more so the massive destruction in Germany during the second half of the war showed construction companies quite plainly that huge tasks awaited them in the post-war period. In 1941, the GB Bau already assumed that the post-war construction volume would still exceed the sector's capacity, which was also seen as the reason why there would still be need for a *Generalbevollmächtigter*.[79]

Again, it must be stressed that all of the mentioned motives were influenced by politically determined framework conditions. No firm could ignore the reality and consequences of Germany's total war. The most significant of these was the *Wehrmacht's* call for soldiers, which soon extended to almost every German man fit for military service. In the construction sector, the GB Bau cancelled projects within the Reich while the demand of the OT and the *Wehrmacht* in occupied Europe increased. Thus, there was significant structural pressure on construction firms to adapt to the state's objectives and entrepreneurial room for manoeuvre was certainly reduced. Normally, however, firms were still able to choose from several options and freedom of contract was maintained. Most importantly, the logic and general aims of entrepreneurial decision-making remained the same. Entrepreneurial decision-making did not become obsolete under some kind of command economy. Nor did National Socialist economic policy, or in this case, the *modus operandi* of the OT change the fundamental aims of businesses as we know them from market economies: to earn money and secure the long-term survival of the firm.

d Elements of Coercion: The Transport Ban and the Concentration of Firms

While the chapter's previous parts have shown that coercion played a minor role during the mobilisation process for the Wiking mission, this part will focus on the situation after the firms had come to Norway. There were two measures in particular that threatened the entrepreneurial autonomy of EW's contractors: the transport ban on construction machinery and the possibility of relocating and even amalgamating small and inefficient firms. I argue, however, that Willi Henne as the GB Bau's

79 BArch, R 2/28093, microfiche 1, Undated memorandum on 'the organisation, tasks and staff of the agency of the GB Bau', probably from October 1941.

representative and head of the OT never exhausted the two measures to a significant degree.

On 19 September 1942, the Reich Commissioner for Sea Shipping (*Reichskommissar für die Seeschiffahrt*, RK See) in Berlin issued a general ban on the transport of German construction machinery from Norway to Germany. Essentially, the measure aimed to reduce transports from Germany to Norway, too, as machinery now had to be reallocated between firms and construction sites in Norway before additional machinery was requested in the Reich. Only the GB Bau's representative in Norway could grant exceptions from the transport ban. The office reserved its right to prohibit German firms from sending machines back to the Reich and could even confiscate machinery.[80]

There are at least two cases of German firms in occupied Norway that were affected by the transport ban. After finishing projects for A/S Nordag, the construction firm Gehlen requested the return of its machinery to the Reich in spring 1943. The firm wanted to take over works in Germany for IG Farben, which had been one of Gehlen's customers for years. In the light of the tense transport situation and the lack of machinery in Norway, however, the GB Bau's Max-Erich Feuchtinger blocked Gehlen's request. Gehlen's alternative wish, to work for IG Farben at the hydropower station in Mår (Telemark), was turned down as well. Instead, the firm was sent to the OT's railway construction sites in Northern Norway.[81] The second example took place in late July 1943, when the *Luftgaukommando Norwegen* in Oslo asked the GB Bau to "preventively confiscate" construction machinery from the firm Lorscheidt & Linge (Kiel). The material had to be transported to an air force construction site in Bardufoss so urgently that an agreement with the firm had not yet been reached. Although the *Luftgaukommando* expected the firm's consent and subsequently the signing of a proper contract, it is of course questionable whether Lorscheidt & Linge could have stopped the transport and revoked the confiscation at that point.[82] The firms on the railway construction sites in Northern Norway would also certainly have preferred to return their machinery to Germany instead of seeing it confiscated by Norwegian authorities in May 1945.

The transport ban was probably the most distinct element of coercion in the relation between EW and the GB Bau on the one side and German industry on the other. If the GB Bau's representative did not give his permission for the return

80 RAO, RAFA-2188/1/E1a-L0004, file 2150, Note by the GB Bau in Oslo on the use of German firms' machinery for construction projects in Norway, 2.8.43.

81 On the entire process with meetings between representatives of the firm, the OT and the GB Bau in July 1943, see RAO, RAFA-2188/1/E1o-L0050, file L1. It should be kept in mind that Feuchtinger and his co-workers at the GB Bau office in Oslo were at the same time members of the OT apparatus. Thus, this case might also be interpreted as a conflict over resources between the OT and IG Farben, with Feuchtinger using his position as GB Bau's representative in Norway to benefit the OT.

82 RAO, RAFA-2188/E1a-L0004, file 2150, Letter *Luftgaukommando Norwegen* to GB Bau, Oslo, 27.7.43; ibid., Letter GB Bau, Oslo, to construction firm Lorscheidt & Linge, 17.8.43.

transport, a firm's machinery was basically trapped in Norway. It seems, however, that the transport ban never became a major issue between EW and its contractors. By late autumn 1944, many of the construction projects in Norway were either finished or had been cancelled. Germany now became the prime area of activity for the OT and with Germany's defeat on the horizon, private firms likewise wished to bring their equipment and workers back home. This overlapping of interests certainly prevented the topic evolving into a major confrontation between state and industry.

In late September, the GB Bau handed the competence to grant exceptions from the transport ban over to EW's Department for Supplies.[83] Lists compiled by the department show the majority of firms was able to send huge amounts of construction machinery and equipment back to the Reich in late 1944 and early 1945.[84] The research literature also provides similar examples from other OT missions. The Teerag-Asdag AG backed out of OT projects in Eastern Europe when the conditions became unbearable due to the constant air strikes of the Soviet air force.[85] Polensky & Zöllner, on the other hand, tried to bring their machinery back home from various places in occupied Europe by stressing that their OT projects in Kaufering and Mühldorf were vital to the war effort.[86]

Nevertheless, according to a joint post-war overview by EW and the *Wehrmacht*, German construction firms had left behind equipment with an overall value of 44 million RM and a weight of 48,000 tonnes in Norway.[87] The information on the value, however, refers to value as new. Given the considerable wear and tear of machines, which were often used around the clock under harsh climatic conditions, the residual value was certainly lower. As the overview also shows, firms were more likely to have lost machinery the farther north they had been deployed. This was probably due to the transport difficulties in the north and the fact that the railway construction in Northern Norway was still being prioritised in the last months of the war. The firms Hochtief in Engan and Sager & Woerner in Trondheim topped the list, as they had lost around 3,000 and 2,500 tonnes of equipment respectively. In occupied France, too, firms lost costly machinery because the OT refused to permit them to evacuate equipment in time during the Allied advance.[88]

83 RAO, RAFA-2188/2/Hfa-L0007, file 27 Regelung des Bauschaffens in Norwegen, Letter Willi Henne to the RK See, Oslo, and to the *Höherer Feldwirtschaftsoffizier* at the WBN, Oslo, 28.9.44.
84 See for example RAO, RAFA-2188/1/E7a-L0025.
85 Lütgenau and Schröck, *Teerag-Asdag*, 56.
86 Raim, *Kaufering und Mühldorf*, 111–16.
87 NHM, FO.II/11.4, box 27, file 56b, Report by (Geb.) AOK 20 (OK/WBN) – *OT Wiking Kommandostab* in Lillehammer, 16.7.45. In contrast to the figure on the value, the overall weight does not include barracks and vessels.
88 Lemmes, 'Arbeiten für das Reich', 404.

Among the firms returning from Norway to Germany was Strabag Weimar. In December 1944, after 31 months' work for *Einsatzgruppe Wiking*, Strabag Weimar's workers and employees left the occupied country together with 400 tonnes of machinery.[89] During their final weeks in Norway, the company had once more proven that the OT was never able or willing to erect a system that would have resembled a command economy. Over the course of the autumn, EW had tried to make Strabag Weimar carry out road works at the Hardangerfjord. For the company, however, its prime objective was to get all equipment back to Germany and not to go into "an assignment that (1.) does not suit us and (2.) where construction machines could get lost."[90] The attempts of EW's representatives to change the mind of Strabag's site manager failed. The firm did not enter into another contract with the OT and left Norway.

The second element of coercion concerned the GB Bau's authority to take action against firms or branches of firms, which lacked the skilled workers and machinery to carry out projects efficiently. As we have seen in chapter III.3.c, the decision on the deployment or closure of construction firms within the Reich and the tasks of the GB Bau lay with the *Amt Bau* and Carl Stobbe-Dethleffsen, that is, with the industry itself. From summer 1944 onwards, however, the OTZ under Xaver Dorsch took over these responsibilities, which has been interpreted as a loss of autonomy for the construction industry and militarisation of the sector. Chapter III.3.c has posed the question whether this change in Germany had any significant effects on OT contractors in occupied Europe.

In contrast to the Reich, the offices of the GB Bau and the OT in occupied Europe had already been closely connected for years. In occupied Norway, for instance, Willi Henne was GB Bau's representative and head of *Einsatzgruppe Wiking* since 1942. Hence, for the vast number of German construction firms working in occupied Europe, nothing really changed in this respect in 1944.

The question is then to what extent the GB Bau/OT in occupied Europe threatened construction firms' autonomy. The Speer decree of 28 October 1943, according to which less efficient firms could be forced to work as subcontractors under larger firms, enter working combines, or – as a last resort – could be merged into larger firms for the time of the war, was introduced in occupied Norway in May 1944.[91] Sæveraas has argued that the decree was a means for EW to increase pressure on its contractors and an important element of the firms' militarisation.[92] Another decree of the GB Bau of 26 September 1944 enabled the leaders of the OT *Einsatzgruppen* to request the closure of very small firms ("*Kleinstbetriebe*") for the duration of the

89 ThHStAW, Straßenbau AG Weimar Nr. 144, fol. 10 ff.
90 Ibid.
91 BArch, R 13-VIII/87, fol. 177–79, Decree by GB Bau's representative in occupied Norway on the concentration of German construction firms in Norway, 20.5.44.
92 Sæveraas, 'Beton macht Geschichte', 168–69; Sæveraas, 'Festung Norwegen', 197.

war.[93] However, it is unclear whether any German construction company in Norway was ever affected by these decrees. EW's contractors were mostly "big players" while smaller firms had already come to Norway as members of working combines, which was no unusual arrangement in construction. To date, there is no documented case of a German construction firm under EW that was broken up due to a lack of workers and machinery or insufficient performance.

2 Contracts and Price Regulation

Earning money remained the single most important aim of entrepreneurial decision-making for construction firms working under the OT in occupied Europe. Negotiations on contract types, their design and the underlying legal regulations thus lay at the heart of the relation between the OT and the construction industry in wartime Europe. No other topic was deemed so crucial by both sides. The OT tried to strike the balance between containing costs and the incentives necessary to attract firms and stimulate their productivity. The German construction firms, on the other hand, tried to ensure that their work under the OT became an economic success, not a risky adventure in the remotest corners of occupied Europe.

Despite their crucial role in the OT system, the contracts between the Organisation and its contractors have only received some attention in recent years.[94] While these earlier contributions provide helpful overviews over the various contract types and their incentive structures, this chapter will try both to provide an in-depth analysis of the contracts and at the same time to broaden the picture. It argues that in addition to a thorough understanding of the contracts themselves, it is crucial to investigate the preceding phase in which the contracts were drafted and negotiated. Furthermore, the focus will once more lie on the actual effects of the contracts and EW's ability to monitor its contractors.

After dealing with attempts at containing construction costs in occupied Norway before EW's arrival in the chapter's first part, part b and c show that EW depended heavily on the construction industry's expertise when drafting its contracts and calculating construction prices from 1942 onwards. In the process, the firms were able to realise considerable information rents, that is, additional earnings derived from

---

93 Circular decree by the GB Bau regarding the closure of businesses and the establishment of *Kriegsgemeinschaftsbetriebe* in construction, 26.9.44, in: *Ministerialblatt des Generalbevollmächtigten für die Bauwirtschaft, des Generalinspektors für das deutsche Straßenwesen, des Generalinspektors für Wasser und Energie und für den Rüstungsausbau, herausgegeben vom Reichsministerium Speer* 2, no. 34 (25.11.44).

94 Seidler, *Die Organisation Todt*, 212–16; Raim, *Kaufering und Mühldorf*, 118–27; Streb, 'Negotiating Contract Types'; Sæveraas, 'Beton macht Geschichte', 155–69; Ingulstad, 'Regulering', 258–66.

having information not provided to the principal. However, the OT was willing to pay this price to mobilise the industry's reserves as fast and widely as possible. The argument is that neither Todt, Speer and Dorsch in Berlin nor Willi Henne in Oslo were "construction dictators" who steered the industry *ad libitum*. I argue that the industry's entrepreneurial scope of action to a considerable degree resulted from their close involvement in the drafting of the contracts. Part c discusses in detail the OT's efficiency-output contract, which must be seen as an attempt to raise companies' performance while limiting their risk in the midst of a total war. Part d will conclude the chapter by focusing on the question of what time pressure and lack of administrative personnel meant for the contract and price negotiations between EW and its contractors.

a The Reich Commissariat's Attempts to Contain Construction Costs

In the last months before the outbreak of the war, the works on the *Westwall* had led to the insight that cost-plus contracts were a convenient choice when complex projects had to be executed as fast as possible and without proper projection. Thus, despite being costly and inefficient in the long run, it did not come as a surprise that the OT resorted to cost-plus contracts during and after the military campaigns all over Europe.

The use of cost-plus contracts in combination with a lack of effective price and wage regulations had contributed to significant price increases in construction, including in occupied Norway. During the first two years of the occupation, the Norwegian regulations and controls of prices and profits had not succeeded in keeping the price spiral in construction under control.[95] The *Wehrmacht* in particular tried to attract Norwegian labour and firms by any means. By the end of 1941, this had led to a situation where the RK's Department for Technology estimated that many construction projects in Norway cost many times more than comparable ones in the Reich.[96] The German construction companies in the country were not subject to the Norwegian regulations. They were answerable to the authorities in Germany for their prices, and there are indications that the Reich Price Commissioner was rather lenient towards German constructions firms in the occupied territories.[97]

The Reich Commissariat had become convinced by autumn 1941 that stricter regulations for German construction projects were necessary to regain control over wages and prices. This would have to include the German firms in particular.

95 Ingulstad, 'Regulering', 259–60; Harald Espeli, 'Incentive Structures and State Regulations', in *Industrial Collaboration in Nazi-Occupied Europe: Norway in Context*, ed. Hans Otto Frøland, Mats Ingulstad, and Jonas Scherner (London: Palgrave Macmillan, 2016), 258–66.
96 BArch, R 13-VIII/264, First report of Thierbach, Oslo liaison office of Wigru Bau, 27.12.41.
97 See chapter V.2.a, fn. 105.

Therefore, on 1 December 1941, the Department for Technology under Heinz Klein issued provisions on the observance of wages and prices in construction, effective from 15 January 1942.[98] They applied to all types of contracts, including cost-plus and fixed-price contracts, and to all of the occupying power's awarding authorities. The rules made it obligatory to keep a daily record of hours worked. A maximum worktime (10 hours per day, or 60 hours per week) was set and overtime pay was only allowed when 48 hours had been worked. Piecework earnings were allowed to exceed the standard wage rates by 40 percent at most.[99] For works executed on a cost-plus base, the department set the maximum rates that German and Norwegian firms were allowed to add for expenses, salaries, risk and profit. For German firms, the permissible overall mark-up on the wage costs was 50 percent, for larger Norwegian firms 38 percent and for small Norwegian firms 30 percent, all including 8 percent for risk and profit.[100] To compensate firms for their administrative work when using subcontractors, they were paid 2 percent of the latter's invoice total. Finally, the regulations demanded transparent price calculation also in the case of fixed-price contracts. The Department for Technology defined certain reference values above which a fixed price was regarded to contain unjustified profits. In practice, this amounted to a limitation of profits for fixed-price contracts.

Read *ex negativo*, the regulations give an idea of what had contributed to exploding costs in the previous years. Companies had inflated their wage costs and increased their profits by billing undocumented hours. The order that overtime pay was only permissible after 48 hours indicates that firms had demanded that their workforce work long hours in order to inflate costs. The *Wehrmacht* had often condoned such practices and awarded contracts with more than generous premiums, not least since the *Wehrmacht* branches competed with each other for firms.[101]

The initiative's immediate effect was not entirely as intended. RK's *Oberbauleitung West* in Bergen complained in early 1942 that the *Wehrmacht* had started to consistently grant its Norwegian contractors the maximum mark-up of 38 percent, even though some firms previously had received lower rates. The maximum rates had *de facto* become the norm, which in some places had even led to price increases. OBL West therefore asked the Department for Technology for permission to grant this rate, too. Otherwise, "[. . .] the good firms would exclusively work for the *Wehrmacht* while

98 RAO, RAFA-2188/1/E1a-L0002, file 2134, Provisions on the observance of wages and prices in construction by the Department for Technology, 1.12.41. The regulations were also made public on 11 December in: *Verordnungsblatt des Wehrmachtbefehlshabers in Norwegen* 2, no. 48 (1941). See also: Ingulstad, 'Regulering', 260–61.

99 Special tariff agreements on piecework in the major Norwegian cities remained unaffected by this regulation.

100 A Norwegian firm was regarded large when it had its own book-keeping office.

101 Korff, 'Norwegens Wirtschaft', 1949, pt. II, 255–56, 259.

OBL West would have make do with the bad firms."[102] In the following months, too, the pricing experts of the RK and the Price Commissioner realised that the OBLs of *Einsatzgruppe Wiking* largely were ignoring the provisions.[103]

Initially, however, the German construction industry was somewhat concerned. For the head of Wigru Bau's liaison office in Oslo, Friedrich Thierbach, the regulations came as a surprise and a real shock. In his first reaction, he called them a "highly alarming" intervention by the agency of the Reich Commissioner.[104] Thierbach virtually apologised to the firms that he had not been involved in the formulation of the rules. While Wigru Bau had to admit that none of the regulations, such as the demand to correctly report hours worked, should actually be a problem for a reputable German construction firm, it promised to discuss them with the Department for Technology and the Price Commissioner in Berlin after consulting the major German firms in Norway.[105] A meeting was called for 12 March 1942 in Berlin. It seems that a significant number of the German construction firms in Norway were present. The attendance list counts 21 firms; another firm (Ludwig Czaikowski, Saarbrücken) did not participate, but had submitted a written statement.[106] They criticised three points of the new price regulations in particular. First, in their view, the granted rates for expenses, salaries, risk and profit in the

102 RAO, RAFA-2174, Ef-L0001, file 7, Letter Thote, OBL West, to the Department for Technology, 7.2.42

103 Ingulstad, 'Regulering', 262.

104 BArch, R 13-VIII/264, Attachment 'Miscellaneous' to the first report by Thierbach, Oslo liaison office of Wigru Bau, 30.12.41.

105 RAO, RAFA-2174/Ef-L0001, file 7, Letter by Thierbach, Wigru Bau, 2.3.42. Wigru Bau itself had contributed to the price increases. In October 1940, the construction company Sager & Woerner had submitted their Norwegian price calculations to the Wigru Bau and asked for advice on whether the mark-up of 73 percent on the wage bill was acceptable and legal. Wigru Bau responded that Sager & Woerner could easily add another 5 percentage points and added that from their experience, the Reich Price Commissioner would most likely approve those rates if the prices were audited: RAO, RAFA-2188/2/He-L0005, file 20 (Dr. Klein Mappe IV), Letter Riedel, Wigru Bau, to Sager & Woerner, 19.10.40. This was significantly higher than the maximum mark-up of 50 percent set in December 1941.

106 22 is certainly just the minimum number of German firms working in Norway before the establishment of *Einsatzgruppe Wiking*, as Strabag Weimar, for example, which had been in the country since 1940, is not on the attendance list. In this respect, Sæveraas has made an interesting observation: Wigru Bau's list of German construction firms in Norway of early July 1942 seems to have been drawn up chronologically, meaning that every new firm arriving in Norway was simply added at the bottom of the list. Around 40 of the firms on top of the list are registered with proper addresses in Norway, among them almost all firms that were present at the meeting on 12 March 1942, while the firms further down are only listed with approximate areas of deployment. We can therefore assume that c. 40 German firms were already working in occupied Norway before the establishment of *Einsatzgruppe Wiking*. See BArch, R 13-VIII/264, Letter by Thierbach, Oslo liaison office of Wigru Bau, to Wigru Bau, Berlin, with attached list on German construction firms in Norway (as of 10 July 1942), 11.7.42. Cf. Sæveraas, 'Beton macht Geschichte', 152.

cost-plus contracts did not reflect the difficult conditions in Norway. Second, a strict cap of piecework earnings to at most 140 percent of the standard pay would destroy the workers' incentive to give their best. Third, the German firms found it unacceptable that awarding authorities could gain insight into the firms' price calculations under fixed-price contracts and, to make matters worse, could withhold payments when they regarded the prices too high.[107] There was also some confusion about the question of whether the mark-up of 50 percent should be added on the wages of the German workforce only. If this were the case, the firms would have had a hard time covering their operating expenses, since they often used local workers.[108]

In the following months, Wigru Bau had intense talks both with the RK's Department for Technology in Oslo[109] and the Reich Price Commissioner in Berlin. In particular, the association commissioned its leading expert on pricing, Gerhard Opitz, to draft a more sophisticated method to calculate the German firms' rates for operating expenses, salaries, risk and profit in Norway.[110] However, when it became clear that most German firms would be working under soon-to-be-established *Einsatzgruppe Wiking*, the negotiations on the December 1941 regulations were soon overshadowed by those on EW's new contract. And indeed, Opitz' work would eventually become a key element of EW's contracts between 1942 and 1945.

b The Wiking-Bauvertrag: Return to the Cost-Plus Principle

The establishment of EW and the deployment of staff, workers and firms to occupied Norway was a hasty undertaking in spring 1942. Firms started working without any contractual basis as the negotiations on EW's so-called *Wiking-Bauvertrag* dragged on until August. During the negotiations, Wigru Bau relied particularly on the expertise of two firms, Sager & Woerner and Strabag, that were widely considered to be "*Norwegenkenner*", that is, experts on working in occupied Norway.

From the very beginning, it was clear that the *Wiking-Bauvertrag* could only be a cost-plus contract, as the projects' number and scope, and even the location of many construction sites, were still uncertain. Only after some months, EW envisaged,

107 BArch, R 13-VIII/264, Minutes of a meeting between Wigru Bau and their member firms active in Norway on 12 March 1942 in Berlin, 7.4.42.
108 Ibid., Letters by the construction firms Habermann & Guckes, Berlin, and Hermann Milke, Soest, to Wigru Bau, 12.3.42 and 11.3.42, respectively.
109 Thierbach pinned his hopes on the new strong man on the Norwegian construction sector, Willi Henne, "[. . .] who will probably realise the absurdity of the regulations [. . .]". See BArch, R 13-VIII/264, Letter Thierbach to Riedel, Wigru Bau, 7.4.42.
110 Cf. the calculations and drafts by Opitz in: BArch, R 13-VIII/264.

would negotiations on a fixed-price contract start. The industry supported this strategy. While considering increased work in the occupied territories in early 1942, Strabag Weimar's Alex Korffmacher had reassured himself that works would start in a risk-free cost-plus contract and made clear that he would only submit a tender for a fixed-price contract when the actual performance under local conditions was foreseeable. Hence, Korffmacher expected "[. . .] that these works could never result in losses [. . .]." Quite the contrary, they would be "beneficial and lucrative".[111] By choosing a cost-plus contract, EW reduced the risks for the involved firms during the mission's first months, but delayed the introduction of a fixed-price contract, which the RK had advocated in late 1941.[112]

Already on 16 January 1942, as a reaction to the Department for Technology's provisions of 1 December 1941, Wigru Bau had forwarded a request of the Reich Price Commissioner, asking Siemens-Bauunion's Gerhard Opitz to elaborate a more sophisticated, but still easy-to-use method for the calculation of the mark-ups for operating expenses and profit under a cost-plus contract in Norway.[113] In the following months, Opitz developed the rates based on empirical values from the previous years. He already delivered a first, simple draft in February, a second one in early April and a final draft in early June 1942.[114] The work on the new contract seems to have taken place primarily in Berlin, where Opitz collaborated closely, and apparently frictionlessly, with the office of the Reich Price Commissioner. Finally, the Reich Price Commissioner drafted lists based on Opitz' calculations, from which firms could easily read the mark-up they were allowed to add to the production costs of their construction site. The key variables were the project's value, the company's size, the share of material costs and, finally, what share of the project was executed by subcontractors. The size of a company was defined as its turnover in the last business year before the Wiking mission. The mark-up decreased the higher a project's value and the bigger the company. It further decreased the more parts of a project were executed by subcontractors and the higher the share of material costs in the project's value.[115]

The OT and Wigru Bau finalised their negotiations on the new *Wiking-Bauvertrag* in early August 1942. EW sent the contract along with additional guidelines both to the companies' headquarters in Germany and the construction sites in Norway in the second half of August.[116] The *Wiking-Bauvertrag* applied exclusively to German

---

111 ThHStAW, Straßenbau AG Weimar Nr. 105, fol. 505–6, Note by Korffmacher on a consultation with *Bauoberinspektor* Dölle, 3.1.42.

112 Ingulstad, 'Regulering', 262.

113 BArch, R 13-VIII/264, Letter Wigru Bau to Opitz, 16.1.42.

114 All three drafts can be found in BArch, R 13-VIII/264.

115 See the *Wiking-Bauvertrag* with attached assessment tables (*Bemessungstabellen*) in RAO, RAFA-2188/2/Haa-L0034.

116 BArch, R 13-VIII/265, Circular no. 7 of the Oslo liaison office of Wigru Bau, 13.8.42, and circular no. 8 of the Oslo liaison office of Wigru Bau, 8.9.42.

firms.[117] It consisted of the contract itself that had to be signed by the firm, the OBL and EW's Department for Contracting and Pricing in Oslo. The three assessment tables for the calculation of the mark-ups were attached to the documents. The mark-ups granted for operating expenses and profit were the very core of the cost-plus contract and the rates were identical to those in Gerhard Opitz' drafts. They normally ranged from 11–14 percent. They could climb to more than 16 percent for mid-sized firms executing works entirely without the help of subcontractors on a very small construction site, with material costs amounting to less than five percent of the project's value. At the other end of the table, a German firm that delegated its entire work to subcontractors received only the bottom rate of six percent. The latter case, however, probably never occurred in reality. German firms were allowed to add 0.3 percentage points to all of these rates as a special reward for their complicated mission in Norway.[118] There are no reactions from summer 1942 indicating that the firms were unhappy with the rates granted in the *Wiking-Bauvertrag*, although they were slightly below the rates granted by the air force, for example.[119] Strabag Weimar expected a mark-up of almost 16 percent.[120] Habermann & Guckes AG, already working in Trondheim before 1942, likewise welcomed Gerhard Opitz' proposal in contrast to the regulations of 1 December 1941, which would have granted them only 6.4 percent for operating expenses and profits.[121]

It is worth noting that the entire method of calculating the profits as a percentage of the construction sum was the OT's concession to the unpredictable conditions and time pressure in the occupied territories. As a lesson from the exploding costs at the *Westwall*, the LSBÖ of 25 May 1940 had demanded that the profit could no longer increase proportionately to the construction sum, but had to be fixed as an absolute sum before construction started.[122] However, the OT was not able to implement this demand in its various contracts in occupied Europe, as the fixing of an

117 For Norwegian firms, EW introduced a separate cost-plus contract (*Wiking-Stundenlohnvertrag für norwegische Baufirmen*) in the second half of 1942.

118 Ibid.

119 The air force seems to have granted a mark-up of 18 percent for operating expenses and profit throughout the occupation, as the construction firms Steffen Sohst, Kiel, and Allgemeine Baugesellschaft Lorenz, Berlin, report in September 1942 and November 1944, respectively: RAO, RAFA-2188/1/H1a-L0008, file Schriftwechsel Kiel No. 1, Letter Steffen Sohst, Kiel, to Steffen Sohst, Bergen, 3.9.42; LAB, A Rep. 250-05-13, Nr. 45, Letter by Allesch, Allgemeine Baugesellschaft Lorenz, Lunde construction site, to the firm's headquarters in Berlin, 27.11.44.

120 ThHStAW, Straßenbau AG Weimar Nr. 107, fol. 74–75, Letter by Strabag's office in Stavanger to the Berlin headquarters, 21.7.42.

121 BArch, R 13-VIII/264, Habermann & Guckes AG, Berlin, to Wigru Bau, 12.3.42.

122 RGBl. 1940, I, 851–59. See also: Streb, 'Preisregulierung', 42.

absolute profit sum would have required at least a basic overview over the projects' scope and costs.[123]

Beyond the mark-ups for operating expenses and profits, the OT reimbursed several costs upon receipt, like operating supplies provided by the firms or costs for transport, insurance and rented machinery. The firms also received allowances to cover the costs of repairs and spare parts for the machinery that they had brought to Norway. If the firm provided tools and minor machinery from its own stock, the OT granted a fee of 3 percent on the wage bill of all workers on the construction site. The *Einsatzgruppe* also reimbursed the bills of Norwegian subcontractors, to which the German main contractor added its mark-up for operating expenses and profit, although this mark-up slightly decreased – as we have seen above – the higher the subcontractor's share in the project was.[124]

When EW and the firms had agreed on an engagement, the OT apparatus tried to smooth out bureaucratic obstacles and minimise the specifically "Norwegian" risks for the contractors. Among other things, the OT agreed to cover the insurance for seaborne transport of machinery between Germany and Norway,[125] as well as all war-related damages and loss of machinery on the construction sites.[126] The machinery was transported as *Wehrmachtsgut*, meaning that its import to Norway and later re-import to Germany was free of duty. As it already has been mentioned in chapter IV.3.b, EW also reimbursed the German firms for all fees and taxes that they might have to pay in occupied Norway. *De facto*, this meant the comparably high Norwegian turnover tax of 10 percent.

For all German companies working under it, the *Wiking-Bauvertrag* overruled the highly criticised regulations of 1 December 1941.[127] As a comparison shows, the German firms had been quite successful in representing their interests. The rates for the firms' mark-up on the wage costs had been replaced by a method that was for the most part drawn up by the industry itself. Moreover, the Wiking contract clarified that the mark-up was granted on the wage bill of all workers on the construction site, including Norwegians, other civilian (forced) labourers and POWs. The provisions of 1 December 1941 remained valid for the construction projects of the *Wehrmacht*, the RK and other German agencies in Norway, although they soon were dismantled bit by bit.

The first to fall was the limitation of piecework earnings to 140 percent of the hourly standard wage. On 29 May 1942, Willi Henne issued a supplement to the

---

123 Walter Daub, 'Zur Gestaltung von Bauverträgen', *Die Bauindustrie* 11, no. 5 (1943): 117–21.

124 RAO, RAFA-2188/2/Haa-L0034.

125 BArch R 13-VIII/265, Note by Freckmann, Wigru Bau, for Mr. Vialon, 21.8.42.

126 RAO, RAFA-2188/2/Haa-L0034, Part 1, paragraph 5, of the terms of engagement of *Organisation Todt – Einsatzgruppe Wiking* for the execution of construction works on cost-plus basis.

127 RAO, RAFA-2188/1/E1a-L0002, file 2134, Draft of a letter by Wick, Department for Labour and Welfare, to the head of the Department for Technology in RK's regional office in Trondheim, 3.6.44.

provisions of 1 December 1941, which aimed to encourage higher performance, as individual workers were allowed to earn more than the maximum amount. The provision now stated that piecework earnings should not exceed 140 percent *on average*.[128] Admittedly, this was not exclusively the result of Wigru Bau's lobbying. The cap on piecework earnings had aroused discontent among the Norwegian workforce, too, and provoked the protest of the Union of the Norwegian Construction Workers (*Norsk Bygningsarbeiderforbund*).[129] Moreover, the GB Bau and the RK's Department for Labour and Welfare agreed on abandoning the paragraphs on the maximum worktime and overtime in spring 1943 as they conflicted with the new labour agreement for Norwegian construction workers of 26 March 1943.[130] As the Department for Labour and Welfare noted, those paragraphs had caused a lot of confusion especially within the awarding offices of the *Wehrmacht*, anyway.[131]

The events of spring and summer 1942 show the industry's ample scope of action and influence on state authorities in the midst of the war. Far from being intimidated, leading German construction firms and their association accomplished significant changes to the rules that aimed to regulate their businesses and contributed to the formulation of the new Wiking contract. Moreover, a committee consisting of several company representatives was established in order to monitor the contract's introduction and discuss potential problems. Based on an agreement between Wigru Bau and Willi Henne, director Karl Pfitzner of Sager & Woerner was appointed Contracts Consultant (*Vertragsreferent*) to support the *Einsatzgruppe* with his expertise in future as well.[132]

What is more, it was possible to see very early the monitoring problems that the state would inevitably face when trying to enforce its regulations. As discussed in chapter I.1.c, the aim of containing costs was in principle not incompatible with cost-plus contracts. However, this required thorough *ex post* controls of the firms' invoices. In early 1942, however, the problematic staffing situation was already looming. The Supply Staff Wiking envisaged that the personnel of the future OBLs would consist mainly of technicians and engineers. Only three men were concerned with the financial aspects of the works, that is, the control of invoices.[133] However,

128 BArch, R 13-VIII/264, Copy of the supplement to paragraph 4 of the provisions on the regulation of the construction industry of 1 December 1941, issued by the Department for Technology, 29.5.42.

129 Ingulstad, 'Regulering', 268.

130 RAO, RAFA-2188/1/E1a-L0002, file 2134, Wick, Department for Labour and Welfare, to Feuchtinger, GB Bau, Oslo, 30.3.43; Feuchtinger to Wick, 13.5.43; Wick to Feuchtinger, 17.5.43.

131 Ibid. It is tempting to speculate to what extent this "confusion" was simply disregard.

132 BArch, R 13-VIII/265, Note by Thierbach, Oslo liaison office of Wigru Bau, 8.9.42, and letter Thierbach to Riedel, Wigru Bau, 9.9.42.

133 HHStAW, Abt. 485, 480, fol. 9–11, Letter by Koester, Supply Staff Wiking, to OBR Frankfurt am Main, 9.3.42. See also: BArch, R 3901/20211, Letter by Koester, Supply Staff Wiking, to Lesch, RAM, 11.3.42.

OBR Frankfurt am Main already reported in mid-March that they were unable to provide those three auditors per OBL.[134] Tellingly enough, when EW presented the *Wiking-Bauvertrag* to the *Wehrmacht* in Norway and asked whether the military, too, wished to introduce this contract, the answer was no: considering the permanent shortage of staff, the introduction of a new accounting method was not practicable.[135] Thus the industry does not seem to have expected to be monitored too tightly. With regard to the requirement to record the hours worked on the remote construction sites on a daily basis, a requirement that also remained in force under the *Wiking-Bauvertrag*, Friedrich Thierbach already asked the obvious question in mid-January 1942: "Who is to control the correctness of the records in the wilderness?"[136]

c Increasing Performance in a High-Risk Setting: The Efficiency-Output Contract

The Wiking contract had laid the contractual foundation for the quick start of construction works in Norway. However, it became clear very soon that under the given circumstances, both EW and the construction industry were increasingly unhappy with it. Most of the – sometimes downright indignant – complaints by the construction industry in autumn 1942 referred to the lack of allocated workers. The firms had brought machinery and skilled workers to Norway, relying on EW to provide them with additional (forced) labourers and POWs in order to operate their construction sites profitably. In the second half of 1942, it dawned on them that EW was not always able to keep its word. Under the cost-plus contracts, the labour shortage severely threatened the companies' profitability. Hence, they came to welcome the OT's fundamental reform of contracts in early 1943.

The OT was unsatisfied, too, as the head of EW's Department for Contracting and Pricing, Josef Fröhler, explained in a meeting with entrepreneurs on 7 February 1943 in Oslo.[137] "The latest visits to construction sites have revealed", he declared, "that the progress of construction works falls far short of expectations." The results were "[. . .] downright alarming and disastrous."[138] He

134 HHStAW, Abt. 485, 480, fol. 33–39, Minutes of a meeting concerning the Wiking mission on 17.3.42 in Berlin, 18.3.42. In late June, the liaison office of Wigru Bau even published a request in which EW asked the German construction firms whether they had free typists and wages clerks that they could place at the disposal of the *Einsatzgruppe*: BArch, R 13-VIII/264, Circular by Thierbach, Oslo liaison office of Wigru Bau, 26.6.42.

135 RAO, RAFA-2188/1/E5f-L0046, Minutes on a meeting of representatives of the RK, *Wehrmacht* and OT at the office of the Chief Quartermaster at the WBN in Oslo, 7.9.42.

136 BArch, R 13-VIII/264, Second report by Thierbach, Oslo liaison office of Wigru Bau, 18.1.42, 5.

137 RAO, RAFA-2188/1/E5f-L0014, Letter by Josef Fröhler to all OBLs with the attached transcript of a speech with the title "The efficiency output contract in Norway", held by Fröhler on 7 February 1943 in Oslo, 23.2.43.

138 Ibid.

identified several reasons for the unsatisfactory situation, such as the lack of workers, building materials and machinery, which was mostly a result of transport problems. Fröhler also criticised the workforce's inadequate performance, although it is unclear whether he was referring to unmotivated Germans and Norwegians or rather to groups like the Soviet POWs, who died in their hundreds in the winter of 1942–43 due to insufficient provisions and medical care on the construction sites of the heavy shore batteries in Northern Norway.[139] The fortification pioneers, too, were unsatisfied with EW's performance in its first year.[140]

In his talk, however, Fröhler focused on what he had identified as the "basic evil of the considerable drop in performance": the cost-plus contract. Under this contract type, firms would have no incentive to increase their workforce's performance. Fröhler estimated the performance during the last months to be only one third of the "normal performance".[141] Thus, the *Einsatzgruppe* wished to replace the *Wiking-Bauvertrag* as soon as possible. The aim was the introduction of a so-called efficiency-output contract (*OT-Leistungsvertrag*). In contrast to the old cost-plus contract, the firms calculated and submitted an offer with a fixed price before construction works started. Put very simply, the calculation consisted of three elements. First, a service and price specification (*Leistungsverzeichnis*) listed typical, predefined work steps. An item on the list could, for example, specify the time necessary to blast one cubic metre of rock or to dig 10 metres of drainage ditch. Second, the items were multiplied by the average hourly wage of the workers on the construction site (*Mittellohn*). Finally, the firm added a mark-up for its profit, thereby arriving at the total estimated cost of the project. This price was to remain unchanged unless major unforeseen circumstances changed the project's design or delayed the works. The idea behind this principle was simple: if a firm managed to increase its efficiency, that is, to carry out a project faster and at lower costs, it could keep the saved costs as an additional profit. Through this EW hoped to implement an incentive to increase performance and accelerate the construction process.

The preparations for introducing the efficiency-output contract in Norway coincided with a general effort to increase productivity and efficiency in the German construction industry. Within the OT's empire, *Einsatzgruppe West* was pioneering the conversion from cost-plus to efficiency-output contracts on the construction sites of the Atlantic Wall in France, Belgium and the Netherlands, where negotiations between *Einsatzgruppe West*, Wigru Bau and a committee of several major contractors had started in May 1942.[142] Most importantly, the firms were to participate in the

139 Sæveraas, 'Festung Norwegen', 200.

140 Sæveraas, 198.

141 RAO, RAFA-2188/1/E5f-L0014, Letter by Josef Fröhler to all OBLs with the attached transcript of a speech with the title "The efficiency output contract in Norway", held by Fröhler on 7 February 1943 in Oslo, 23.2.43.

142 On the negotiations with *Einsatzgruppe West* in 1942, see Bilfinger Archive, A 367.

formulation of official service and price specifications that would become the core of the new contract.

In Norway, too, the OT relied heavily on the expertise of private industry during the negotiations on the *Leistungsvertrag*. Drafting the service and price specifications required experience and knowledge about the Norwegian conditions. *Einsatzgruppe Wiking* was forced to admit that it did not possess the necessary expertise and therefore commissioned private industry to draft service and price specifications for the construction of concrete bunkers and bunkers blasted into rock, as well as for the construction of roads, railways, tunnels, piers and barracks. On 31 October 1942, Wigru Bau's liaison office in Oslo contacted construction firms that had gained first-hand experience in Norway and asked them to submit specifications by 5 December. The drafts would then be discussed and harmonised at the firms' monthly meeting on 7 December in Oslo.[143]

The specifications for EW's efficiency-output contract differed from those in the occupied West in one absolutely crucial point: they neither contained any fixed prices for the various work steps, nor did they specify how much time the execution of the work steps should take.[144] The specifications only provided a guideline for the firms. Factors such as soil conditions, the availability of infrastructure and supplies and the climate were factors that were far too diverse in Norway's far-stretched territory for mandatory prices to be fixed. Contractors were still to set their prices for the various items individually. They then negotiated the overall prices for every project with the respective OBL and EW's Department for Contracting and Pricing.[145] Thus, the only advantage of the "official" specifications for various types of works was that they made comparisons easier and helped EW to detect particularly overpriced items. Specifications can be found in the archives at least for the construction of fortifications and railways.[146] Whether price and service specifications existed for other activities also is uncertain.

The German construction industry steered clear of dictating the prices set by the state in occupied Norway. EW still depended on private industry's expertise, and the latter continued to profit from information asymmetries after 1943. After several months in occupied Norway, and years in other regions of occupied Europe, it was primarily the firms who possessed the knowledge on how expensive the construction of different types of military infrastructure actually was. As mentioned above, the firms discussed their offers on their meeting on 7 December 1942 in Oslo,

143 BArch, R 13-VIII/265, Draft of the letter from Thierbach to construction firms regarding the service and price specifications for the efficiency-output contract in Norway, 31.10.42.
144 RAO, RAFA-2188/1/F3a-L0011, file Leistungsvertrag, Letter by Thierbach to Wigru Bau, Berlin, 1.1.43.
145 See the numerous price and service specifications attached to the contracts in RAO, RAFA-2188/1/E5e.
146 RAO, RAFA-2188/1/E5d-L0002, file Leistungsverzeichnisse.

1. Bunkers (concrete)
Straßenbau AG, Weimar
Schwerm & Lange, Cologne
Carl Brandt, Bremen
Arge Nordmark, Hamburg
Wayss & Freytag, Berlin
Diehl Hoch-Tief-und Betonbau, Essen

2. Bunkers (rock)
Straßenbau AG, Weimar
Hermann Milke, Soest
Boersch Strassenbau, Kassel
Siemens-Bauunion, Berlin

3. Road construction
Straßenbau AG, Weimar
Franz Bodmann, Göttingen
Boersch Strassenbau, Kassel
Braun & Fleckenstein, Ludwigshafen
Gaiser, Thiele & Nufer, Furschenbach

4. Railway construction and tunnelling
Philipp Holzmann, Hanover
Universale Hoch- und Tiefbau, Vienna
Grün & Bilfinger, Trondheim
Polensky & Zoellner, Munich
Peter Bauwens, Cologne
Carl Rose, Berlin

5. Piers
Hermann Möller, Kiel
Ohlendorff'sche Baugesellschaft, Hamburg
Heinrich Butzer, Dortmund
Henry Dehning, Bremen

6. Barracks
Arge Sachsen, Dresden
Christoph & Unmack, Bonn
Ostsee Holzindustrie, Stettin
Arge Pommern, Stettin

Figure 17: German construction firms asked to prepare service and price specifications for Einsatzgruppe Wiking's efficiency-output contracts, November 1942. Source: BArch, R 13-VIII /265, Note by Thierbach regarding the draft of price and service specifications for the new efficiency-output contracts in Norway, 3.11.42. Note the remarkable fact that the construction firm Carl Rose Tief- und Betonbau GmbH was asked to prepare a draft, even though the firm's owner had been arrested after audits in connection with the construction of the *Westwall*: BArch, R 2301/7187, fol. 41.

and finally, the head office of Wigru Bau in Berlin harmonised the proposals before sending them to EW. What looks like price-rigging on the part of the industry was not illegal according to current law, however. As long as the awarding authority had given its consent, construction firms were allowed to agree on their prices.[147]

When drafting the preliminary construction design, the firms multiplied the costs for the various work steps by the *Mittellohn*, that is, the average hourly wage

147 Bauch, 'Die Durchführung der Baupreisverordnung'.

earned by the workers on the construction sites. As we will see in the following chapter, the latter value was adjustable depending on how many workers of which nationality were actually working on the project. In doing so, the OT made allowances for the at times considerable fluctuation of the workforce in the occupied territories. With regard to the construction price, this meant of course that the price offer a firm submitted to the *Einsatzgruppe* was only a preliminary one. If the wage costs increased during the construction works, the project's price did, too. It is therefore important to note that the *OT-Leistungsvertrag* was not a fixed-price contract in the narrow sense of the term.

One important feature of the contract was the fact that the sum that the firms received for operating expenses, risk and profit was still calculated as a percentage of the production costs, not an absolute sum. The firms thus still profited from higher costs. The rates worked out for the *Wiking-Bauvertrag* in 1942 remained the basis for the mark-ups in the efficiency-output contract, too. They were taken from the same table, granting 6–8 percent of the production costs for operating expenses and 6.5 percent for risk and profit. The additional mark-up of 0.3 percent for Norway remained in effect, too. Moreover, the firms were allowed to add a 2.5 percent risk premium under the efficiency-output contract, where a worse performance could theoretically result in lower profits. Taken together, the firms added a fee of approximately 15–18 percent on the estimated production costs of the construction site. In a meeting in late December 1942, EW proposed to the firms that this fee should not be "hidden" in the total construction price, but visible as a distinct item on the invoice. Thus EW hoped bothersome checks by the German Court of Audit could be avoided. EW stuck with this proposal even when representatives of the industry countered (with remarkable frankness!) that they could still hide excess profits in the prices for the single work steps instead of the final mark-up. Apparently, EW was afraid that audits would yield unpleasant results not only for the firms, but for the *Einsatzgruppe*, too.[148]

Firms do indeed seem to have used fixed prices to cover price elements that were illegally high, as an example of Berlin-based Allgemeine Baugesellschaft Lorenz from 1944 shows. Lorenz was constructing a pier in Lunde (near Farsund in Vest-Agder) on the southern tip of Norway. As in the *Wiking-Bauvertrag*, German firms received a mark-up on the invoices of their subcontractors. Thus, the higher the bill of the subcontractor, the higher was the German firm's profit. To lighten the burden on its own administration, the OT gave the monitoring of local subcontractors over to the German construction firms. A letter from Lorenz's construction office in Lunde to the firm's headquarters in Berlin contains a revealing sentence: "[. . .] it became necessary to calculate fixed rates for all items in order to cover up the wages of the firm Karlsen, which contain [illegal, S.G.] bonuses." Lorenz's site manager

148 RAO, RAFA-2188/1/F3a-L0011, file Leistungsvertrag, Minutes on a meeting between representatives of EW, Wigru Bau and construction firms in Oslo, 1.1.43.

concluded, pleased, that this very special form of German-Norwegian cooperation "[. . .] redounds very much to our advantage [. . .]".[149]

As in the *Wiking-Bauvertrag*, firms were compensated for wear and tear of their machinery under the efficiency-output contract, and again, EW reimbursed them for the 10 percent Norwegian turnover tax.[150] The industry had managed to make modifications to several points of the contract's initial draft. The head of Sager & Woerner's office in Oslo, Karl Pfitzner, seems to have played a particularly influential role during the negotiations in early 1943.[151] Most importantly, the firms got their way with delivering the construction materials to the construction sites by themselves.[152] In this case, Gustav Klumpp from the liaison office in Oslo noted, the firms were allowed to add a fee of approximately 15–18 percent to the prices of building materials. This was an additional way of earning money and constituted "[. . .] a considerable reserve for the entrepreneur."[153]

EW officially introduced the *OT-Leistungsvertrag* as of 1 April 1943, first and foremost for the construction of fortifications. Here, a considerable part of the works consisted of rock blasting and excavation, activities that depended less on construction materials supplies and thus could be projected more easily for the firms under wartime conditions.[154] However, as had happened one year before with the *Wiking-Bauvertrag*, firms complained that they had actually not received the new contract and the price and service specifications in time.[155] Most of the efficiency-output contracts were in fact signed months later and entered into force retroactively.[156] Even the abovementioned strategy reported from D.B.H.G. in the late 1930s of simply submitting the invoices after works were finished and calling them tenders, can be observed in Norway, for example in the case of the Norwegian

149 LAB, A Rep. 250-05-13, Nr. 149, Letter by Allesch, Allgemeine Baugesellschaft Lorenz, Lunde construction site, to the firm's headquarters in Berlin, 29.10.44. The local representative of the OT was informed about this circumvention of regulations: see below, chapter V.3.d, fn. 329.

150 RAO, RAFA-2188/1/E5f-L0014, *Einsatzgruppe Wiking*: Guidelines for the price formation under the efficiency-output contract in Norway, 5.3.43.

151 RAO, RAFA-2188/1/F3a-L0011, file Leistungsvertrag.

152 Cf. the different drafts of the *OT-Leistungsvertrag* and the industry's modifications between early January 1943 and the version of 6 February 1943 in: RAO, RAFA-2188/1/F3a-L0011, file Leistungsvertrag.

153 RAO, RAFA-2188/1/F3a-L0011, file Leistungsvertrag, Report by Klumpp on the transition to the efficiency-output contract in Norway, 29.10.43.

154 We know, however, that even in late November 1943, no firm participating in the construction of coastal batteries and fortifications under ABL Liinahamari, one of *Einsatzgruppe Wiking*'s remotest construction sites in the north, had switched to the new contract yet: RAFA-2188/1/F3a-L0009, file Vertrauensmänner, Letter by construction firm Gottfried Hallinger, *Einsatz Wiking*, to the Oslo liaison office of the German building trades, 30.11.43.

155 ThHStAW, Straßenbau AG Weimar Nr. 139, fol. 33, Korffmacher, Strabag Weimar, to OBL Süd, Kristiansand, 11.4.43.

156 See for example RAO, RAFA-2188/1/E5e.

contractor A/S Betongbygg.[157] However, it appears that EW did manage to switch to the new contract type at least in the fields of fortifications and road construction. By autumn 1943, most German firms were working under efficiency-output contracts in those fields.[158]

For many other types of works, however, the transition to efficiency-output contracts came much later or even had to be called off completely. As early as February 1943, EW's Josef Fröhler doubted that it would be possible to introduce the new contract for the construction of piers and on the railway construction sites.[159] In the following years, too, the efficiency-output contract was never introduced comprehensively.[160] In the case of railway construction, the *Einsatzgruppe* even negotiated on a modified *Wiking-Bauvertrag* in spring 1943. For the firms working on the *Nordlandsbane*, working under an efficiency-output contract was out of question, and thus they submitted their demands regarding a modified cost-plus contract to Wigru Bau in late April. On 29 April 1943, the organisation then met with representatives of the OT in Berlin to come to an agreement. The modified contract addressed the main problem on the Northern Norwegian construction sites, which was EW's inability to guarantee the firms sufficient numbers of forced labourers. Therefore, the fee for operating expenses, risk and profit was no longer based on the wage bill of all workers on the site, but on those factors that the firms themselves could influence. According to the supplementary agreement to the contract, the railway construction firms were to receive a mark-up of at least 9.5 percent on the fourfold sum of their German employees' salaries, the wages of their German workers (both including social contributions) and the reimbursement for their machinery.[161] The effect of this was that the firms were better able to predict how much they could expect to earn on the *Nordlandsbane*. Subsequently, EW made a last effort to replace cost-plus contracts in railway construction in early 1944, but it seems that the conditions in the north were still too insecure for the firms to accept an efficiency-output contract.[162]

---

157 RAO, S-3138/0026/Db-L0005, dommer 75/45, trial against A/S Betongbygg, Kristiansand, document 4, Report by controller Kurdøl, 7.7.45.

158 RAO, RAFA-2188/1/F3a-L0011, file Leistungsvertrag, Report by Klumpp on the transition to the efficiency-output contract in Norway, 29.10.43.

159 RAO, RAFA-2188/1/E5f-L0014, Letter by Josef Fröhler to all OBLs with the attached transcript of a speech with the title "The efficiency output contract in Norway", held by Fröhler on 7 February 1943 in Oslo, 23.2.43.

160 Ingulstad, 'Regulering', 263.

161 RAO, RAFA-2188/1/G2b-L0023, Undated copy of a note on a meeting between representatives of Wigru Bau and OT in Berlin on 29 April 1943.

162 In 1943, EW had plans to introduce the efficiency-output contract on the *Nordlandsbane* as of 1 May 1944. Grün & Bilfinger wanted to prevent this at any cost, as the preconditions for this type of contract were not met in Northern Norway. The firm lacked skilled German workers in particular: Bilfinger Archive, A 14, Report by the Hamburg branch office for 1943, 31.12.43. The firms seem to

In light of increasing transport problems and Soviet troops already having crossed the Norwegian border, EW's Department for Contracting and Pricing realised in autumn 1944 that cost-plus contracts were sometimes the better – because more flexible – choice, especially in Northern Norway.[163] Around the same time, an overview from OBL Alta's contracting office listed several firms with which no agreement had been reached and which were thus working under cost-plus contracts.[164] In Southern Norway, too, EW does not seem to have been able to force construction firms into efficiency-output contracts against their will. In September 1944, the German firm Hermann Wegener refused to operate a quarry at the Gardermoen airport under such a contract. Instead, EW agreed a cost-plus contract with the firm.[165] Hence, despite EW's general preference for the latter, cost-plus contracts and efficiency-output contracts continued to coexist in occupied Norway until the end of the war.

By and large, however, the construction firms were not against the introduction of the new efficiency-output contract in principle. This was due to the fact that EW's proposal satisfied additional motives that were important to the firms. Most importantly, the contract guaranteed that the entrepreneurs retained full and exclusive control over their firms. The OT had no right to split up a firm or withdraw parts of its permanent workforce.[166] The same applied to machinery. In spring 1943, Steffen Sohst pressed for the signing of the new contract with the OT, because "[. . .] the air force in Drontheim and Andalsnes [sic] is still using our tugboats and barges for other works, something that would stop immediately when taking over works under the efficiency-output contract."[167] Secondly, EW reassured the firms that they would be compensated for additional costs in the event of unforeseen changes or delays

have prevailed in this conflict. However, some firms that were blasting tunnels on the *Nordlandsbane* worked under efficiency-output contracts from the summer of 1943 onwards. From EW's point of view, however, these works were not "railway construction" in the narrow sense, but rock cavern construction. Accordingly, EW seems to have used the efficiency-output contracts developed in the Fortifications Department: RAO, RAFA-2188/1/E5d-L0003, file Felshohlbau, Survey on rock cavern construction by Dr. Wiedemann in the vicinity of OBL Drontheim and OBL Mo i Rana between 20 July and 1 August 1943.

163 RAO, RAFA-2188/1/E5a-L0029, Letter by Fröhler to *Einsatz Polarbereich*, Narvik, regarding the introduction of the efficiency-output contract for the firms Arge Metze & Leiterer/Strauss and Gebr. v. d. Wettern, 20.10.44.

164 RAO, RAFA-2188/1/G2b-L0023, Undated list by OBL Alta on the status of works in contracting, presumably from September 1944.

165 RAO, RAFA-2188/1/H1b-L0014, file tysk korrespondanse, Letter construction firm Hermann Wegener, *Einsatz Wiking*, to its head quarter in Adelebsen, 13.9.44. Cf. the copy of a note by *Einsatzgruppe Wiking* on the agreement with Hermann Wegener, 9.9.44, in: ibid.

166 RAO, RAFA-2188/1/E5f-L0014, Letter by Josef Fröhler to all OBLs with the attached transcript of a speech with the title "The efficiency output contract in Norway", held by Fröhler on 7 February 1943 in Oslo, 23.2.43.

167 RAO, RAFA-2188/1/H1a-L0010, file Schriftwechsel Kiel No. 2, Letter by Steffen Sohst's office in Bergen to the firms' headquarters in Kiel, 6.4.43.

that the firms were not responsible for.[168] Josef Fröhler also explicitly flagged the option to return to a cost-plus contract if conditions should change fundamentally.[169] Thirdly, as convenient the old cost-plus contracts may have been, the firms had to submit compilations of their costs to the OBLs to get their expenses reimbursed. Through this, the firms feared, the organisation would over time gain insight into the actual production costs and their price formation. In other words, they would lose the advantage of information asymmetry.

Finally, the German construction industry did not see their profits threatened by the new contract. The *OT-Leistungsvertrag* rewarded higher efficiency and in this respect was no different from a classical fixed-price contract. In order to earn more money, the firms had to raise the work pace on the construction sites. But as long as the contract included additional risk premiums, which the *OT-Leistungsvertrag* in fact did with its 0.3 percent mark-up for Norway and the 2.5 percent risk premium, most firms were not afraid of the challenge.The big players in the industry in particular saw the advantages of an environment that would be more competitive than under the cost-plus regime as it allowed them to outrival smaller competitors. Jochen Streb already concluded in the case of construction firms under the OT's *Einsatzgruppe West* that they "had obviously fast recognised that the economic risk of building standardised bunkers and gun emplacements in occupied countries could be more than compensated by a high-risk premium [. . .]."[170] Accordingly, statements of German construction firms document that they indeed preferred the *OT-Leistungsvertrag* as soon as they deemed the risks calculable. Strabag Weimar stated in October 1942 that they "would be most grateful" to work under this contract as long as EW involved the firms in its formulation and took the difficult conditions in Norway into account.[171] Gelsenkirchen-based *Tiefbau* firm Gottfried Hallinger, working on the heavy coastal battery *Peter* (also called *Suomi*) in the vicinity of Petsamo, reported in late 1943 that it saw no reason why the efficiency-output contract should not be introduced even under the region's sub-polar

168 Ibid. See also: RAO, RAFA-2188/1/F3a-L0011, file Leistungsvertrag, Note on a meeting between Fröhler and representatives of several construction firms in Berlin, 8.1.43.
169 Ibid.
170 Streb, 'Negotiating Contract Types', 375. Already at the *Westwall*, the German Court of Audit had noticed "considerable excess profits" under fixed-price contracts, because the firms included every imaginable risk in their prices. Put in the terms of PA theory, they realised high information rents: BArch, R 2301/3007, Memorandum of the President of the German Court of Audit on the Reich budget account for the budget year 1939, 19.4.43, fol. 261.
171 ThHStAW, Straßenbau AG Weimar Nr. 109, fol. 154–61, Statement by Korffmacher on works under efficiency-output contracts, Stavanger, 13.10.42. Another example is the construction firm Steffen Sohst, working in Western Norway, which already indicated in June 1942 that it would prefer an efficiency-output contract: RAO, RAFA-2188/1/H1a-L0009, file Schriftwechsel Kiel No.1 [forts.], Letter Steffen Sohst, Kiel, to its office in Bergen, 11.6.42.

conditions.[172] Similar positions are documented in the case of the OT's *Einsatzgruppe Italien*, where the construction firm Hermann Klammt pushed for the introduction of an *OT-Leistungsvertrag* in the summer of 1944.[173] At the time, the firm was already working under this contract in Norway and Denmark. In his argument, Hugo Klammt, the firm's owner, appealed to the spirit of the true entrepreneur who did not fear the challenges and risks of being fully responsible for a construction project: "Someone who possesses neither the skills to overcome such unforeseen challenges, nor the courage to put himself and his wealth at stake, should not become an entrepreneur, but a civil servant, and if he is technically talented, a building officer."[174] Admittedly, Klammt's preference for the efficiency-output contract had some rather mundane motives, too. He assumed that monitoring agencies would have no choice other than to accept additional profits from increases in productivity under this type of contract as "justified".[175]

Internal calculations from construction firms survive sporadically in the archives and give an impression of how information asymmetries translated into excess profits. In 1943, EW had commissioned Strabag Weimar with two tunnelling projects in Byberg, southwest of Stavanger. The firm was to blast large chambers for gun emplacements into the hillside. From 12 April 1943, the works were conducted under the new *OT-Leistungsvertrag*. The fixed price at Byberg was calculated as follows: based on the *Mittellohn* earned by the workers on the building site, Strabag Weimar was allowed to add 16.78 percent as a fee for operating expenses, risk and profit. The sum was then multiplied with the estimated time the blast operations would take. Finally, 20 RM per cubic metre were added for material costs, mainly explosive agent and fuel. The information asymmetry in the company's favour consisted of the fact that only Strabag Weimar could realistically assess how much time it would take to blast one cubic metre of solid rock. In the negotiations leading to the fixed-price contract of April 1943, Strabag Weimar had convinced the OT authorities that the workload at Byberg would amount to 55 hours per cubic metre. This was a vast exaggeration. As the internal Strabag documents reveal, it took the company just half that time to finish one cubic metre between April and September 1943. After renegotiations, EW and Strabag Weimar agreed on a revised service and price specifications on 6 October 1943, in

172 RAFA-2188/1/F3a-L0009, file Vertrauensmänner, Letter by Gottfried Hallinger, *Einsatz Wiking*, to the Oslo liaison office of the German building trades, 30.11.43. Although on Finnish territory, ABL Liinahamari and the works on the Peter battery were subordinate to OBL Kirkenes and thus *Einsatzgruppe Wiking*.

173 On this, see the conflict between Hermann Klammt and the liaison office of the German building trades for *Einsatzgruppe Italien*, which did not support the replacement of the existing cost-plus contracts: BArch, R 13-VIII/85, fol. 29 ff.

174 BArch, R 13-VIII/85, fol. 69–71, Letter Klammt to Knecht, Wigru Bau, 30.6.44.

175 Ibid., Letter Klammt to Trapp, liaison office of the German building trades for *Einsatzgruppe Italien*, 10.8.44.

which the hours/m³ rate and hence the fixed price were reduced significantly. However, the rates were still high enough for Strabag's site manager Korffmacher to be perfectly pleased with the outcome of the negotiations.[176] The actual profit was even higher, since Strabag's material costs in fact amounted to 17.46 RM per cubic metre, whereas EW paid 20 RM.

Table 19 shows the striking discrepancy, especially from April to October 1943, between the prices on which Strabag Weimar and EW agreed and the actual costs of the works in Byberg.[177] In this period, the prices for the three construction projects in Byberg exceeded the real costs by 101 percent, 115 percent and 703 percent, respectively. Prices in the Far North were significantly higher. In June 1944, OBL Alta still regarded 188 RM per cubic metre in rock cavern construction (*Felshohlbau*) as acceptable.[178] Of course, we must not forget that it was the efficiency-output contract's express intention to reward the entrepreneur's efforts to increase efficiency by allowing for correspondingly higher profits. However, the presented figures leave no doubt that this policy got out of hand during certain phases, not to mention that it thwarted the general National Socialist price stop policy.[179]

Table 19: Approved fixed prices (in RM, including 20 RM material costs) and hours of work, per cubic metre, Byberg building site, April and October 1943, versus actual costs (including 17.46 RM material costs) and hours of work, per cubic metre, 12 April – 20 September 1943.*

Building site	Type of work[1]	Hours/m³ approved by OT, Apr. 43	Price per m³, approved by OT, Apr. 43	Hours/m³ approved by OT, Oct. 43	Price per m³, approved by OT, Oct. 43	Actual hours/m³ needed	Actual costs per m³
Byberg I	a	55	109.70	28	66.00	23.6	54.46
Byberg I	b	55	445.50	33	74.10	31	55.46
Byberg II	a	55	109.70	28	66.00	22.6	50.96

[1]a = "*Gang- und Kammersprengungen*", b = "*Schleppschachtsprengungen*".
*ThHStAW, Straßenbau AG Weimar Nr. 126, fol. 148 ff., Service and price specifications for the construction of fortifications in Norway for the contract of 1 April 1943, 15.10.43.

176 ThHStAW, Straßenbau AG Weimar Nr. 126, fol. 148 ff., Service and price specifications for the construction of fortifications in Norway for the contract of 1 April 1943, 15.10.43.

177 Note that the "actual costs" still contain the mark-up of 16.78 percent, that means that Strabag Weimar would have earned money even at this price.

178 RAO, RAFA-2188/1/G2b-L0023, OBL Alta to several of its ABL, 7.6.44.

179 Another example is Grün & Bilfinger, which was constructing a pier for the German navy in the harbour of Trondheim under a efficiency-output contract. The firm managed to undermatch the agreed prices by almost 20 percent for the production of reinforced-concrete piles and by almost 50 percent for driving the piles into the ground, while the production of one cubic metre of ferro-concrete slabs became c. 13 percent more expensive than planned. Below the line, Grün & Bilfinger seems to have been satisfied with the earnings: Bilfinger Archive, A 14, Building report up to 31 December 1943, 11.2.44.

After the introduction of the *OT-Leistungsvertrag*, EW put considerable effort into analysing its effects and gaining an idea of the real level of productivity in rock cavern construction. EW requested detailed information from the OBLs on their firms' productivity on a regular basis. EW also carried out two additional surveys in the summer of 1943 on bunker construction sites in Southern Norway and around Trondheim as well as on the *Nordlandsbane*, where some tunnelling firms were now working under efficiency-output contracts, to gain further information on the effects of the transition.[180]

In spring 1943, the *OT-Leistungsvertrag* had helped to increase the productivity in rock cavern construction and remedy the most extreme excesses of the cost-plus regime. One survey on the *Nordlandsbane* focused on the tunnelling firms. They had worked extremely inefficiently and their Norwegian workforce had earned extraordinary wages. On one construction site, after the new contract had been introduced the performance in tunnelling works increased from 150 h/m^3 to 26 h/m^3 while at the same time half of the former workforce were sent home.[181] In the following one and a half years, OBLs all over Norway had to gather detailed information from their firms,[182] and EW did indeed register significant increases in productivity between spring 1943 and autumn 1944. After sifting through the reports, EW estimated that the average time required for one cubic metre in rock cavern construction were up by 30.2 hours until the end of September 1943, by 22.4 hours between October 1943 and April 1944 and by 17.2 hours between May and July 1944.[183]

Besides eliminating the incentive to increase earnings by working slowly and inefficiently, which had existed under the old cost-plus contracts, a major reason for the increases in productivity was that the new contract rewarded learning effects. In the summer of 1943, EW's surveys still found that most German firms in rock cavern construction were not working according to accepted standards in the field. According to the *Einsatzgruppe*, courses for blasters, the provision of better drill steel and detailed instructions by the producer of the drill steel had increased performance.[184] Only under the new efficiency-output contract did these learning effects and the improvement of work routines actually lead to higher profits and were thus in the firms' own interest. In this respect, the *OT-Leistungsvertrag* was a success.

180 RAO, RAFA-2188/1/E5d-L0003, file Felshohlbau, Surveys on rock cavern construction by Dr. Wiedemann, conducted between 13 July and 19 July 1943 in Southern Norway and between 20 July and 1 August 1943 around Trondheim and in Northern Norway.

181 Ibid., Survey on rock cavern construction by Dr. Wiedemann in the vicinity of OBL Drontheim and OBL Mo i Rana between 20 July and 1 August 1943.

182 Ibid., Head of EW's Fortifications Department, Köbele, to all *Einsätze* and OBLs of EW's Fortifications Department, 21.6.43.

183 RAO, RAFA-2188/1/E1a-L0002, file 2132, Head of EW's Fortifications Department, Köbele, to all *Einsätze* and OBLs, 22.9.44.

184 Ibid.

Results are more mixed when it comes to cost reduction. The surveys and reports showed EW that the first efficiency-output contracts still led to excess prices. Very similarly to the situation a few years before at the *Westwall*, calculations had been based on the very low performances under the previous cost-plus contract regime. Hence, EW had to initiate multitudinous renegotiations in autumn 1943 to reduce prices. The renegotiations with Strabag Weimar that led to the reduced prices of October 1943 also have to be seen in the context of larger attempts on the organisation's part to gain an overview over the real production costs in the construction of fortifications.

As a comparison with Table 19 shows, the prices that Strabag Weimar and EW agreed on in October 1943 conformed exactly to the average value of 30.2 hours that seems to have been EW's state of knowledge at that point in time. But as Table 19 also shows, the firm was already much more productive then. Moreover, there are no indications of further renegotiations in the case of Strabag Weimar after October 1943, even though EW was generally aware of the continued increases in productivity. Obviously, the costs of monitoring the firms' performance and initiating several rounds of renegotiations would have exceeded the gains at some point.

Between 1933 and 1945, the state's difficulties in monitoring prices were intensified by the shift towards complicated *Tiefbau* projects and growing importance of large-scale engineering projects. This was especially the case for the OT with its focus on complex, singular projects. Price calculations in *Tiefbau* traditionally contained far more uncertain elements due to different soil conditions.[185] Furthermore, construction sites in the occupied territories were often scattered over large areas, making transport more unpredictable and expensive. Finally, exceptional projects like the railway in Northern Norway were difficult to standardise. All this contributed to the fact that information asymmetries were particularly high under the OT's wartime projects. After 1939, for example, the management of Allgemeine Baugesellschaft – A. Porr AG specifically pursued the strategy of abandoning residential construction and pure earth moving works and focusing on complex civil engineering works. The reasoning behind the decision was that in the latter field, it was easier to justify high prices to the awarding authorities based on highly complex calculations.[186]

The development described in this part of the chapter contributed to rising prices in the construction of fortifications in occupied Europe. In 1938, the German Court of Audit's delegate at the *Westwall* had already criticised that prices had risen from 65 to 200 RM per cubic metre of concrete within only a few years.[187] In March 1943, the OTZ calculated that the Organisation's fortifications would cost 200 RM in France, 220 RM in Denmark and 250 RM per m^3 in Norway in the accounting year 1943.[188] While the OTZ was able to keep the price in Denmark stable, at the end of the accounting year it

185 Plum, 'Abschlüsse', 7.
186 Matis and Stiefel, *Porr*, 417.
187 Cf. chapter III.2.b.
188 BArch, R 3/1808, fol. 179–80, letter by Panka, OTZ, to the OKH, 26.3.43.

assessed that costs in France had risen to 280 RM per m^3 and in Norway to 400 RM m^3.[189] The price explosion in Norway was explained by the fact that the OT had to build many smaller, separate installations, which often lay in remote regions.

It is impossible to say to what degree these rising prices translated into higher profits for German construction firms working under the OT in occupied Europe. While prices for construction materials used by the OT seem to have remained stable throughout the war, transport costs certainly played a paramount role in why construction in Norway became particularly expensive. Moreover, the conditions under which firms had to carry out their projects were of course much more dangerous and complicated than during peacetime. However, it seems fair to assume that construction firms at least did not earn less under the OT as the war went on.

As has been shown in this part of the chapter, the effort involved in controlling German construction firms threatened to get out of proportion even in rock cavern construction, where monitoring and measuring productivity as well as the formation of prices were relatively easy. And as the following section will show, problems were no smaller in other fields of EW's activities.

d Information Asymmetries and Industrial Self-Responsibility

In the course of 1943, EW also introduced efficiency-output contracts for many firms in the northernmost part of Nazi-occupied Europe. Under OBL Alta, German firms built fortifications, piers, roads and snow fences under the harsh conditions of a short, arctic summer and months of darkness during the winter. Although the OBL had its own contracting office under *Oberbau-Inspektor* Kulike, negotiations seem to have been centralised for the most part under Josef Fröhler in EW's Department for Contracting and Pricing in Oslo. In early October, Fröhler took on the arduous journey to Alta – around 1,300 kilometres as the crow flies – to negotiate on prices for road and snow fence construction. Kulike was dissatisfied with the results of Fröhler's negotiations and he laid his concerns down in two extensive notes. They give a striking insight into how the OT concluded contracts in the occupied territories in the midst of the war.

The negotiations took place on 8 October 1943. Fröhler, who, according to Kulike, "appeared to be overworked", conferred with several firms in a marathon meeting that took the whole day, lasting till after midnight. Kulike complained that he had tried to convince his superior several times using detailed calculations that the prices claimed by the firms were far too high. He estimated, for example, that the offer of Allgemeine Baugesellschaft Lenz & Co, Berlin, was 100 percent over the prices of the construction firm Krull, even though the latter was working under a cost-

189 Ibid., fol. 189–92, 237, 290–91.

plus contract. In some cases, firms agreed to reduce their prices. According to Kulike, however, Berlin-based road construction firms Frericks and Zech & Voigt had initially claimed such "absurdly high" prices that they were able to reduce them by up to 40 percent without further ado and still realise excess profits. Kulike thought the reduced prices were still outrageous. A representative of Frericks himself openly admitted that the firm had realised a profit of 100,000 RM on a construction site with a project value of 260,000 RM. Fröhler, however, perceived his colleague's objections and post calculations of his co-worker as "indecent towards the firms". Obviously working under high time pressure, he waved through more or less all offers that the firms presented to him.[190]

The firms took advantage of the specific conditions in occupied Norway when they drew up their offers. First, the German firms present in Norway had scarcely any competition. Inviting offers from firms outside Norway, which had no knowledge of the actual conditions, was impracticable, not to mention the time and effort that would have been necessary to transport their equipment and workers northwards. Moreover, the number of firms that had the size and know-how to execute the demanding projects in Norway and that had some idle capacity in 1942–43 was probably rather limited. Under huge pressure to conclude efficiency-output contracts as soon as possible, EW had no realistic option other than to rely on the firms that were present in occupied Norway. This put the latter in a powerful bargaining position. Second, most contracts seem to have been signed months after the works had already started.[191] This minimised the firms' risk of losses and meant a considerable advantage during price negotiations. In the case of the abovementioned firms Frericks, Zech & Voigt and several others, which were building snow tunnels north of Alta, price negotiations likewise took place when a large part of the works was already finished. Their offers were thus in fact based on interim calculations and sometimes almost final accounts, as OBL Alta noted.[192] Third, EW's priority was speed and efficiency. Even though various officials in the ranks of EW were obviously in the picture about the massive information rents that the firms were realising under the *OT-Leistungsvertrag*, budgetary concerns were hardly ever raised. And even if EW had tried to regain control over the prices and excess profits, it is doubtful whether it would have been able to conduct the complicated and tedious price monitoring. The number of staff available for these tasks was simply too low.[193]

---

190 RAO, RAFA-2188/1/G2b-L0023, Two notes by *Oberbau-Inspektor* Kulike on price negotiations in Alta, 11.10.43 and 28.12.43.

191 See for example numerous efficiency-output contracts from OBL Kristiansand, which were effective from 1942–43, but were not signed before autumn 1944: RAO, RAFA-2188/1/E5e-L0001.

192 RAO, RAFA-2188/1/G2b-L0023, Note by *Oberbau-Inspektor* Kulike on price negotiations in Alta, 28.12.43.

193 Cf. chapter IV.2.c.

This is illustrated by a letter from Fritz Köbele to EW's Staff Department of 25 February 1943. Fritz Köbele was head of EW's Fortifications Department and was at the same time the superior of Josef Fröhler (Department for Contracting and Pricing) and Fritz Motsch (Department for Accounting). In other words, besides leading EW's most important field of construction activities, Köbele was responsible for the entire transition to the new *OT-Leistungsvertrag*, that is, its formulation, price negotiations, monitoring and accounting. For this task, he had at his disposal only three typists in the Fortifications Department, two in the Department for Contracting and Pricing (one of which was absent due to illness) and one in the Department for Accounting. Now he complained that the Staff Department was intending to withdraw even this last typist in the Department for Accounting. According to Köbele, the situation was so severe that it threatened the introduction of the *OT-Leistungsvertrag*.[194] The tight staffing situation is further illustrated by the fact that Josef Fröhler personally concluded all efficiency-output contracts in the summer of 1943 between Southern Norway and Narvik. He travelled along the coast in a small cutter and often only found time to sleep aboard the vessel on his way to the next construction site.[195]

EW was dependent on the construction firms not only while designing the *OT-Leistungsvertrag*, but also after its introduction. In 1943, EW and Wigru Bau agreed that the latter would appoint members of construction firms as Delegates for Pricing (*Vertrauensmänner für die Preisbildung*), one for every ABL. They usually came from one of the bigger construction firms working in the respective area.[196] "Regardless of their own person or firm", they were to mediate personally between the *Einsatzgruppe* and the construction firms.[197] They were to work in the field of contracting and pricing, assist EW when transferring firms, workers and machinery between construction sites or when firms and machinery were returned to Germany towards the end of the occupation.[198] Most importantly, they supported the firms in drawing up offers, since many firms lacked employees in their accounting offices. EW, however, lacked both the expertise and the personnel to assess whether these offers were fair. The *Einsatzgruppe* therefore asked the industry whether the Delegates for Pricing could also advise its local officers that checked the offers. What the OT was proposing here would indeed have been a new high point of industrial self-responsibility in public procurement: the firms were basically supposed to monitor themselves. It was the firms, however, that

194 RAO, RAFA-2188/1/E5a-L0003, Letter Köbele to Minuth, Staff Department, 25.2.43.

195 RAO, RAFA-2188/1/F3a-L0011, file Leistungsvertrag, Report by Klumpp on the transition to the efficiency-output contract in Norway, 29.10.43.

196 See for example the list of delegates as of May 1943 in RAO, RAFA-2188/1/E5a-L0009, Circular no. 18 by the Oslo liaison office of the German building trades, 17.5.43.

197 RAO, RAFA-2188/1/F3a-L0009, file Vertrauensmänner, Letter Klumpp to Dipl. Ing. Schneider, Wayss & Freytag, on the occasion of Schneider's appointment as Delegate for Pricing in the vicinity of OBL Bergen, 22.12.44.

198 Ibid.

refused to take over this task, as it would have been too time-consuming for them to visit and stay in touch with all contractors in the widespread country.[199]

This chapter has focused on contracts as a means to understand the relation between the *Organisation Todt* and the German construction industry. As has been shown, the German firms were to a high degree able to participate in the formulation of contracts and to safeguard their interests in the process. What is more, the Organisation relied heavily on the expertise and self-responsible work of private business. Given the fact that the German administration was becoming increasingly thinned out as the war progressed, drawing on the administrative resources of private industry was a necessity. At the same time, however, the OT's leadership always held the belief that industrial self-responsibility would yield better results than a system based on commands, threats and force could ever do. The fact that private business realised high information rents was not a failure of the OT's *modus operandi*, it was its very nature.

EW's strategy to increase productivity at almost any financial cost did in fact produce results. Besides the failed plan to build the Polar Railway, a project that never seems to have had the wholehearted backing of EW's leadership anyway, the organisation did to a large extent manage to deliver what it had been asked for. Most notably, the task given in 1942 to fortify the Norwegian coast with a line of bunkers and gun emplacements was by and large fulfilled when the occupation ended.[200] Detecting excess prices and prohibiting war profiteering, on the other hand, were not the priority of the OT in the occupied territories. Hence, when it comes to price monitoring, it seems justified to see the division between principal and agent not so much between the firms and the OT, but between an alliance of private industry and OT on the one side and German financial authorities like the Court of Audit or tax authorities on the other side.

When EW bent the rules because it prioritised speed while the Court of Audit struggled with controlling EW's accounting and expenses, the risk that deceptive business acts would be uncovered of course declined. Referring to the OT, Franz-Otto Gilles has already suggested that "[. . .] the entanglement of state and private administrations, with their unclear allocation of competencies [. . .], left the doors wide open for corruption and excess profits to become more or less the norm, facts that were hardly compatible with the traditional image of an orderly, thrifty and economical administration."[201] Nevertheless, Gilles judges the role of EW's pre-auditing activities as a positive example of the Court of Audit's work in occupied Europe. The *Wehrmacht* appreciated EW's efficiency. "This efficiency rested to a large extent on an effective administration and accounting control, which were

199 RAO, RAFA-2188/1/F3a-L0011, file Leistungsvertrag, Report by Klumpp on the transition to the efficiency-output contract in Norway, 29.10.43.

200 Sæveraas, 'Beton macht Geschichte', 114.

201 Gilles, *Effizienz ohne Ethik?*, 14–15.

supported substantially by the '*Vorprüfung*' [. . .]."[202] Gilles' verdict rests mostly on the judgement of Norwegian historian Helge Paulsen.[203] In the light of what has been said throughout this thesis and based on documents from the OT archives in Oslo, however, the degree to which the Court of Audit really managed to keep tabs on the *Einsatzgruppe* and its entrepreneurs seems questionable. For most of the occupation, the pre-auditing office under Fritz Motsch consisted of no more than three to five auditors for the entire country.[204] Once again, the state had to appeal to private industry's decency and self-responsibility, as *Einsatzgruppe West* did in a circular of 19 February 1943, which can be found in Motsch's papers and arguably also reflected the position of EW:

> The German Court of Audit and the pre-auditing office of the OT have complied with a maximum degree of simplification of the accounting process, but highlighted specifically that the entrepreneurs have to be aware of their self-responsibility more than ever when drawing up their invoices. A significant simplification of the accounting process at the OBLs is possible when the audit of firms whose invoices were deemed factual and calculative correct in the past is limited to spot checks.[205]

One crucial problem for the auditors was that they could not keep step with the ever-changing structure of EW. In March 1944, the pre-auditing office had to audit several OBLs in Northern Norway that did not even exist anymore. Their firms had been relocated to other construction sites long ago.[206] Finally, after August 1944, the combing out of the German administration for soldiers forced the Court of Audit to abandon pre-audits altogether.[207]

If productivity and speed were the OT's top priority, the question was of course how to reach this goal. Facilitating learning effects and the acquirement of know-how have been mentioned. However, the most important thing was to raise the productivity of the workforce and, in particular, to urge on the forced labourers, often beyond the point of total exhaustion. The firm's role and economic motives in this regime of forced labour form the topic of the following final chapter of this thesis.

---

202 Gilles, 16.
203 Helge Paulsen, 'Norwegen', in *Verwalteter Beutepartikularismus. Finanz-, Verwaltungs- und Wirtschaftskontrolle und nationalsozialistische Besatzungspolitik in den von Deutschland besetzten Gebieten: Ein Tagungsbericht*, ed. Franz-Otto Gilles and Gerhard Otto (Berlin: Freie Universität Berlin, 1991), 37–44.
204 RAO, RAFA-2188/1/E5a-L0007, file Prüfdienst OT. Zentrale, Note by Department V5 (pre-auditing office) at OTZ, 19.6.43; ibid., Letter by Nolte, head of the Department V5 (pre-auditing office) at OTZ, to Berger, Department P5, 24.3.44.
205 Ibid., Circular by *Einsatzgruppe West* to all OBLs and firms, 19.2.43. In fact, Willi Henne mentioned after the war that the EW had only performed random spot checks of the firms' invoices: Henne, 'Organisation Todt in Norwegen', pt. II, 4.
206 RAO, RAFA-2188/1/E5a-L0007, Letter Motsch to Nolte, head of the Department V5 (pre-auditing office) at OTZ, 10.3.44.
207 Paulsen, 'Norwegen', 44.

3 Labour

Labour was one of the German war economy's major bottlenecks. As for any other business in wartime Germany, protecting its workforce from draft calls and gaining access to additional (forced) labour became prime motives for construction companies. In view of the rising inflation, this motive rather gained than lost importance over the course of the war. Accordingly, one of the incentives that the EW could offer its contractors was to give workers a protected status and provide the firms with additional labour.

Part a of this chapter will focus on these incentives before providing an overview over the various groups of workers under *Einsatzgruppe Wiking* in Norway between 1942 and 1945.

Part b and c will analyse OT's pay system and its implications for prices and entrepreneurial profits. For first time, the financial aspects of the deployment of Soviet POWs, the largest group of forced labourers in occupied Norway, will be investigated systematically. Studies on forced labour in the Third Reich have described the basic financial modalities around the work deployment of POWs.[208] However, special regulations existed for the construction sector, for firms working under the OT and for occupied Norway. Earlier findings can therefore not always be readily applied to the case of POWs working under EW. A major question will be to examine under which contractual conditions and to what degree German construction firms profited financially from the use of POWs. Furthermore, I argue in part c that the OT's attempts to increase productivity on its construction sites in occupied Europe increased the pressure on the workforce, rather than the firms. The organisation set incentives that motivated firms to raise the work pace on the construction sites and make additional profit from the work of forced labourers in particular.

Finally, part d will describe OT's system of performance factors. Only cursorily mentioned by earlier research, I regard this system as a pioneering attempt to make the exploitation of forced labour manageable under the unpredictable conditions of war and occupation. Once more, the crucial role of the construction firms in designing and implementing this system will be highlighted.

---

208 Hans Pfahlmann, *Fremdarbeiter und Kriegsgefangene in der deutschen Kriegswirtschaft 1939–1945* (Darmstadt: Wehr und Wissen, 1968), 180–82; Christian Streit, *Keine Kameraden. Die Wehrmacht und die sowjetischen Kriegsgefangenen 1941–1945* (Stuttgart: Deutsche Verlags-Anstalt, 1978), 214–15, 280–85; Spoerer, *Zwangsarbeit unter dem Hakenkreuz*, 164–66; Rolf Keller and Silke Petry, eds., *Sowjetische Kriegsgefangene im Arbeitseinsatz 1941–1945. Dokumente zu den Lebens- und Arbeitsbedingungen in Norddeutschland*, Schriftenreihe der Stiftung Niedersächsische Gedenkstätten 2 (Göttingen: Wallstein Verlag, 2013), 24–26.

a Einsatzgruppe Wiking's workers

One of the major incentives for working under EW was to secure workers from being drafted. The construction industry had earned good money during the boom of the previous years and could probably compensate a period with substantially reduced activities, but idle workers would lose their protected status and be lost to the *Wehrmacht* or other war-related parts of the industry. In the months before Strabag Weimar managed to take over works for EW in occupied Norway in early 1942, the firm's future site manager in Norway, Korffmacher, suggested "hiding" workers in one of Strabag's workshops in Romania, for example.[209]

Managers, skilled workers and salaried employees in particular with their experience and expertise could hardly be replaced with forced labourers. Kiel-based construction firm Steffen Sohst reported in May 1942 that it had successfully fought against their home region's labour office, the *Gauleiter* and GB Bau's regional representative in order to free up their workers for the mission in Norway.[210] *Einsatzgruppe Wiking*, too, was interested in occupying firms that had sufficient numbers of competent employees at their disposal. As Steffen Sohst's site manager in Bergen was able to report after a meeting with Willi Henne and Fritz Köbele in June 1942, the two officials had enquired whether he was in danger of being drafted into the *Wehrmacht*. They also offered to advocate that the firm's owner Steffen Sohst, who was serving at the eastern front at the time, was exempted from military service.[211] The German construction industry managed to keep the number of "owners, managers and salaried workers" relatively stable throughout the war. Labour statistics registered 59,600 men and women as working for the member firms of Wigru Bau in May 1939, and in July 1944 there were still 52,117 (87 percent). The picture was bleaker among construction handwork firms, which were more frequently affected by closedowns. Here, the numbers were 165,452 in June 1939 and 109,073 (66 percent) in May 1943.[212]

Accepting orders from EW not only secured the companies' existing workforce, but promised access to additional, allocated workers. The machinery investments in the previous years and the fight for their skilled workers would have been meaningless if construction companies now lacked the hands to operate their construction sites profitably. EW's promise to provide these additional workers thus became a major incentive for firms. By 30 June 1942, firms mobilised for EW's fortifications

209 ThHStAW, Straßenbau AG Weimar Nr. 105, fol. 505–6, Note by Korffmacher on a consultation with *Bauoberinspektor* Dölle, 3.1.42.

210 RAFA-2188/1/H1a-L0009, file Schriftwechsel Kiel No.1 [forts.], Letter by construction firm Steffen Sohst, Kiel, to construction site manager Behrens, Bergen, 9.5.42.

211 RAFA-2188/1/H1a-L0009, file Schriftwechsel Kiel No.1 [forts.], Letter by Steffen Sohst's site manager in Bergen to Steffen Sohst, 24.6.42. There is no evidence, however, that they were successful.

212 *The Effects*, 236–37. Figures are for the territory of pre-war Germany.

programme had brought 2,454 of their own regular workers to Norway, while EW's Supply Staff had provided them with additional 2,557 Germans. Further 1,920 German workers were already standing by to be sent to Norway.[213] After years of fierce fighting for German workers, the labour offices all over the Reich were now combing out construction sites on behalf of EW and its firms. The GB Bau had even issued a command that claimants who hitherto had enjoyed protected status were obligated to release up to 5 percent of the workers on their construction sites in favour of the Norwegian mission.[214] Later in the war, Grün & Bilfinger's branch office in Hamburg noted in a report to its head office: "We repeat that our earlier standpoint, to accept only as many orders as workers are available is not correct. Rather, the allocation of workers results from the acquisition of orders and by that, the orders on hand increase."[215] As an example for the use of this strategy, the company explicitly referred to its construction site in Fauske, that is, EW's railway project in Northern Norway.[216]

When it comes to foreign labour, there could have never been any doubt among the German firms that the OT would provide them with forced labour. In Norway, this included conscripted Norwegians and other foreigners as well as POWs and concentration camp inmates. Generally, the German industry would have probably preferred to retain their German workforce over using forced labourers.[217] But the draft calls of the *Wehrmacht* defined the scope within which any entrepreneurial decision-making with regards to labour took place between 1939 and 1945. It is therefore not entirely wrong that, after the war, the allocation of forced labourers was depicted as an inevitable consequence of the war economy.[218] However, this is still only half of the story. Under the given circumstances, German construction firms proactively tried to gain access to forced labour by taking OT orders in occupied Europe. In 1942, this was one of the main incentives the OT could offer them. The firms knew that being hired in Norway would rely heavily on the use of forced labour and at no point did any of the German construction firms expect to carry out its orders merely with its German workforce.

The construction industry's proactive role can be documented throughout the war. During the first month of the war, when German authorities put Polish POWs to work on German farms in order to secure the first harvest of the war, authorities had already felt impelled to inform an obviously impatient industry that the deployment of POWs beyond agriculture was not yet possible.[219] In late September,

213 HHStAW, Abt. 485, 480, fol. 114–22, Report on the activities of Supply Staff Wiking for the time from 1 March 1942 to 30 June 1942, 15.7.42.
214 Ibid.
215 Bilfinger Archive, A 14, Report by the Hamburg branch office for 1943, 31.12.43.
216 Ibid.
217 Mark Spoerer, 'Profitierten Unternehmen von KZ-Arbeit? Eine kritische Analyse der Literatur', *Historische Zeitschrift* 268, no. 1 (1999): 70.
218 Knechtel, *Arbeitseinsatz im Baugewerbe*, 23.
219 *Die Bauindustrie* 7, no. 38 (1939), 1222.

rumours that the industry, too, could soon expect the allocation of POWs, had "provoked vivid interest, not least in the ranks of the construction industry."[220] In occupied Norway, logistical problems fuelled a severe conflict between construction firms and EW in autumn 1942, when the organisation was unable to provide the firms with the labour it had promised. Many complaints reached Wigru Bau from Northern Norway, particularly the Narvik area, where valuable machinery was standing idle due to the lack of auxiliary workers. Under the cost-plus principle of the *Wiking-Bauvertrag*, idle time did not directly result in losses, but there was nothing to be earned either.[221] The construction firm Heinrich Lenhard from Saarbrücken initially even refused to sign the contract for railway construction works in Northern Norway in October 1942. Among other things, Lenhard criticised that "[. . .] prisoners of war and other workers are allocated relatively late. The accounting method of the *Wiking-Bauvertrag* does not make allowance for this, thus, the capacity provided by the firms is not appropriately valued and paid for."[222]

In Southern Norway, too, Strabag Weimar requested the allocation of additional workers. The firm estimated in October 1942 that it could easily employ 20 Norwegians for every German worker on its bunker construction sites around Stavanger. At the present time, however, their German workers were either sitting around or executing simple tasks that Strabag's site manager deemed unworthy of skilled German workers: "By all means, our German personnel on the construction sites should and has to serve as supervisory personnel, not as workers."[223] In this respect, Strabag's opinion conformed completely with the "*Führer*" order of 8 September 1942, in which Hitler had forbidden the use of German construction workers in the occupied territories for heavy and simple tasks like breaking stones or carrying cement bags: "Such a deployment of German labour, whose withdrawal from its previous workplace meant in all cases a sacrifice for the homeland, is

220 *Die Bauindustrie* 7, no. 39 (1939), 1238.

221 Complaints by the firms Jos. Hoffmann & Söhne (Ludwigshafen) and Arge Bold-Doll (Achern/Sasbachwalden) in: RAO, RAFA-2188/1/F3a-L0002, Minutes of the monthly meeting between German construction firms and the Oslo liaison office of Wigru Bau on 7 November 1942 in Oslo, 8.11.42. Complaint by Arge Preusse-Weiss (Braunschweig) in: BArch, R 13-VIII/265, Letter to Einsatzgruppe Wiking, 26.10.42. See also: ibid., Letter Thierbach, Oslo liaison office of Wigru Bau, to Riedel, Wigru Bau, Berlin, 5.9.42.

222 BArch, R 13-VIII/265, Letter by the construction firm Heinrich Lenhard, Saarbrücken, to *Einsatzgruppe Wiking*, OBL Skipagurra, 22.10.42. Lenhard eventually signed a contract and worked on the *Nordlandsbane* between Mo i Rana and Fauske. As two plans from September 1943 and March 1944 show, EW in fact managed to allocate the agreed number of workers to Lenhard's construction sites in the following years: RAO, RAFA-2188/2/Hfa-L0027, map of the railway Mo i Rana – Korsnes, 16.9.43; RAO, RAFA-2188/1/E9a-L0029, map of the railway Mo – Fauske – Korsnes, 3.3.44.

223 ThHStAW, Straßenbau AG Weimar Nr. 109, fol. 150–53, Letter by site manager Korffmacher to Strabag's headquarters in Berlin, 14.10.42; ibid., fol. 162–66, Letter by site manager Korffmacher to *OT-Bauleitung* Stavanger, 13.10.42.

unjustifiable and moreover completely incompatible with the reputation of the German in the occupied territories."[224] In February 1944, Strabag Weimar very politely requested the allocation of 60 Soviet POWs for its construction site in Hundvåg.[225] When the firm Hermann Wegener took over construction works for EW on Gardermoen airport in autumn 1944, it was the firm itself that requested the allocation of ten *Ostarbeiter* (civilian forced labourers from the Soviet Union) on 8 September 1944.[226] The firm wanted to use more manual labour to reduce the wear and tear of its stone-crushing machinery. EW granted the request and the ten men were working for the firm in the following months, as the time sheets from the construction site prove.[227] The apologetic claims by Erhard Knechtel, according to which "a firm had no influence whether it was provided with German or foreign, voluntary or involuntary [labour, S.G.]" and "[. . .] practically no room for manoeuvre was left to the firms" are thus not supported by historical evidence.[228] Although it ultimately was the OT that assigned forced labourers to the firms, the allocation was usually based on previous negotiations and often on requests from the firms. Knechtel's report, published by the Federation of the German Construction Industry in 2000 as a reaction to the public debate on compensation for former forced labourers, neither reflected the state of research in the late 1990s, nor is it supported by the findings presented in this thesis.[229]

Throughout the occupation, scarce labour would remain one of the limiting factors for EW's construction projects, although the organisation at least managed to overcome the most severe problems that had bred the conflicts with the German firms in autumn 1942. While the number of EW's German workers remained relatively stable in the following years, for most of the time somewhere between 6,000 and 7,500, forced labourers were the backbone of the OT's construction activities in Norway (see below, Figure 18). Approximately 130,000 forced labourers from all over Europe were brought to Norway during the occupation, around 90,000 of which had to work on EW's construction sites. Among these were also women who mostly worked in camp facilities such as kitchens and laundries. In late January 1945, EW occupied 1,315 female workers in Norway.

---

224 Moll, *Führer-Erlasse*, 280–81.
225 ThHStAW, Straßenbau AG Weimar Nr. 147, fol. 682, Letter by Strabag Weimar, Stavanger, to BL Stavanger, 28.2.44.
226 RAO, RAFA-2188/1/H1b-L0014, file tysk korrespondanse, Letter construction firm Hermann Wegener, *Einsatz Wiking*, to its head quarter in Adelebsen, 13.9.44.
227 See the time sheets and invoices in: RAO, RAFA-2188/1/H1b-L0014, file tysk korrespondanse.
228 Knechtel, *Arbeitseinsatz im Baugewerbe*, 23.
229 In the preface of the report, Knechtel thanks Franz W. Seidler for "[. . .] the critical revision of the manuscript and for his encouragement to the publication of the report [. . .]". At the time, Seidler was already the target of criticism for releasing apologetic books with far-right publishers.

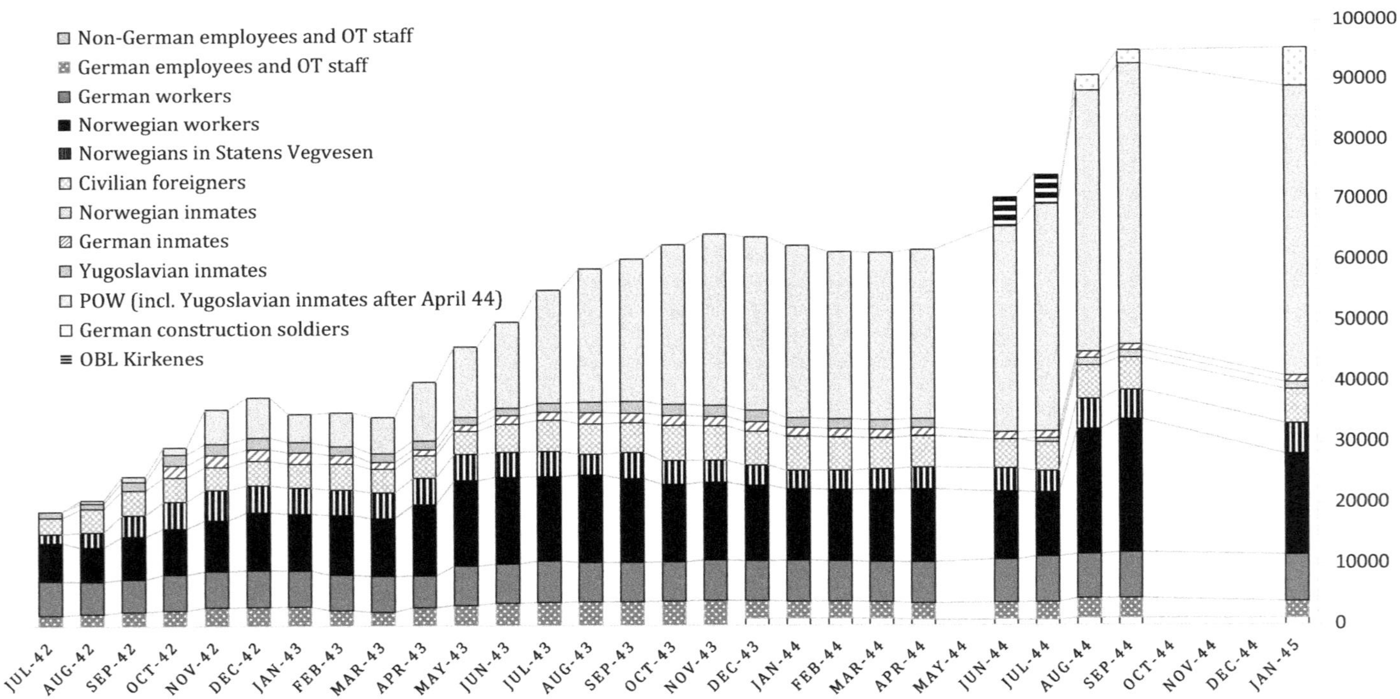

Figure 18: Workforce of Einsatzgruppe Wiking in Norway (from August 1944 with Finland), 1942–45. Source: appendix, Table A9.1.

Only 24 of them were German. The largest groups were Norwegians (950) and women from Russia and Ukraine (270).[230]

The workers' chances of survival to a considerable degree depended on their status in the National Socialist racial hierarchy. "Beyond the murder of the Norwegian Jews," Hans Otto Frøland has noted, "it is this extensive forced labour that ties Norway most clearly to German racial policy during the Second World War."[231] The numbers given in Figure 18 show the number of workers at work on EW's construction sites at certain dates. Hence the figure does not reflect numbers of workers who died. Accordingly, many more than the approximately 95,000 men and women registered in late 1944 had worked under EW in Norway at some point in time, whether they were Germans, Norwegians, or foreigners, whether forced or voluntary. According to a recent estimate by Gunnar D. Hatlehol, their number easily exceeded 100,000.[232]

Einsatzgruppe Wiking's projects put enormous pressure on the Norwegian labour market. Together with the *Wehrmacht's* construction activities, they are the reason that Norway became the only occupied country to never send a significant number of workers to the German Reich. Norwegians constituted a considerable part of EW's workforce. In the summer of 1943, they numbered almost 15,000. Furthermore, around 4,000 Norwegians worked for the Norwegian Public Roads Administration (*Statens Vegvesen*) on projects ordered by EW. The number of Norwegians increased massively in August 1944, when the organisation took over the air force construction sites. Altogether, up to 30,000 Norwegians may have worked under EW between 1942 and 1945. Most of them were conscripted. Acting under the pressure of the German occupiers, Norwegian authorities introduced increasingly extensive measures for the conscription of labour, resembling the provisions in the Reich. From autumn 1940, unemployed Norwegians could be conscripted for up to three months, and after March 1941, in 23 branches of the economy, the termination of a labour contract depended on the approval of the labour offices. From June 1941, employed Norwegians could also be conscripted for up to six months, and finally, in February 1943, for an indefinite time.[233]

It is important to note, however, that EW's labour statistics probably did not include tens of thousands of Norwegians who worked voluntarily on the construction sites. They were handicraftsmen or members of the infamous work gangs (*Akkordkolonnen*) that often worked for Norwegian subcontractors that had only

230 RAO, RAFA-2188/2/Hh-L0001, file 90-I, Chart A1d on the number of EW's staff members and workers, 25.1.45.

231 Frøland, 'Organisation Todt i Norge', 169.

232 Gunnar D. Hatlehol, 'Einsatzgruppe Wiking og kampen om arbeidskraften 1942–1945', *Historisk Tidsskrift* 97, no. 3 (2018): 240–55.

233 Gunnar D. Hatlehol, 'Tvangsstyringen av arbeidslivet under hakekorset 1940–1945: Diktat og kollaborasjon', *Arbeiderhistorie* 22, no. 1 (2018): 49–71.

been established to hire out workers to German firms.[234] In its fight for scarce labour, the OT ignored existing law, hiring and paying high wages to these more or less self-employed workers.[235]

EW's forced labourers included inmates of the Emsland camps, most of them German soldiers who had been convicted by military courts. Approximately 2,600 German inmates were sent to Norway, the first 2,000 arriving in September 1942. In the following years, however, around 1,000 were returned to Germany, most of them because they were too enfeebled to work. Of those who had to stay, hundreds perished.[236] In the course of the war, the OT together with the SS also organised the transport of more than 4,000 Yugoslavian camp inmates to Norway. Most of them were Serbian resistance fighters.[237] The death toll among these men was dreadful, and over 60 percent of them did not survive.[238] This is why the number of Yugoslavs actually working for EW never exceeded 2,000. In addition, a mix of civilian labourers from more than 20 nations worked on EW's construction sites. Major groups came from Denmark, Poland, France, Belgium, the Protectorate of Bohemia and Moravia and the Soviet Union. In addition, many Dutch sailors worked in the *Transportflotte Speer*. They are not included in Figure 18, since the *Transportflotte Speer* was an independent organisation.[239] As research on forced labour has repeatedly shown, workers' living conditions differed significantly depending on where in Europe they came from. Often, the transition from voluntary work to forced labour was fluid, for example when EW made it difficult for workers who had formally taken up work voluntarily to terminate their contracts and return to their home countries.[240]

In the following months and years, prisoners of war would become the biggest group of forced labourers in Norway. Besides 1,700 Polish POWs, the prisoners came from the Soviet Union. Their number grew particularly from early 1943 on and

---

234 Ingulstad, 'Regulering', 271.

235 It is probably impossible to quantify this group of workers. However, we can at least gain an idea of the scope of the phenomenon when we compare EW's labour statistics of September 1944 (that is, including all Norwegian workers of former *Wehrmacht* construction sites) with the survey of the Norwegian Statistical Office of February 1944. The former registered less than 30,000 Norwegian workers and employees on EW's construction sites. The latter survey, however, had estimated that almost 90,000 Norwegians were working on German construction sites in February 1944: *Statistisk-økonomisk utsyn*, 221.

236 Hatlehol, 'Norwegeneinsatz', 387–418. The exact number of German inmates that perished in Northern Norway is unknown. Hatlehol estimates their number somewhere between 358 and 647: Hatlehol, 401.

237 Hatlehol, 'Norwegeneinsatz', 261–66.

238 Trond Risto Nilssen, 'Jugoslaviske fanger i Norge under andre verdenskrig', in *Sotavangit ja internoidut. Prisoners of War and Internees. A Book of Articles by the National Archives*, ed. Lars Westerlund (Helsinki: Kansallisarkisto, 2008), 166–81.

239 RAO, RAFA-2188/F3a-L0012, file Korrespondez, Letter by Fröhler to the *NSDAP Auslandsorganisation* in Oslo, 16.2.44.

240 Hatlehol, 'Norwegeneinsatz'; Frøland, Hatlehol, and Ingulstad, 'Regimenting Labour', 14–15.

again in mid-1944, when EW took over the construction sites of the *Wehrmacht* and incorporated *Einsatz Finnland*. It can been estimated that around 90,000 Soviet POWs were brought to Norway during the Second World War and between 45,000 and 50,000 of them worked on EW's construction sites at some point in time between 1942 and 1945.[241] Between 10,600 and 10,700 of the Soviet POWs lost their lives on Norwegian soil.[242] A recent tentative estimate assumes that approximately 40 percent of them perished on construction sites under the responsibility of EW.[243]

b Einsatzgruppe Wiking's pay system 1942–43

When investigating what traces these various groups of workers have left in the archival material on EW and its contractors, questions concerning the workers' performance, the level of wages and the type of pay, that is, whether workers should be paid per hour or according to their output, stand out. As has been shown in the previous chapter, wages remained a crucial factor in EW's contracts and the basis for the firm's calculations of their profits. As is the case for the entire German war economy, gaining an overview over the numerous wage regulations for the various categories of workers under EW can be difficult and confusing.[244] For our purpose, however, it is sufficient to outline the main features of the OT's pay system and discuss what they meant for the construction firms.

All civilian workers were paid by the private construction firms, regardless of whether they were Germans, Norwegians or from other countries. The worker's earnings consisted of two parts. An hourly wage was paid in reichsmark from the German firm's office in Germany directly to the worker's German account. Wage accounting was particularly difficult in the case of the German workers: their wage depended on the standard wage rate of their home region in Germany, leading to the situation that workers doing the same job on the same construction site did not necessarily earn the same amount of money. A foreman from Pomerania earned less than an unskilled worker from a city like Hamburg, where wage rates were particularly high. Norwegians were of course paid in Norwegian kroner, while for all other civilian foreigners, the firms paid the wages in RM to special accounts from which

241 Cf. Hatlehol, 'Norwegeneinsatz', 214.
242 Soleim, *Sovjetiske krigsfanger*, 2009, 96.
243 Hatlehol, 'Norwegeneinsatz', 243. A further 3,000 Soviet prisoners of war drowned off the Norwegian coast when Allied aircrafts sank the M/S Palatia in October 1942 and the M/S Rigel in November 1944.
244 For occupied Norway, Gunnar D. Hatlehol's work provides the most comprehensive overview over the topic: Hatlehol, 'Norwegeneinsatz'. For the German Reich in general, the chapter on wages in Mark Spoerer's study on forced labour can serve as a starting point: Spoerer, *Zwangsarbeit unter dem Hakenkreuz*, 151–66. Furthermore, the early study by John H. E. Fried, already published in 1945, still provides valuable information: Fried, *Exploitation*.

the money was transferred via clearing to the workers' home countries. Table 20 shows the hourly earnings for various groups of workers in the first summer of EW's mission in Norway. It depended on the workers' qualifications and their position in the National Socialist racial hierarchy, resulting in a significant wage differential within the group of German workers, but most importantly, between the Germans and the foreign workers. For workers from Poland, even the low wages given in Table 20 were further reduced by the *Sozialausgleichsabgabe*, a discriminatory 15 percent tax.

Table 20: Hourly earnings in RM, including daily allowance and free food and accommodation, for unmarried civilian workers under *Einsatzgruppe Wiking* in Norway, summer 1942.*

	Germans (Hamburg)	Germans (Pomerania)	Norwegians	Danes	Other foreigners	Poles	Poles (GG)
Foremen	1.35	1.07	1.15	.	1.04	0.67	0.51
Skilled	1.31	1.06	1.04	1.21–1.44	0.93	0.56	0.45
Semi-skilled	1.29	0.99	0.97	.	0.86	0.49	0.32
Unskilled	1.10	0.92	0.82	1.04	0.79	0.42	0.20

*Wage rates for German workers: RAO, RAFA-2188/1/E5a-L0040. For Norwegians, according to the national Norwegian wage agreement of 1941: RAO, RAFA-2188/2/Hca-L0001; see also: Ingulstad, 'Regulering', 266–71. For Danes, other foreigners (including among others Dutch, Frenchmen, Italians or Poles from the annexed part of Poland) and Poles from the General Government: RAO, RAFA-2188/2/Haa-L0034, 'Guidelines' for the *Wiking-Bauvertrag*. For the conversion to RM, I have used the official, fixed exchange rates.

The second element of the pay was a daily allowance (*Auslösung*) that the workers received on the construction sites in Norwegian kroner. According to a RAM ordinance of 22 November 1941, it was 2.60 kroner per day for unmarried and 4.40 kroner per day for married workers.[245] Polish workers did not receive this allowance and even had to pay 2.50 kroner per day for accommodation and food. Danes who had signed a work contract before 1 April 1942 were an exception as they received an allowance of three kr. per day.[246] These allowances are also included in Table 20.[247]

In summer 1942, there was no specific OT pay scale for the so-called *Ostarbeiter*, that is, workers from the Soviet Union. They were paid according to tables published by the Council of Ministers for the Defence of the Reich (*Ministerrat für die Reichsverteidigung*) in June 1942.[248] Their wage was significantly lower than

245 RABl. 1941, I, 525–26.

246 See the printed "Guidelines" for the *Wiking-Bauvertrag* in RAO, RAFA-2188/2/Haa-L0034.

247 The daily allowance of the German and foreign workers has been converted to hourly rates assuming a worktime of eight hours. Category "Danes" includes the allowance of three kr. per day, also assuming a worktime of eight hours. Subtracted from the wages of the Poles are the 2.50 kr. per day, that unmarried Poles had to pay for food and accomodation.

248 RGBl. 1942, I, 419–24.

that of a German worker in a comparable position. An *Ostarbeiter* doing work for which a German would earn 8 RM per day only received 3.50 RM, of which 1.50 RM were deducted for accommodation and food.[249] In Norway, the infamous *Ostarbeiterabgabe*, an additional tax that German firms had to pay in order to prevent them replacing German workers with the cheaper *Ostarbeiter*, went partly to the German (75 percent) and to the Norwegian (25 percent) treasury.[250]

Beyond the differing wage rates, discrimination found expression in varying regulations for leave, overtime pay, performance bonuses, social contributions, taxes and the like.[251] For example, the labour offices supported German workers who had been conscripted based on the *Dienstpflichtverordnung* with an additional allowance plus a loyalty bonus after 12 months in the new firm.[252] These elements further increased the differences between Germans, Western and Eastern European workers. Finally, exchange rates that were fixed in favour of the Reich cheated foreign workers of parts of their earnings when they transferred them to their home countries.

Receiving pay that only remotely corresponded to their work effort was something the Soviet prisoners of war could only dream about.[253] The financial modalities of their deployment deserve closer attention as they became the largest group of forced labourers on EW's construction sites. Furthermore, the existing studies on Soviet POWs in occupied Norway have hardly touched upon the topic.[254]

In occupied Norway, all POWs were under the administration of the *Kriegsgefangenen-Bezirkskommandant Norwegen* (KBN), an institution established in August 1942, and were detained in the four *Stalags* or one of their subcamps.[255] The KBN placed the POWs at the disposal of EW, which again distributed them among the construction firms. The POWs never became employees of the firms, a contractual relationship only existed between the firm and the OT. The *Wehrmacht*

249 On this topic, see Spoerer, *Zwangsarbeit unter dem Hakenkreuz*, 151–66.

250 *Verordnungsblatt für die besetzten norwegischen Gebiete*, no. 9 (1942).

251 'Die arbeits- und sozialrechtlichen Bestimmungen für ausländische Arbeiter und Kriegsgefangene in der Bauwirtschaft nach dem Stande vom 1. Dezember 1942', *Die Bauindustrie* 10, no. 26 (1942): 690–91.

252 The two payments could amount to more than 40 RM per month, which added to the income of a conscripted German OT worker: RAO, RAFA-2188/1/E5a-L0040.

253 The following explanations deal with the Soviet POWs. Regulations for the Yugoslavian inmates, often just called "Serbs" in the sources, were identical: RAO, RAFA-2188/1/E5a-L0040, KBN to EW, Department for Contracting and Pricing, 20.3.43. Slightly different rules applied to Polish POWs, see RAFA 2188/1/E5f-L0005, file Durchschriften 1943–1944, Letter by Motsch, EW, to OBL Kristiansand, 29.9.43. For the German inmates from the Emsland camps, *Einsatzgruppe Wiking* paid 70 RM per month and inmate, via *Strafgefangenenlager Nord* in Alta, to the administration of the Emsland camps in Papenburg: RAO, RAFA-2188/1/E5a-L0040, telegram by Autenrieth, OBL Alta, to Fröhler, EW Oslo, 11.4.43; and ibid., telegram by OTZ, Berlin, to Fröhler, EW Oslo, 11.3.43.

254 Hatlehol, 'Norwegeneinsatz', 237–38.

255 Hatlehol, 214–16.

provided the construction sector with Soviet POWs at a particularly low rate. As the RAM had specified in July 1940, the work of POWs was valued at 60 percent (time pay) or 80 percent (piecework pay) of the standard wage of a comparable German worker. In construction, however, the working hour of an unskilled POW was worth only 0.33 RM.[256] The rate in occupied Norway was even lower, only 0.32 RM per hour and 0.46 RM for skilled POWs.[257] Under efficiency-output contracts, however, the work of POWs seems to have been uniformly registered in the time sheets and invoices at 3.60 RM per day. Of these marginal sums, only 0.20 RM per day were actually paid out to the Soviet POW, however, and not in cash to prevent escapes. While in the Reich, the sums were provided as "camp money" that could only be used within the camps to buy articles like soap or unrationed foodstuffs,[258] the POWs in occupied Norway received their pay in kind.[259] The *Wehrmacht* retained the rest of the wage, which meant that it was *de facto* lost from the POW's perspective.

As the war progressed, German authorities realised that they hardly could expect the POWs to increase their performance as long as the pay was so low. Out of economic, rather than moral considerations, the OKW agreed in May 1942 that the POWs could purchase articles from the camp magazines up to a value of 0.20 RM per day. The sums spent were deducted from the retained part of the POW's pay, which means that this particular measure did not lead to higher costs for the state.[260] In January 1944, the sum was raised to 0.35 RM. The allocation depended on the POW's performance.[261]

It has been argued that Soviet POWs were not generally cheaper than unskilled German workers when the significantly lower performance of the former is taken into consideration.[262] In the case of construction, and particularly for the OT in occupied Europe, this statement has to be qualified, The Reich Price Commissioner estimated in 1942 that the average performance of a Soviet POW in construction lay

---

256 Hans Küppers and Rudolf Bannier, *Einsatzbedingungen der Ostarbeiter sowie der sowjetischen Kriegsgefangenen*, Sonderveröffentlichung des Reichsarbeitsblattes (Berlin: Geschäftsstelle des Reichsarbeitsblattes, 1943), 128, 139.

257 RAO, RAFA-2188/1/E5f-L0005, file Durchschriften 1943–1944, EW to all *Einsätze* and OBLs regarding the wage accounting for POWs working under the OT, 2.10.43. These rates applied to the Yugoslavian inmates, too, and under the efficiency-output contract also to the Polish POWs. See RAFA 2188/1/E5f-L0005, file Durchschriften 1943–1944, Letter by Motsch, EW, to OBL Kristiansand, 29.9.43.

258 Spoerer, *Zwangsarbeit unter dem Hakenkreuz*, 165.

259 RAO, RAFA-2188/1/E1m-L0047, Head of the OKW to the Chief Quartermaster at the WBN, 11.5.42.

260 Ibid.; and ibid., Chief Quartermaster at the WBN to Stalag 303 and 330, 6.4.42.

261 Ibid., Letter Feuchtinger, EW, to all *Einsätze* and OBLs, 18.1.44.

262 Spoerer, *Zwangsarbeit unter dem Hakenkreuz*, 188; Johann Custodis, 'Employing the Enemy: The Economic Exploitation of POW and Foreign Labor from Occupied Territories by Nazi Germany', in *Paying for Hitler's War: The Consequences of Nazi Hegemony for Europe*, ed. Jonas Scherner and Eugene N. White (New York: Cambridge University Press, 2015), 78.

between 40 and 50 percent of a German worker's performance.[263] As we will see below, *Einsatzgruppe Wiking* used the estimate of 50 percent.[264] And indeed, the gross hourly wage of a Soviet POW under EW (0.32 RM) was 42 percent of that of an unskilled worker from Pomerania (0.73 RM). But while social contributions and insurance were approximately on the same level for both groups,[265] the German OT worker among other things received the additional daily allowance of 2.60–4.40 kroner,[266] bonuses for overtime work, work on Sundays and holidays, or an additional 0.05 RM per hour if he was a *Stammarbeiter*. At the end of the day, a Soviet prisoner of war on a construction site of EW was significantly cheaper than a German worker. To a lesser degree, this holds true for the civilian forced labourers, too.

Under EW's cost-plus contracts, however, the construction firms did not benefit from the cheaper work of the Soviet POWs, in the sense that the lower wage costs would have resulted in higher profit margins. The reason for this was that the firms initially did not pay for using the POWs. The KBN rented the POWs out to EW, which again distributed them between the firms. Since under the *Wiking-Bauvertrag* EW reimbursed all wage costs to the firms anyway and at the same time paid for the construction projects, it was unnecessary to establish a payment arrangement between EW and the firms. There was no monetary flow between EW and the firms regarding the Soviet POWs.[267]

Although the firms initially used the Soviet POWs for free, they still kept records of the hours that the prisoners worked on the construction sites. The first reason for this was that records were needed to know how much money EW owed the KBN for the use of the POWs.[268] The OBLs transferred the sums monthly to Stalag 303 in Lillehammer, regardless of where in the country the POWs were deployed.[269] These monthly payments only stopped in November 1944, when the KBN decided to

263 BArch, R 3/1446, fol. 23–27, Circular decree 13/42 by the Reich Price Commissioner regarding the deployment of POWs in construction, 2.3.42.

264 Cf. chapter V.3.d.

265 Mostly because the client had to pay 10 percent extra tax on the gross wage of the POW to the Stalag: Küppers and Bannier, *Einsatzbedingungen*, 188.

266 Converted to reichsmark and based on an 8-hour work day, this meant 0.18 RM and 0.31 RM per hour for an unmarried and a married worker, respectively.

267 Cf. RAO, RAFA-2188/1/G1c-L0006, file OTW, Circular no. 33 (= tax circular no. 5) by the Oslo liaison office of the German building trades, 12.8.44.

268 RAO, RAFA-2188/2/Fab-L0080, Circular by *Einsatzgruppe Wiking* to all *Einsätze* and OBLs, 12.2.43. From September 1943 at the latest, the settlement between the *Wehrmacht* and EW was simplified. EW no longer paid for the use of Soviet POWs and Yugoslavian inmates under cost-plus contracts: RAO, RAFA-2188/1/E5f-L0005, file Durchschriften 1943–1944, Decree 485/43 of Einsatzgruppe Wiking, 7.9.43.

269 Ibid., EW to all *Einsätze* and OBLs regarding the wage accounting for POWs working under the OT, 2.10.43.

no longer demand compensation from EW for the use of POWs. In so doing, the KBN followed a decision by the OKW of July of that year.[270] Since the *Wehrmacht* funded EW's projects anyway, this rearrangement had no practical consequences. The second reason was that the construction firms calculated their profit as a percentage of the wage bill on their site. Hence, the firms also needed a record of the hours worked by the Soviet POWs, which they then multiplied by 0.32 RM and 0.46 RM for unskilled and skilled POW, respectively, in order to have a calculation basis.[271]

As mentioned above, firms had an interest in higher wage bills as long as they were working under a cost-plus contract. Against this background, it does not come as a surprise that Strabag Weimar was not entirely happy with the deployment of Soviet POWs on its construction sites around Stavanger in November 1942. A day's work by a Soviet POW was valued at 3.60 RM, while a Norwegian worker received a sum equivalent to around 8 RM. As the wage bill on the construction site was the basis for calculating Strabag's profit, the firm preferred the expensive Norwegians. However, "because the further employment of Norwegian workers will be difficult, our workforce will presumably be reinforced with Russian prisoners only. By that, the workload is increased, sales and profit, on the other hand, only to a limited extent."[272] A similar motive was at the bottom of Arge Stallmann's creative 1943 attempt to relocate its head office *pro forma* from Gelsenkirchen to Berlin, in order to profit from the capital's higher wage rates for its German workers.[273] Apart from the case of the construction industry, historians so far have only sporadically referred to the connection between cost-plus contracts and the industry's interest in higher wage bills in the National Socialist war economy.[274] It must be stressed, however, that German firms of course profited from forced labour insofar as the allocation of forced labourers was the basic prerequisite for executing orders in the German war economy. In Norway, it would have been unthinkable to realise the same turnovers and profits only using the scarce Norwegian labour.

270 RAO, RAFA-2188/1/E1a-L0002, EW to all *Einsätze* and OBLs, 16.11.44.

271 Some of these records have survived in the archives, for example in RAO, RAFA-2188/2/Hda-L0031 and RAO, RAFA-2188/2/Faa-L0062.

272 ThHStAW, Strassenbau AG Weimar Nr. 108, fol. 200 ff., travel report Stavanger 4 November – 12 November 1942, Weimar, 25.11.42.

273 RAO, RAFA-2188/1/E5a-L0040, Letter to construction firm Wilhelm Stallmann, Gelsenkirchen, 28.7.43. Under the *Wiking-Bauvertrag*, the wage of a German worker depended on the payscale effective in the region where the firm's head office was located.

274 Streb, 'Preisregulierung', 41; Eichholtz, *Kriegswirtschaft II/2*, 513. Beyond the construction sector, it has been mentioned, albeit not analysed systematically, in the case of the Volkswagenwerke, Daimler-Benz and Siemens: Hans Mommsen and Manfred Grieger, *Das Volkswagenwerk und seine Arbeiter im Dritten Reich* (Düsseldorf: Econ Verlag, 1996), 564; Barbara Hopmann et al., *Zwangsarbeit bei Daimler-Benz* (Stuttgart: Franz Steiner Verlag, 1994), 334–35; Feldenkirchen, *Siemens*, 206, 550–51.

c Raising the Pressure: Einsatzgruppe Wiking's Efficiency Reforms of 1943

In the course of 1942, the OT prepared a major reform of its pay and contract system. The Organisation launched changes in four fields. Although they were introduced independently of each other, these changes must be understood as complementary. As described in the previous chapter, EW introduced the efficiency-output contract in spring 1943. For the workforce, the reforms brought the introduction of a new pay scale, the transition from time to efficiency pay (*Leistungslohn*), as well as group-based performance factors. The guiding idea behind the reforms was to increase the pressure on contractors and particularly on their workers to boost the work pace on the construction sites. Higher efficiency was to be rewarded in order to mobilise the last reserves of the construction sector for Germany's final victory. Three weeks before Goebbel's total war speech at the *Sportpalast*, Price Commissioner Fischböck answered the rhetorical question "What does improved performance in industry mean?" in a speech in front of his agency's staff. It meant bringing the utmost out of the available workforce and material, Fischböck explained.[275]

In early 1943, the OT introduced new pay scales for most of their civilian workers in Europe. In the case of the German workers, a standardisation of wage rates primarily aimed to create some order in the jumble of payment principles. The fact that the same work did not result in the same wage, but depended on where in Germany the construction firm came from, had aroused discontent among the men. The OT's Special Trustee for Labour (*Sondertreuhänder der Arbeit für die Organisation Todt*), the Hessian Trustee for Labour Fritz Schmelter, issued the new pay scale for German OT workers, the *OT-Frontarbeiter-Tarifordnung*, on 1 October 1942.[276] It became effective from 1 February 1943.[277] The pay scale introduced consistent hourly wages for all German OT workers in occupied Europe (Table 21). The daily allowance was replaced by a monthly pay (*OT-Frontarbeitersold*) in local currency, based on the pay that the soldiers of the *Wehrmacht* received. In the Norwegian case, the pay was 120 kroner for assistant overseers, 111 kroner for foremen and 96 kroner for all other workers. Married workers could apply for family benefits (*OT-Familienbeihilfe*) if their new wage lay significantly below the old one. At the same time, however, the OT cancelled

275 Hans Fischböck, 'Sparsame Wirtschaftsführung – Kommerzialisierung der Preisüberwachung – Rede des Reichskommissars für die Preisbildung Staatssekretär Dr. Hans Fischböck auf dem Gefolgschaftsappell des Preiskommissariats am 29. Januar 1943', *Mitteilungsblatt des Reichskommissars für die Preisbildung Teil 2* 7, no. 3 (1943): 32–37.

276 RABl. 1942, IV, 1194. On Fritz Schmelter, see Botzet, 'Generalbevollmächtigter', 127, fn. 58.

277 RAO, RAFA-2188/1/E5a-L0040, OTZ, Main Department for Work Deployment and Welfare (*Hauptabteilung Arbeitseinsatz und Sozialpolitik*), to all *Einsatzgruppen* of the OT, 17.12.42.

Table 21: Hourly earnings in RM, including pay and free food and accommodation, for unmarried civilian workers under *Einsatzgruppe Wiking* in Norway, summer 1943.*

	Germans	Norwegians	Other foreigners	Poles
Assist. overseer	1.29	1.25	1.16	0.96
Foremen	1.22	1.00	1.10	0.92
Skilled	1.06	0.91	0.98	0.80
Semi-skilled	0.98	0.80	0.88	0.72
Unskilled	0.91	0.68	0.81	0.65

* For wages for Germans, including *OT-Frontarbeitersold*, according to the *OT-Frontarbeiter-Tarifordnung* of 1 October 1942, see RAO, RAFA-2188/1/E5f-L0014; for Norwegians according to the amendment of the national Norwegian wage agreement of 26 March 1943, see ibid.; for foreigners, now including Danes, according to the OT's *Ausländer-Bautarif Nord* of 18 January 1943, see ibid. On the monthly pay under the *Ausländer-Bautarif Nord*, cf. RAO, RAFA-2188/2/Fab-L0081, file Tarifordnungen, Letter by the Special Trustee for Labour for the OT, Schmelter, to the associations in the construction sector, 28.1.43; for the Polish workers according to the *Polen-Auslands-Bautarif*, see RAO, RAFA-2188/2/Fab-L0081.

the labour offices' supplementary allowance for conscripted workers and their loyalty bonuses. The 10 percent performance bonus, which potentially all workers could receive under the old regulations, could now only be given to at most one quarter of the workers on a construction site. Below the line, the new OT pay scale meant a pay cut for the majority of German OT workers in occupied Europe. Although for some groups of workers from rural regions, the new pay scale would in fact bring a slight pay rise, most of the big players active in the occupied territories came from larger German cities with higher wages. Especially single men working for firms from high-wage regions saw their earnings severely reduced.[278]

Norwegian workers likewise had to accept lower wage rates in the revised national Norwegian wage agreement of 26 March 1943.[279] However, as Ingulstad has shown, the wage agreement never covered the entire sector. It did not apply to stone, earth moving and concrete works and special regulations applied to different regions and handwork branches. Furthermore, many workers banded together in the aforementioned work gangs. As such, they were regarded as self-employed and not subject to the agreement.[280]

Finally, the civilian foreign labourers were now covered by the so-called *Ausländer-Bautarif Nord* of 18 January 1943.[281] Their wages were lower than those of German workers, while the monthly pay (in their case called *Einsatzgeld*) was

278 See the examples in RAO, RAFA-2188/1/E5a-L0040.
279 RAO, RAFA-2188/1/E5f-L0014.
280 Ingulstad, 'Regulering', 269–71.
281 RAO, RAFA-2188/1/E5f-L0014 and RABl. 1943, IV, 57.

at the same level for married workers. Unmarried workers received 60 percent of the pay.[282] For Polish workers, the *Polen-Auslands-Bautarif* was issued on 11 January 1943.[283] Under this pay scale, the OT no longer differentiated between Poles from the General Government and other Poles. Unmarried Poles did not receive the monthly *Einsatzgeld*. Again, Poles still had to pay the 15 percent *Sozialausgleichsabgabe* on their wages shown in Table 21. Both the *Ausländer-Bautarif Nord* and the *Polen-Auslands-Bautarif* became effective from 1 February 1943.

The industry's reaction to the new pay scales in autumn 1942 was sceptical. In a statement, of which Wigru Bau's liaison office in Oslo received a copy, the head of Sager & Woerner's office in Oslo, Karl Pfitzner, demonstrated to the firm's office in Berlin the consequences of the new *OT-Frontarbeiter-Tarifordnung*. A married machine operator from Berlin faced a wage cut of almost 12 percent, an unmarried machine operator of even 25 percent. Pfitzner feared a "huge drop in performance" among the workers.[284] The firm Lohse & Berger from Chemnitz among other things resented the new limitation according to which only the most hardworking quarter of the German workers on a site could receive the 10 percent performance bonus. All of their 18 German workers in Norway had leading positions on the construction sites and had received the performance bonus for years.[285] For obvious reasons, the firms refrained from mentioning another very evident downside of the new pay scales to the OT: under cost-plus contracts, lower wages resulted in reduced profits.

In a five-hour meeting between more than 60 firm representatives, as well as clients like EW, the RK, the *Wehrmacht* and A/S Nordag under the direction of Wigru Bau's Friedrich Thierbach on 7 December 1942 in Oslo, however, the entrepreneurs abandoned their reserves.[286] Concordantly, the industry warned against the "disastrous" and "intolerable" consequences of the new pay scale for German workers. Wigru Bau and the Reich Association of the Building Handwork Guilds were also targets of indignation, as they had not prevented the reform. It was criticised quite openly that the lower wages would reduce the basis on which the firms calculated their mark-ups under the *Wiking-Bauvertrag*.

The conflict eased over the following months, however. To understand why, we have to consider another change in the OT's pay and contract system that was implemented about the same time: the transition from time pay to efficiency pay. Within the Reich, efficiency pay in construction had already been introduced on

---

282 Under the *Ausländer-Bautarif Nord*, the level of the pay (*Einsatzgeld*) did not depend on the workers' nationality, as Hatlehol assumes. Cf. Hatlehol, 'Norwegeneinsatz', 350–51.

283 RAO, RAFA-2188/2/Fab-L0081 and RABl. 1943, IV, 109.

284 RAO, RAFA-2188/1/F3a-L0003, Copy of a letter by Pfitzner, Sager & Woerner Oslo, to Sager & Woerner Berlin, 7.10.42.

285 Ibid., Letter by Lohse & Berger to the Special Trustee for Labour for the OT, Schmelter, 17.11.42.

286 RAO, RAFA-2188/1/F3a-L0002, Minutes of the monthly meeting between German construction firms and the Oslo liaison office of Wigru Bau, on 7 November 1942 in Oslo, 8.11.42.

2 June 1942 with a national wage scale (*Reichstarifordnung über den Leistungslohn im Baugewerbe,* RTOLL) that was to be in effect as of 1943.[287] The wage scale was based on ergonomic surveys that the DAF had conducted during the 1930s. In the RTOLL, the state defined the normal performance that could be expected from workers for certain work steps. Every worker's individual performance was measured and then multiplied by the hourly standard wage. Workers could achieve their hourly standard wages when they finished their task within the given time, while a better or slower performance would raise or lower their earnings, respectively.[288] The RTOLL, however, remained piecemeal. The factors used to calculate the workers' hourly wages could be changed if local conditions required it, and it was often the firms themselves that were supposed to measure the performance of their workers.[289] The administrative burden was also high. Until the end of the war, 13 supplements to the RTOLL were published, containing detailed specifications for various construction activities.[290]

Outside of the borders of the Reich, *Einsatzgruppe West* was first to base their bunker construction contracts at the Atlantic Wall on the RTOLL performance values.[291] As the conditions in the occupied West were more difficult than in the Reich, Fritz Schmelter's representative in Paris had to make some concessions, however. The firms' site managers were allowed to factor in obstacles like transport problems, the higher average age of the workforce, or negative effects of enemy attacks, communication problems, and the like. Hence, the expected "normal" performance was lower in the occupied West than in Germany.[292] For the firms, this once more opened up an opportunity to capitalise on information asymmetries.

In occupied Norway, efficiency pay was introduced on 1 April 1943, that is, two months after the new OT pay scales. After the new pay scale had reduced their hourly wages, the workers were now presented with the "solution": by working harder and faster, both German and foreign workers were to bridge the gap between their new wages and the previous wage level. At the same time, the OT got the construction firms on board by introducing the efficiency-output contract, which financially rewarded higher performance. Thus the firms, too, were now motivated to urge on their workforce. As much as high wages and low performance had increased entrepreneurial profits under the cost-plus contracts and had helped to

287 RABl. 1942, IV, 827.
288 Wolfgang Franz Werner, *'Bleib übrig!' – Deutsche Arbeiter in der nationalsozialistischen Kriegswirtschaft* (Düsseldorf: Schwann, 1983), 234–35; Siegel, *Leistung und Lohn*, 142; Streb, 'Preisregulierung', 43–45.
289 Werner, *Bleib übrig!*, 234–35.
290 *Tarifordnung über den Leistungslohn im Baugewerbe nebst den bisher erschienenen Anhängen 1 bis 13* (Düsseldorf: Droste Verlag, 1953).
291 Cf. also BA-MA, RH 8/4579.
292 BArch, R 50-I/366, OTZ to *Einsatzgruppe West* regarding efficiency pay, 20.4.43.

attract workers, the development still had been a double-edged sword. The construction firm Sager & Woerner stated after the war that it had been interested in maintaining a reasonable ratio between performance and wage, "[. . .] because we realised early that it would be difficult to accomplish a reorientation towards the principle of efficiency after the war."[293] In early 1943, Strabag Weimar welcomed the introduction of the efficiency pay for Norwegian workers. The new pay scale for Norwegians would reduce their earnings by 15–20 percent, and thus give the Norwegians an incentive to work harder and put an end to their previous "sluggish performance".[294] Much of the pressure of the OT's efficiency reforms in the winter of 1942–43 thus rested on the shoulders of the workforce, rather than on the firms' management. The efficiency pay was of course unpopular and the – already low – support for the Norwegian collaboration regime under Vidkun Quisling's *Nasjonal Samling* (NS) dwindled away among the Norwegian workforce, as German intelligence observed in spring 1943.[295] Efficiency pay was a strain on the ever-expanding group of older, unfit or forced labourers in particular, because it did not provide for a certain hourly minimum wage as traditional piecework pay had done.[296]

d Making Exploitation Manageable: OT's System of Performance Factors

The fourth element of the OT's reforms concerned the varying levels of performance of the different groups of labourers. Many groups of forced labourers did not possess the skills or were simply too enfeebled to achieve the same performance as German workers. Whenever fixed-price contracts were used, this became a problem both for construction firms and the awarding authorities. For the firms, the fact that they had to submit offers, though without being able to predict the composition of their workforce during the entire construction process was a major element of uncertainty. How to handle the constant transfers of forced labourers between construction sites, plants and even entire industries was a matter of major debate not only in construction, but in the German war economy as a whole. How should the state react when firms realised excess profits because they had been provided with forced labourers who proved to be more productive than expected? On the other hand, could firms be expected to bear the financial losses when they were provided with POWs or concentration camp inmates who were so sick and weak that they

293 Sager & Woerner AG, *Sager & Woerner 1864–1964*, 2: Sammlung-Einsatz-Krise-Erstarkung, 1920–1945:173.
294 ThHStAW, Straßenbau AG Weimar Nr. 139, fol. 34 f., Note by Strabag's site manager Korffmacher, Stavanger, 21.4.43.
295 Ingulstad, 'Regulering', 270.
296 Ingulstad, 269.

achieved scarcely half of a German worker's productivity? Questions of this kind were particularly relevant for the OT and its contractors after the introduction of the efficiency-output contracts in the occupied West in the summer of 1942 and in Norway in spring 1943. Working under ever-changing conditions and a flexible organisational structure in the occupied territories, considerable fluctuations in the workforce were a daily occurrence. In occupied Norway, the discussion concerned especially the deployment of Soviet POWs, as they were not only the biggest group of forced labourers, but also the group whose performance deviated most from that of healthy and well-nourished Germans. It is needless to say that the forced labourers themselves were degraded to not much more than cost factors in the negotiations between the firms and the OT.

In spring 1942, Reich Price Commissioner Fischböck realised that POW deployment in construction had become a widespread phenomenon, which "had a great effect on prices due to varying wages, performances and social contributions". Because there was some confusion among customers and entrepreneurs regarding the deployment of POWs, Fischböck issued clarifying guidelines on the topic on 2 March 1942.[297] This decree presented a solution to the problem of the varying performance levels of the constantly relocated groups of foreign forced labourers and introduced so-called performance factors (*Leistungsfaktoren*). These factors were assigned to different groups of workers in order to allow for their lower performance compared to German workers, who were given the factor 1. When drawing up an offer, the firms could now multiply the hourly wage of a group with the number of workers in this group and their performance factor. They did so for every category of workers on their construction site and thus arrived at an average hourly wage (*Mittellohn*) of all workers on a construction site. If the composition of the workforce changed over time, the variables could be easily adjusted. When a project was finished, the effective *Mittellohn* was determined in an adjustment procedure.[298] For example, if a firm had calculated for German and Norwegian workers, but the latter were replaced with Soviet POWs after some weeks, the concluding adjustment took into account both the lower production costs and the lower productivity of the Soviet POWs by multiplying their wage with their lower performance factor.

297 BArch, R 3/1446, fol. 22–29, Circular decree 13/42 by the Reich Price Commissioner regarding the deployment of POWs in construction, 2.3.42.

298 RAO, RAFA-2188/1/E5f-L0014, Letter by Josef Fröhler to all OBLs with the attached transcript of a speech with the title "The efficiency output contract in Norway", held by Fröhler on 7 February 1943 in Oslo, 23.2.43. For a detailed example of an adjustment procedure (*Berichtigungsverfahren*) for a construction project in the area of Sandnessjøen of the construction firm Lohse & Berger, see RAO, RAFA-2188/2/Faa-L0010. *Einsatzgruppe Hansa* also gives a comprehensive explanation of the procedure for its "Ruhrvertrag L" in 1944: BArch, R 50-I/70, fol. 105–12.

It seems that group-based performance factors were first introduced on a large scale by *Einsatzgruppe West* in France, the Netherlands, Belgium and the Channel Islands in the summer of 1942 together with the efficiency-output contract. *Einsatzgruppe Wiking* followed in April 1943. The factors certainly reflected National Socialist racial thinking, particularly in their attempt to measure the performance not of a single worker, but of entire ethnic groups.[299] Nevertheless, the factors presented in Table 22 clearly show a genuine effort to realistically assess the performance of different groups of forced labourers on the OT's construction sites: groups that were on the same step in the Nazi racial hierarchy could have different performance factors, while the performance of the same group of workers was sometimes assessed differently depending on their area of deployment.

Table 22: Performance factors for the OT's workers under the efficiency-output contract.*

	EG Wiking (Norway, 1943–45)	**EG West (Mainland France, 1942)**	**EG West (Channel Islands, 1942)**	**EG VI (Weingut bunkers, 1944–45)**
Germans	1.0	1.0	1.0	1.0
Norwegians	1.2			
Dutch	1.2	1.6	1.8	
French	1.2	1.5	1.5	
Danes	1.3			
Ostarbeiter	1.3			
German construct. soldiers	1.3			
Poles	1.4	1.5	1.5	
Czechs	1.4			
Norwegian inmates	1.5			
French POWs	1.2			
Belgian POWs	1.3			
Polish POWs	1.5			
Yugoslavian inmates	1.5			
British POWs	2.0			
Soviet POWs	2.0		2.5	
Belgians		1.4	1.4	
Spanish inmates		1.4	1.4	
Foreigners (w/o Italians)				1.2

299 This was the main difference to the concept of the RTOLL, which tried to measure the performance of individual workers to arrive at a "fair wage".

Table 22 (continued)

	EG Wiking (Norway, 1943–45)	EG West (Mainland France, 1942)	EG West (Channel Islands, 1942)	EG VI (Weingut bunkers, 1944–45)
Italians and POWs				1.4
Concentration camp inmates				1.7
Hungarian and Latvian Jews				3.0

*Figures for *Einsatzgruppe Wiking* from: RAO, RAFA-2188/1/E5f-L0014, Letter by Josef Fröhler to all OBLs with the attached transcript of a speech with the title "The efficiency output contract in Norway", held by Fröhler on 7 February 1943 in Oslo, 23.2.43. Figures for *Einsatzgruppe West*: BArch R 50-I/272, fol. 109–12, Circular no. 351 by *Organisation Todt*, OBL Normandie, 5.10.42. Figures for *Einsatzgruppe VI*: Raim, *Kaufering und Mühldorf*, 126.

The Soviet prisoners of war in Norway had a performance factor of 2.0, meaning that EW assumed that two POWs could do the work of one German worker. This corresponded roughly with the benchmark values given by the Reich Price Commissioner in his circular decree of 2 March 1942. In their price calculations under the efficiency-output regime, the German firms inserted the number of Soviet POWs they expected to work on the project and multiplied it with the number of days the POWs would be deployed. Every workday of a POW was normally valued at 3.60 RM.[300] The sum was then multiplied by the performance factor 2.0. Thus the firms were safeguarded against losses resulting from the lower performance of "their" forced labourers. There was still no money flow from the firms to EW or the Stalags under the efficiency-output contracts. The wage bills for Soviet POWs (and Yugoslavian inmates) that were part of the calculations were later simply deducted from the invoices when EW paid the firms.[301]

The OT's performance factors have hardly been mentioned in the literature on the OT or forced labour in the Third Reich in general. The studies of Raim and Sæveraas are exceptions in this regard, although they do not investigate the

300 It seems that the firms no longer differentiated between skilled and unskilled POWs in their calculations for the efficiency-output contracts.

301 In this process, EW still had to rely on the construction firms to correctly register the POW hours. This held the danger of fraud. Allgemeine Baugesellschaft Lorenz, for example, convinced EW in autumn 1944 that the Soviet POWs on its construction site had worked 5,905 hours and documented this with a neat time sheet attached to the invoice. Accordingly, the price of 5,905 POW hours was deducted from the invoice. However, as internal documents prove, the Soviet POWs had actually been forced to work 8,900 hours on Lorenz's construction site, which meant an additional profit for the firm: LAB, A Rep. 250-05-13, Nr. 45, Letter by Allesch, Allgemeine Baugesellschaft Lorenz, Lunde construction site, to the firm's headquarters in Berlin with attached invoice no. 6 and time sheets, 7.12.44.

significance and consequences of this system more closely. Also Seidler mentions the performance factors in his study on the OT.[302] The German war economy, and not least the construction sector, relied heavily on the mass deployment of forced labour.[303] Groups of workers were frequently relocated across the borders of countries and sectors alike. This situation produced significant inefficiencies in the field of labour deployment. Procurement agencies and businesses were in a constant tug-of-war over the price of forced labour. Under these conditions, the performance factor system of the OT's *Einsatzgruppen* must be understood as an attempt to fit the realities of the German war economy, and of a (forced) labour market that spanned a whole continent, into a manageable contractual system.

But where did these performance factors come from? Once again, industrial self-responsibility came into play. In January 1943, the OTZ sent a draft of guidelines for the conclusion of efficiency-output contracts to all *Einsatzgruppen* leaders, explaining the role of performance factors in the price calculation among other things. The guidelines stated that the leader of the *Einsatzgruppe* would fix the performance factors after hearing the committee of entrepreneurs, which needed to consist of at least two representatives of firms in the area.[304] What this meant in reality could be observed in the occupied western territories, where *Einsatzgruppe West* had already introduced efficiency-output contracts and performance factors in autumn 1942. On 12 December 1942, *Einsatzgruppe West* contacted the representatives of the committees of entrepreneurs under every OBL and admitted that neither the firms nor the OT possessed empirical values on the performance of their foreign labourers. Therefore, the initial values, which seemed to have been determined rather arbitrarily, should be reassessed. However, performance surveys organised by the OT itself would be so time-consuming and labour-intensive that *Einsatzgruppe West* regarded it as "[. . .] a task of the self-responsibility of the firms [. . .]" to provide it with correct empirical values. The representatives of the committees of entrepreneurs were to ask firms in their area to observe the performance of the workers on their construction sites and convey the values to the *Einsatzgruppe*.[305]

In occupied Norway, the role of business circles in the formulation of the performance factors can be traced even more directly. In early 1943, the construction industry and EW met regularly for negotiations on the efficiency-output contract for

302 Seidler, *Die Organisation Todt*, 214; Raim, *Kaufering und Mühldorf*, 126; Sæveraas, 'Beton macht Geschichte', 164–65.
303 Rauh-Kühne, 'Hitlers Hehler'.
304 BArch, R 50-I/106, fol. 28–31, Draft of guidelines for the conclusion of efficiency-output contracts under the OT by OT's Department for Contracting and Pricing.
305 BArch, R 50-I/286, fol. 5, Letter by *Einsatzgruppe West* to the representatives of the committees of entrepreneurs at the various OBLs, 12.12.42.

the construction of fortifications. At that point in time, two drafts in particular were subject to the discussion, one by EW and one by the firm Carl Brandt from Bremen.[306] After meetings on 7 and 8 January, EW presented a new draft contract including performance factors. Friedrich Thierbach from the liaison office asked a group of firms to give their view on the proposed factors before negotiations proceeded in February.[307] The construction firms' strategy was to fix the performance factors at as high a level as possible during the negotiations (that is, assessing the expected performance as low as possible), before pushing the workers' performance to the level of regular German workers. It was in this margin between the price and expected performance laid down in the contract on the one hand and the actual performance of the forced labourers on the other hand where the additional profit lay.[308] On 6 February, the industry already presented a counterproposal in which the performance factors had been significantly changed. For nearly every group of forced labourers, the performance factors had been adjusted upwards, that is, in favour of the construction firms. For Soviet POWs, for example, they were changed from 1.55 to 2.0, for the Yugoslavian inmates from 1.35 to 1.5, for Norwegians from 1.1. to 1.2, for Danes from 1.0 to 1.3, and so on.[309] These were exactly the factors that EW would eventually use in the new contract, as displayed in Table 22.

The reforms of 1943 had implemented a system that rewarded the German construction firms for increases in efficiency. The prime victims of this policy were the forced labourers, especially exposed and defenceless groups like the Soviet POWs or Yugoslavian inmates. The OT's German workers and the firms held a crucial position in this system of exploitation, as the OT made publicly clear: "[. . .] Reich Minister Speer [. . .] instructs the entrepreneur not to acquiesce to the hazards of a drop in performance, but to fight them actively, to counter them effectively with the skill of his management and leadership."[310]

306 RAO, RAFA-2188/1/F3a-L0011, file Leistungsvertrag, Letter Thierbach to Riedel, Wigru Bau, 1.1.43.

307 Ibid., Invitation by Thierbach to a meeting on 12 February 1943, 20.1.43. The invitation was sent to firms that were constantly involved in the formulation of the efficiency-output contract during these months: Sager & Woerner, Ohlendorff'sche Baugesellschaft, Hochtief, Hermann Milke, Arge Nordmark, Arge Sachsen, Carl Brandt, Boersch Strassenbau, Strabag and August Prien.

308 Example: During negotiations, the performance factor of Soviet POWs is set to 2.0, meaning that the performance of a Soviet POW is expected to be 50 percent of a German worker's performance. The firm suffers no financial loss from the fact that Soviet POWs are less productive than Germans as long as their performance does not drop below this 50 percent. The firm can now increase its profit by raising the performance of the POWs, be it by means of incentives or brute force. Cf. Raim, *Kaufering und Mühldorf*, 126.

309 In the files of the Oslo liaison office of the German building trades, there is even an example of EW's draft contract in which the performance factors are altered and further comments added by hand. The new performance factors seem to have been inserted by Friedrich Thierbach (grey pencil), other changes by his co-worker Joachim Schlimper (red pencil): RAO, RAFA-2188/1/F3a-L0011, Undated preliminary remarks on the efficiency-output contract.

310 Daub, 'Zur Gestaltung von Bauverträgen'.

The *Wehrmacht*, EW and the firms apparently regarded food and sales articles from the *Wehrmacht* camp magazines as the most promising leverage to increase the pressure on the workforce. The measures, however, always remained a process of trial and error rather than to developing into a coherent policy. The army's General Command LXX in Southern Norway already noted in August 1943 that neither fines nor arrests had yielded results when it came to "work-shy" Norwegians. For those Norwegians who received their provisions from the *Wehrmacht*, the army therefore recommended the withdrawal of tobacco rations and a reduction of the weekly ration of meat by 100 grams. EW seems to have adopted the measure on its own construction sites in September. The firms were to report the names of "work-shy" men: "The firms have now got a means of taking action against those unwilling to work."[311] In late 1943, complaints from the ranks of EW reached KBN Kurt Wuthenow, according to which the POWs in Norway were too well-nourished in relation to their work performance. The critics wished the introduction of performance-based food rations, a measure that Wuthenow refused.[312] Though, he agreed with EW that tobacco rations, unrationed foodstuff and articles that the POWs could obtain in the camp magazines, were distributed based on their work performance.[313] In June 1944, regulations were further tightened when the KBN agreed that parts of the warm meals should be distributed depending on the POWs' work performance.[314] This measure primarily targeted the Soviet POWs on the railway construction sites in Northern Norway. It was the *Wehrmacht* that gave out the rations in the camps, but it was the German foremen on the construction sites who were now registering the work performance during the day in a so-called performance book (*Leistungsbuch*).[315] The vast literature on forced labour has repeatedly demonstrated the potentially devastating effects of performance-based food rations (*Leistungsernährung*) for POWs in all parts of the German war economy.[316] The decision on whether an already weakened POW would be punished for his lower performance lay with the employees of private German businesses. Their superior position held the potential to make them masters over life and death. According to the liaison office of the German building trades in Oslo, "the decrees met the general wish of the firms to be provided with further means to increase the

311 ThHStAW, Straßenbau AG Weimar Nr. 130, fol. 88, *Frontführer* Macher, BL Stavanger, to all German firms, 9.9.43.

312 RAO, RAFA-2188/1/E1m-L0047, Note by KBN Wuthenow, 15.12.43.

313 Ibid., Letter by Feuchtinger, GB Bau, Oslo, to all *Einsätze* and OBLs, except Denmark and Tønsberg, 18.1.44.

314 Ibid., Order by KBN Wuthenow, 3.6.44.

315 Ibid. On the topic, see also Hatlehol, 'Norwegeneinsatz', 238–40.

316 See for example: Herbert, *Fremdarbeiter*, 319–20; Walter Naasner, *Neue Machtzentren in der deutschen Kriegswirtschaft 1942–1945* (Boppard am Rhein: Harald Boldt Verlag, 1994), 147–50.

performance of the captive Russians."[317] It seems, however, that the implementation of performance-based food rations did not go as expected. It was unpopular among many of the local *Wehrmacht* camp commanders and the liaison office noted with some disappointment that some construction firms, too, "[. . .] lacked the necessary understanding [. . .] for this severe measure."[318] After Soviet POWs in some locations had gone on hunger strike, the KBN retracted his order on performance-based food rations in the late summer.[319]

With regard to the actual performance of the workforce on EW's construction sites, we must assume that private business stood to benefit from information asymmetries.[320] However, the *Einsatzgruppe* was in no way ignorant of the discrepancy between reality and the contractually fixed performance factors. Fritz Köbele estimated in March 1944 that the work performance of Soviet POWs had risen from 50 percent to 60–70 percent compared to a German worker.[321] In the High North, in December 1943 OBL Alta was fully aware of the fact that firms in the area had doubled their profits by insisting on using the performance factor 2.0 for Soviet POWs during contract negotiations.[322]

The findings presented in this chapter make it necessary to qualify some of the statements that earlier research has made about the financial aspects of POW deployment.[323] As life and working conditions were extremely diverse within the various categories of forced labourers, it goes without saying that all conclusions drawn can only be approximations of this highly complex historical phenomenon. While Spoerer and Custodis have argued that Soviet POWs were not cheaper than comparable German workers,[324] the material analysed in this chapter suggests that in construction, their productivity in fact exceeded the costs of their deployment. This even holds true in the case of Northern Norway, a region where the living conditions of the Soviet POWs were certainly harsh. This supports the position of

317 BArch, R 13-VIII/87, fol. 70–75, Circular no. 32 by the Oslo liaison office of the German building trades, 13.7.44.
318 Ibid., fol. 9–28, here 16, Undated report by the Oslo liaison office of the German building trades for August, September and October 1944.
319 Ibid.
320 The strategy of understating the actual performance of forced labourers during contract negotiations with the state has been documented in the case of POWs in the coal mining industry: Streit, *Keine Kameraden*, 395, fn. 150.
321 RAO, RAFA-2188/2/Haa-L0029, Circular by Köbele to all *Einsätze* and OBLs, 29.3.44.
322 RAO, RAFA-2188/2/G2b-L0023, Note by OBL Alta, 28.12.43.
323 The findings apply to all *Einsatzgruppen* of the OT both in occupied Europe and within the Reich as far as they used the *OT-Leistungsvertrag*. However, it seems that exceptions were possible. Marc Buggeln writes that during the construction of the naval bunker "Valentin" in Bremen, the wages of foreign forced labourers were excluded from the calculation: Buggeln, *Valentin*, 63.
324 See above, chapter V.3.b, fn. 263.

Cornelia Rauh-Kühne insofar as she concludes that some groups of workers were indeed cheaper. However, both Spoerer and Rauh-Kühne doubt that the state would have been willing to provide the industry with forced labour "below market price", since it was the state itself, as customer of armaments and construction projects, that had to foot the bill. While Spoerer assumes that the lower wage levels already reflected the lower productivity more or less correctly, Rauh-Kühne argues that the state skimmed off the saved wage costs with tight monitoring and *ex post* price reductions.[325]

In contrast, this chapter has shown that the OT mostly relied on information provided by the construction industry. Still, it would be wrong to speak about an "information deficit" on the side of the OT. There is much to suggest that the Organisation was aware of its contractors' excess profits. In Alta, for example, EW knew that some firms had used the rate of 0.40 RM per hour for an unskilled Soviet POW, instead of 0.32 RM, in their price calculations. EW's Josef Fröhler, however, refrained from renegotiating the offers.[326] OT's main priority was to secure the cooperation of private industry and to raise performance, not to contain costs. Rigorous price monitoring and the collection of excess profits would even have run counter to the attempts of increasing the firms' efficiency. When it came to POW work, EW even ceded "a moral right" (sic!) to the firms to make an (additional) profit from exploiting prisoners of war.[327]

Finally, there are examples of EW simply ignoring its own pay scale and regulations, as in the case of Allgemeine Baugesellschaft Lorenz. The firm was in a strong bargaining position, as it was a highly specialised expert on pile driving works. In 1944, Lorenz was dissatisfied with the wage level on its construction site. As usual, the profit was calculated as a percentage of the wage bill, and the firm argued that the wages were too low to yield a profitable result for a specialised firm like Lorenz. The concessions made by EW during the following negotiations were remarkable: the Norwegian workers of Lorenz's subcontractor received performance bonuses that meant they were paid for 12 hours work when actually only having worked 10 hours. The OT's justification was that the Norwegians had higher living costs as they had to buy products on the black market. Lorenz's German foremen protested against this preferential treatment of the Norwegians and thus also received 12

325 Spoerer, 'Profitierten Unternehmen von KZ-Arbeit?', 70; Rauh-Kühne, 'Hitlers Hehler', 5, 50. Note that Spoerer and Rauh-Kühne discuss the deployment of concentration camp inmates. The "problem", however, is the same. The difference between my findings and those of Spoerer and Rauh-Kühne is of course one of degree rather than categorical. Some of it may be explained by the fact that the preconditions for effectively monitoring firms were particularly unfavourable in occupied Norway. Moreover, construction was a particularly fragmented sector with little standardisation.
326 RAO, RAFA-2188/2/G2b-L0023, Note by OBL Alta, 28.12.43.
327 Ibid.

hours' wage for their 10-hour workdays. Moreover, the work of the Soviet POWs on the construction site was valued much higher in the price calculation than the official regulations permitted, which of course did not mean that the POWs actually received this higher pay.[328] Below the line, Lorenz could be very satisfied with its earnings in Norway.

328 LAB, A Rep. 250-05-13, Nr. 45, Letter by Allesch, Allgemeine Baugesellschaft Lorenz, Lunde construction site, to the firm's headquarters in Berlin, 4.11.44. Cf. also LAB, A Rep. 250-05-13, Nr. 158, Note by Allesch, Allgemeine Baugesellschaft Lorenz, Lunde construction site, on a meeting with the OT on 23 September 1944, 28.9.44.

VI Conclusion

This thesis has enquired into the relation between the state and the construction industry in the Third Reich, and – more specifically – between the OT's *Einsatzgruppe Wiking* and its contractors in occupied Norway. I have argued that this relation was characterised by two prominent features.

Firstly, whereas workers in the GdSt's and the OT's construction sites faced harsher working conditions and increasing militarisation, Fritz Todt did not have to use force to secure industry's cooperation. Favourable contracts and a "long leash" policy were regarded as the most promising strategy to mobilise construction firms. Thus the relation between the German construction industry and the "Todt apparatus" were quite cordial by the late 1930s. The divide was rather between the GdSt/OT and the financial authorities, which saw rising prices and profits with great concern. When the war broke out, construction bans, rationing and the *Wehrmacht's* draft calls massively changed the framework conditions under which construction firms were working. Moreover, the state relocated much of the construction activity to the occupied territories. There was considerable structural pressure on firms to adapt to these new realities, however, the *Wehrmacht* and OT's orders were very attractive as well. The use of force thus never became a major factor in this state-business relation.

Accordingly, there is little in the archival material to suggest that companies stopped functioning as independent economic entities under the OT, or that the basics of their entrepreneurial decision-making changed fundamentally during the Third Reich's 12-year reign. Most of the time, their relation to the OT was based on a cooperative spirit. The construction sector under the National Socialist rule was of course subject to state regulation and structural pressures, particularly during the war and in the occupied territories. The National Socialist economy was certainly "steered" and "deformed"; however, I have argued that the construction industry's ample scope of action resulted from the specific structure of the construction sector. Apart from being highly fragmented, the sector was characterised by high information asymmetries. On the one hand, these information asymmetries lay in the sector's nature, as construction works are mostly unstandardised. On the other hand, the sheer mass of projects and the vast size of the area in which the OT carried them out made the close monitoring of these projects very difficult.

When trying to place these findings on the OT and the construction sector within the broader debate on entrepreneurial freedom in the Third Reich, it is important not to compare National Socialist economic policy to some ideal type of an unregulated "anything goes" market economy. The National Socialists certainly interfered with private business. However, neither in the other belligerents' economies nor in pre-1933 Germany was state regulation of the economy something unheard of. Jochen Streb has illuminatingly pointed out that the National Socialist

https://doi.org/10.1515/9783110694291-006

economic system was not an atypical reaction to the crisis of capitalism in the early 1930s. British and American economic policy in fact showed similarities with the approach of the National Socialists, albeit later than in Germany, that is, during the war.[1] With regard to Germany, Dieter Ziegler has noted that also "[. . .] the Weimar Republic had anything but a liberal free market economy."[2] One might add that construction projects are highly dependent on political decision-making processes and authorisation procedures and are subject to manifold regulations in today's market economies as well.

Secondly, the analysis has shown how dependent *Einsatzgruppe Wiking* was on its contractors' expertise, experience and administrative manpower. Be it the drafting of contracts, price calculations, the monitoring of the works or the evaluation of forced labourers' performance – none of this was possible without private industry's support and self-responsibility. The OT has often been described as an organisation that took over and absorbed the German construction industry. This thesis has provided a different interpretation: it seems more accurate to speak of a National Socialist state authority that was at many levels taken over by private business, albeit within an increasingly regulated framework. This holds true for the Organisation's activities in the occupied territories in particular. At the highest level, without doubt, political decision-makers and high-ranking officials set the course. Below that, however, little could be done without the support and against the will of the Organisation's contractors. In occupied Europe, the OT's answer to an increasingly fierce total war was not less, but more industrial self-responsibility.

Jochen Streb's research on public procurement contracts in construction, presented in chapter I.1.c, has formed the theoretical background for this thesis. As a result of my analysis, Streb's thoughts can be modified in some points – or rather, certain aspects should be accentuated with respect to the OT's activities in the occupied territories. First, Streb has pointed out overlapping interests of construction firms and the administration *vis-à-vis* superior state authorities, interests which found expression in lax monitoring or even corruption.[3] I have argued that in occupied Europe in particular, personal and institutional entanglements between firms and *Einsatzgruppen* were another main characteristic of the OT that must be taken into consideration. Second, when trying to make sense of the OT's strategies and

1 Streb, 'Wirtschaftssystem', 63, 79–83.
2 Ziegler, 'Regulated Market Economy', 140. Similarly, Horst Dreier has pointed out the tradition of the "[. . .] German interventionist and economic state; in the 1920s, the *Hermann-Göring-Werke* could have been established as the *Friedrich-Ebert-Werke*, without at the time being perceived as a breaking with own traditions [. . .]": Horst Dreier, 'Rechtszerfall und Kontinuität. Zur asynchronen Entwicklung von Staatsrecht und Wirtschaftssytem in der Zeit des Nationalsozialismus', in *Wirtschaftskontrolle und Recht in der nationalsozialistischen Wirtschaft*, ed. Dieter Gosewinkel (Frankfurt am Main: Vittorio Klostermann, 2005), 70.
3 Streb, 'Preisregulierung', 34–35, 45–47.

conduct, I consider *time* to be a crucial factor. Both at the *Westwall* and later in occupied Europe, the Organisation was working under massive time pressure, which often outweighed budgetary concerns during negotiations.

The OT's *modus operandi* came at a price. German construction firms exploited the considerable information asymmetries and realised sizeable profits from their projects. However, it would be wrong to characterise the OT as a "failed organisation". In the face of time pressure, a lack of administrative personnel and monitoring problems, letting the industry take over yielded results. With the typical mixture of astonishment, admiration and revulsion that can occasionally be observed in the Allied experts who interrogated leading Nazis and tried to understand the German war economy after 1945, British intelligence concluded that the OT "has carried out in the space of a little over five years, the most impressive building programme since Roman times."[4] In this respect, the OT was a success – a success, of course, that was achieved on the back of hundreds of thousands of forced labourers and with disastrous consequences for Germany's and the occupied countries' budgets.

Furthermore, this thesis has contributed with new findings to the history of occupied Norway. I have argued that the overall cost of the German occupation of Norway was considerably higher than hitherto assumed. Based on an analysis of the OT's funding mechanisms, I have shown that the massive German construction investments in particular were not exclusively financed by Norway, but to a considerable degree through the German war budget. Besides gaining for the first time a better understanding of the financial aspects of the OT's mode of operation, these findings help to explain Norway's rapid post-war economic upswing. With regard to the importance of German investments and financial contributions, Norway was probably an exception among the occupied territories. However, future research will have to assess the OT's impact on other "occupied economies", too. Finally, the thesis has argued that the German demand for construction works was primarily an additional demand. It did not entirely suppress local Norwegian construction activities and thus left an economic room for manoeuvre for Norwegian construction businesses that tried to avoid economic collaboration.

*

Every thesis raises at least as many new questions as it tries to answer. Some of these questions promise important insights into the *Organisation Todt*, its relation to the German construction industry, and the construction sector in the Third Reich in general. However, it would have gone beyond the scope of this thesis to address them.

Firstly, the question of whether it is possible to readily transfer the findings of this thesis and other research on the OT in Western and Northern Europe to the

---

4 *Handbook of the Organisation Todt*, 1.

eastern theatre of war remains open. Second World War historians have repeatedly highlighted the fundamental differences between German warfare in Western and Eastern Europe, the latter being the scene of a war of annihilation. More research is thus necessary on the specific *modus operandi* of the OT on Polish and Soviet territory. There, German firms much more often operated as mobile units close to the front zone. It may well be possible that the military element in the state-business relation was stronger under these conditions and entrepreneurial room for manoeuvre more limited than in Western and Northern Europe. It has already been mentioned that this perspective would have to include questions on the participation of OT units in war crimes and genocide.[5]

A second aspect touched upon only briefly in this thesis is the relation between the German construction firms and their local subcontractors. Did the German firms use the OT apparatus to establish contacts and even use it to annex their subcontractors' machinery and workers? Or was it rather the OT that let German contractors recruit local firms on its behalf? Such a study would help to cast light on the conditions of local economic collaboration in wartime Europe and the role that German private industry played in mobilising the occupied economies. In the Norwegian case, such a study could use the extensive material of the *Landssvikarkiv* as a starting point.[6]

A third aspect calling for a deeper analysis concerns the post-war careers of the members of the former OT apparatus. The boom of the *Wirtschaftswunder* years secured plenty of work not only for German construction firms, but also for the former co-workers of Todt and Speer, whether in the West German building administration or in private industry. In general, the Allies treated former members of the building administration rather leniently. Experts were needed everywhere to manage the reconstruction of (West) Germany.[7] Xaver Dorsch returned to the private sector and founded the Dorsch Consult Ingenieurgesellschaft, a consulting firm for construction projects, in 1951.[8] The former leader of EW, Willi Henne, advanced to become the head of the Hessian road administration after his return from Soviet captivity. He retired in 1972.[9] Henne's right-hand man as GB Bau's representative in Norway, Max-Erich Feuchtinger, founded a consulting firm, too.[10] Günther Schulze-Fielitz from the GB Bau joined the construction firm Hochtief in 1948 and became a

5 See above, chapter IV.1.b, fn. 68.
6 For the time being, Sæveraas' chapter on the Norwegian contractors under *Einsatzgruppe Wiking* remains the most important contribution in this respect: Sæveraas, 'Beton macht Geschichte', chap. 4.
7 Fuchs and Everts-Grigat, *Züblin*, 77.
8 Lemmes, 'Arbeiten für das Reich', 413.
9 Hatlehol, 'Norwegeneinsatz', 1–2.
10 See above, chapter IV.4.a, fn. 305.

member of the management board and later supervisory board.[11] In addition, many men who had been delegated to the OT returned to their former positions in private industry. A thorough network analysis could possibly reveal and quantify the extent of the links between the National Socialist building administration, particularly the staff of the OT, GB Bau and GdSt, and the private construction industry both during and after the war.

11 Botzet, 'Generalbevollmächtigter', 119, 137.

Appendix

A1 Total German Construction Volume

The construction volume by and large reflects the turnover of the German construction industry and handicraft firms. The figure shows how much money could be earned in the sector and is a first indicator of the businesses' economic situation. Four institutions gathered statistical data on the construction sector during the Third Reich: The *Reichs-Kredit-Gesellschaft* (ERKA), the *Deutsche Bau- und Bodenbank*, the SRA and the office of the GB Bau.

The official publications of the SRA would be a natural starting point for assessing the total output of the construction sector. However, the data in their statistical yearbooks and in the 1949 *Statistical Handbook* of Germany are patchy, and specify construction investments for different sectors only up to 1934, transport investments up to 1938 and investments in residential building up to 1940.[1] Moreover, as the figures of the SRA only refer to investments, they do not include the spending on repairs and maintenance. One document, however – a confidential report from September 1941 – reveals the SRA's estimate on total German construction output.[2] The SRA relied on information on the aggregate wages of *Hochbau* and *Tiefbau* workers as reported by the employers' liability insurance associations. Assuming that the share of labour costs in *Hochbau* was 33 percent from the mid-1930s and 40 percent in *Tiefbau*, the SRA could estimate the total production volume by this indirect method.[3] As will be discussed in appendix A2, the method had some shortcomings that led the SRA to underestimate the construction volume.

The ERKA and the *Deutsche Bau- und Bodenbank* also estimated the German construction output. Their total numbers are higher than those of the SRA, but follow the same trend. When the office of the GB Bau, after its establishment in late 1938, began to gather material on the construction sector, it obviously relied on the estimates of the *Deutsche Bau- und Bodenbank* and the ERKA, as it can be seen from Table A1.1.[4]

The GB Bau then continued the statistics during the war. The result was a compilation of statistical data issued by the GB Bau, entitled *Die deutsche Bauwirtschaft im Kriegseinsatz*. So far, this survey appears to be lost, but it has left traces in numerous

1 *Statistisches Handbuch von Deutschland 1928–1944*, 609–11.
2 BArch, R 3102/2701, fol. 3–10, confidential report on construction turnover 1929–1940, 10.9.41. The document was discovered and used by Scherner, 'Investment Pattern'.
3 The construction volume calculated on the basis of the wage bills did not include *Regiearbeiten*, that is, works that public authorities carried out with their own workforce. According to the SRA, this applied largely to work on the tracks of the National Railways. This is only a minor problem, though, since we are primarily interested in the sums earned by private construction firms.
4 BArch, R 2301/5342, Deutsche Bau- und Bodenbank, 'Die Entwicklung der deutschen Bauwirtschaft im Jahre 1937'.

https://doi.org/10.1515/9783110694291-007

Table A1.1: Estimates of German construction volume 1928–44, million RM, current prices, including repairs and maintenance. For area covered, see table footnote.*

	German Statistical Office	Dt. Bau- und Bodenbank	Reichs-Kredit-Gesellschaft	GB Bau
1928	.	8,500	.	8,500
1929	8,550	8,300	8,300	.
1930	.	6,500	.	.
1931	.	3,500	.	.
1932	2,370	2,300	.	2,300
1933	3,100	3,200	3,200	3,200
1934	.	5,200–5,700	5,700	5,500
1935	6,430	7,000–7,500	7,500	7,200
1936	.	8,500–9,000	9,000	8,700
1937	8,950	9,000–10,000	10,000	9,500
1938	10,720	.	12,000	12,000
1939	12,370	.	.	13,200
1940	10,420	.	.	8,800
1941	.	.	.	9,100
1942	.	.	.	7,200
1943	.	.	.	6,640
1944	.	.	.	6,364

*German Statistical Office: BArch, R 3102/2701, fol. 3–10, confidential report on construction turnover 1929–1940, 10.9.41. ERKA: Reichs-Kredit-Gesellschaft, *Deutschlands wirtschaftliche Lage in der Jahresmitte 1939* (Berlin: Reichs-Kredit-Gesellschaft, 1939), 6. *Deutsche Bau- und Bodenbank*: Deutsche Bau- und Bodenbank, 'Die Entwicklung der deutschen Bauwirtschaft im Jahre 1937', 5. GB Bau: see Table A1.2. For 1928–1938, the area covered is Germany with its pre-1938 borders, that is, not including Austria and the Sudetenland. From 1939, figures refer to Greater Germany with its shifting borders. Therefore, the growth in production volume between 1938 and 1939 stems mostly from the increase in the area covered. The GB Bau gives also a figure for 1938 including Austria and the Sudetenland: 13 billion RM. The SRA, in turn, provides figures for Germany with pre-1938 borders also for 1939 (11.4 billion RM) and 1940 (9.5 billion RM).

documents and publications. First, the figures on the construction output from the survey's April 1943 edition were used by *Oberregierungsbaurat* Joachim Steffens from the GB Bau office. His name is frequently mentioned in wartime documents on the topic and he is likely to have been instrumental in compiling the GB Bau survey.[5] Presumably in 1944, Steffens finished an unpublished manuscript on the history of the building trades that includes the GB Bau estimates on building production.[6] To allow for price changes, Steffens used the GB Bau's own construction price index that registered a considerable rise in construction prices especially after the outbreak of

5 The Statistical Office, for example, compare their September 1941 calculations with the figures of Steffens: BArch, R 3102/2701, fol. 10. On Steffens, see above, chapter III.1.d, fn. 73.
6 BArch, R 3/Film 77815, Steffens, 'Bauwesen'.

the war.[7] The same output figures and price index were then used in the *Report on the German Economic Situation 1943/44* (*Bericht zur deutschen Wirtschaftslage 1943/44*) that was prepared by the Planning Office in the Armaments Ministry in spring 1944.[8] The data of this report became the basis for all post-war accounts of the construction sector.[9]

The advantage of the GB Bau material is that it provides a comprehensive set of data from the late 1920s until 1944 for the construction output within the German borders (Table A1.2). More importantly, the GB Bau also calculated a construction price

Table A1.2: GB Bau's estimate of German construction turnover 1928/32–44, including repairs and maintenance, million RM, current area of "Greater Germany" (from 1938 including Austria and Sudetenland), and construction price index of GB Bau/Scherner.*

	Current prices	1938 = 100	GB Bau/Scherner construction price index, 1938 = 100
1928	8,500	6,600	129
1932	2,300	2,370	97
1933	3,200	3,460	93
1934	5,500	5,690	97
1935	7,200	7,540	96
1936	8,700	9,000	97
1937	9,500	9,600	99
1938	13,000	13,000	100
1939	13,200	12,800	103
1940	8,800	8,265	107
1941	9,100	6,855	133
1942	7,200	4,671	154
1943	6,640	4,000	166
1944	6,364	3,700	172

*Steffens, 'Bauwesen', fol. 40. For 1943–44, see *The Effects*, 55; Wagenführ, *Deutsche Industrie im Kriege*, 160. Construction price index for 1943–44 as estimated by Scherner, 'Preparation', 466. Note that Steffens' manuscript contains a typing error assuming for 1933 a construction volume in current prices of 3.7 billion RM instead of 3.2 billion RM.

7 For the construction price index see Table A1.2.

8 Jonas Scherner, 'Bericht zur deutschen Wirtschaftslage 1943/44. Eine Bilanz des Reichsministeriums für Rüstung und Kriegsproduktion über die Entwicklung der deutschen Kriegswirtschaft bis Sommer 1944', *Vierteljahreshefte für Zeitgeschichte* 55, no. 3 (2007): 499–546.

9 For example: *The Effects*, 55; Wagenführ, *Deutsche Industrie im Kriege*, 56; Eichholtz, *Kriegswirtschaft II/2*, 378; Burton H. Klein, *Germany's Economic Preparations for War* (Cambridge/Mass.: Harvard University Press, 1959), 105. Note that Klein's figures are in 1938 prices instead of 1939 prices, as he wrongly indicates.

index through 1942 that displayed the price increases for construction works more realistically, especially after the outbreak of the war. For 1943–44, the price trend has been estimated by Scherner.[10] The higher nominal prices to a certain degree covered up how fast the real output of the sector decreased between 1939 and 1945. Some caution is advised, however, as the GB Bau figures cover a shifting terrain. Between 1939 and 1941, the Greater German Reich expanded significantly, while in 1944, territory was lost to the Allied forces.[11] Therefore, the GB Bau figures presumably underestimate the contraction of the construction sector that firms within the *Altreich* (Germany with its pre-1938 borders) faced between 1939 and 1941, while some of the decrease between 1943 and 1944 has to be attributed to the shrinking territory of Germany.[12]

A2 Hochbau, Tiefbau

There are two documents that contain information on *Hochbau* and *Tiefbau* output in the 1930s. The first is the SRA's 1941 confidential report on construction turnover, which contains a time series on *Hochbau* and *Tiefbau*.

The figures reflect the growing importance of the *Tiefbau* sector over the course of the 1930s. The share of *Tiefbau* production in total output, however, seems questionable. Can we really assume that in the late 1930s, when works in the transport sector, including the *Autobahn*, on the *Westwall* construction sites and for the *Wehrmacht* were in full swing, the *Tiefbau* share was only 23 percent? And that it was not more than 25 percent in 1940, when residential construction had already been cut back substantially? These figures seem unlikely and contradict other evidence available on the development of *Hochbau* and *Tiefbau* during the 1930s. The reason for this false estimation might be that the SRA relied on the wage bills as reported by the 12 employers' liability insurance associations of the handwork sector (*Baugewerks-Berufsgenossenschaften*) and the respective national association for the *Tiefbau* sector (*Tiefbau-Berufsgenossenschaft*). The SRA assumed that workers covered by the former associations were engaged in *Hochbau*, and those covered by the *Tiefbau-Berufsgenossenschaft* worked in *Tiefbau*. In contrast to earlier times, this assumption was no longer valid under the National Socialist building boom. As

10 Scherner, 'Preparation', 466.

11 Cf. Tooze, 'No Room', 444.

12 Another set of statistics was compiled by the American industrial statistician Edward O. Bassett in a USSBS Special Report in 1946. Bassett was able to draw upon data collected by the peak organisation of the German industry, *Reichsgruppe Industrie*. However, the *Reichsgruppe* took into account only larger construction industry firms and did not include smaller firms and the considerable handicraft branch, both of which suffered most from the outbreak of the war. Thus, Bassett underestimated both the total construction volume in Germany and the rapid contraction of the construction sector between 1939 and 1945. See Tooze, 454; Edward O. Bassett, *Industrial Sales, Output and Productivity. Pre-War Area of Germany 1939–44 (Special Paper No. 8)*, United States Strategic Bombing Survey, 1946.

Table A2.1: SRA's estimate of German construction turnover 1929–40, million RM, current prices, including repairs and maintenance, with pre-1938 borders.*

	Tiefbau	Tiefbau %	Hochbau	Total
1929	1,400	16	7,150	8,550
1932	320	14	2,050	2,370
1933	500	16	2,600	3,100
1935	1,280	20	5,150	6,430
1937	1,800	20	7,150	8,950
1938	2,500	23	8,220	10,720
1939	2,630	23	8,790	11,420
1940	2,380	25	7,100	9,480

*BArch, R 3102/2701, fol. 7.

has been mentioned on several occasions throughout this work, many members of handwork firms reacted to the rapidly changing demand and engaged in *Tiefbau* works, but they obviously remained members of their original associations. One of the leading figures of the *Tiefbau* industry, Adolf Mast, already stated in early 1934 that the reported figures on wage bills in *Tiefbau* were too low, because "[. . .] a considerable share of construction firms, which executed both *Tiefbau* and *Hochbau* works (so-called *Mischbetriebe*), are not members of the *Tiefbau-Berufsgenossenschaft* and pay the taxes on their wage bills to the employers' liability insurance associations of the handwork sector."[13] And their Bavarian branch, for example, reported in 1938 that many of its members were actually working at the *Westwall*, a *Tiefbau* project.[14] Hence, the SRA method underestimated the increasing output of the *Tiefbau* sector after the National Socialists seized power. This may also be the reason why the SRA's estimate of the total German construction output was too low (see above, Table A1.1): the SRA assumed that wages made up 40 percent of the production value in *Tiefbau* construction and 33 percent in *Hochbau* construction. It then multiplied the wage bills with these percentages to arrive at the total German construction output. Thus, the underestimation of the *Tiefbau* production with its higher share of wage costs reduced the SRA's overall estimate of the German construction output.

Another time series can be derived from newly discovered documents from the National Archives in Oslo. They make it possible to reappraise the *Hochbau*

13 Adolf Mast, 'Das Tiefbaugewerbe an der Jahreswende', *Deutsche Bauzeitung* 68, no. 1 (1934). As long as the share of their *Tiefbau* works did not exceed 40 percent of their total turnover, firms were allowed to insure all of their workers with one of the employers' liability insurance associations of the handwork sector. Cf. Gorgass, *Baugewerbe*, 141.

14 Bayerische Baugewerks-Berufs-Genossenschaft, Munich, business report 1938.

and *Tiefbau* shares in the German construction volume. They provide figures on construction output up to 1941, split up into different areas of building production.[15] The provenance of this statistical material is somewhat unclear. It is attached to the minutes of the second meeting of the "Community of Experience for Construction Volume and Capacity" (*Erfahrungsgemeinschaft für Bauvolumen und -kapazität*) on 8 May 1941 in the Armaments Ministry in Berlin. The purpose of this *Erfahrungsgemeinschaft* was to find ways to improve the statistical coverage of the construction sector. It was led by Hans Meyer-Heinrich, board member of the Philipp Holzmann AG, a committed National Socialist, and the company's most important link to the Armaments Ministry, OT and GB Bau.[16] The participants included representatives of different organisations and offices like the SRA, Wigru Bau and the GB Bau. It is not clear which of the participants prepared the statistical material for the meeting, neither is it possible to say how it ended up in Oslo. However, when we compare it with the figures of the GB Bau, it becomes clear that the figures on construction output provided in the *Erfahrungsgemeinschaft* documents are obviously based on the GB Bau.[17] Another piece of evidence for this assumption is the fact that the author of the *Erfahrungsgemeinschaft* documents used the GB Bau's construction price index.

15 RAO, RAFA-2188/2/Hfa-L0007, file 26 Statistikk og utredninger.

16 On Meyer-Heinrich, see above, chapter II.1.b, fn. 52.

17 Note the difference in the GB Bau construction output for 1938 between Table A2.2 and Table A1.2. The reason for this is that the value for 1938 in Table A2.2 does not include Austria and the Sudetenland. The *Erfahrungsgemeinschaft* documents show a slightly lower construction output from 1939 on, most likely because they refer to the *Altreich*. For 1938, where also the GB Bau figure in Table A2.2 refers to the *Altreich*, the values are almost identical.

Table A2.2: Total German construction output according to the Erfahrungsgemeinschaft (pre-1938 borders) and to the GB Bau (pre-1938 borders, area of Greater Germany from 1939); estimate of Hochbau und Tiefbau production. In million RM, constant 1938 prices.*

	Hochbau	Tiefbau	Total (Oslo doc.)	Total (GB Bau)
1928	4,072	2,522	6,594	6,600
1929	3,865	2,991	6,157	.
1930	3,469	1,588	5,057	.
1931	2,119	1,145	3,264	.
1932	1,552	1,030	2,582	2,370
1933	1,898	1,599	3,497	3,460
1934	3,595	2,230	5,825	5,690
1935	3,680	3,479	7,159	7,540
1936	4,872	4,306	9,178	9,000
1937	4,810	5,029	9,839	9,600
1938	5,483	6,366	11,849	12,000
1939	5,146	6,019	11,165	12,800
1940	2,991	4,486	7,477	8,265
1941	1,955	4,586	6,541	6,855
1942	.	.	4,403	4,671
1943	.	.	.	4,000
1944	.	.	.	3,700

*For total construction output (GB Bau), see Table A1.2. For total construction output according to the *Erfahrungsgemeinschaft* documents, see RAO, RAFA-2188/2/Hfa-L0007, file 26 Statistikk og utredninger. 1940–42: Figures for *Kriegswirtschaftsjahre*. 1942 estimated by the document's author. My estimate of the *Hochbau* and *Tiefbau* shares was calculated as follows: The *Erfahrungsgemeinschaft* documents provide direct values for 1939–41. For 1928–38, total construction output is split up into "residential buildings", construction of "administration and commercial buildings", "traffic infrastructure" and "further construction works". While the two former categories can be attributed to *Hochbau* and the construction of traffic infrastructure to *Tiefbau*, we do not know the *Hochbau*/*Tiefbau* ratio in the category "further construction works". However, for 1928, 1936 and 1938, the document gives values for *Hochbau* and *Tiefbau*. The *Tiefbau* share in total German construction output was 38.2 percent in 1928, 46.9 percent in 1936, and 53.7 percent in 1938. Hence, we can calculate the *Hochbau*/*Tiefbau* ratio in the category "further construction works" for these three years. The *Tiefbau* share in "further construction works" was 82.9 percent in 1928, 81.4 percent in 1936 and 79.2 percent in 1938. I assume a linear decrease in the *Tiefbau* share between 1928 and 1936 (-0.1875 points per year) and between 1936 and 1938 (-1.1 points per year), respectively. Since the production figures for the four categories mentioned above are in current prices, I used the construction price index of the GB Bau/Scherner to convert them to 1938 prices (GB Bau's price index for the years 1929–31 can be reconstructed from the *Erfahrungsgemeinschaft* documents). It is possible to cross-check this estimate of the *Hochbau* and *Tiefbau* share, since the *Erfahrungsgemeinschaft* documents provide an aggregate figure for the years 1933–38 (50 percent *Tiefbau*, 50 percent *Hochbau*). With the estimation method applied here, one arrives at a very similar result (48.6 percent *Tiefbau*, 51.4 percent *Hochbau*).

A3 Public Construction

Table A3.1: Production volume in private construction, public construction and housing, 1924–38. In billion RM, current prices.*

	Public construction	Private construction	Housing construction . . .	. . . of which privately financed
1924	1.1	1.2	1.0	0.4
1925	1.8	2.2	1.7	0.7
1926	1.9	1.8	1.9	0.7
1927	2.4	2.5	2.9	1.5
1928	2.7	3.0	2.8	1.4
1929	2.7	2.7	2.9	1.6
1930	1.7	2.4	2.4	1.3
1931	1.0	1.3	1.2	0.7
1932	0.9	0.6	0.8	0.6
1933	1.7	0.6	0.9	0.7
1934	3.5	0.8	1.4	1.1
1935	4.9	1.0	1.6	1.2
1936	5.4	1.4	2.2	1.9
1937	6.1	1.8	2.1	1.7
1938	7.9	2.1	2.0	1.7

*Reichs-Kredit-Gesellschaft, *Deutschlands wirtschaftliche Lage*, 6; Deutsche Bau- und Bodenbank, 'Die Entwicklung der deutschen Bauwirtschaft im Jahre 1937', 5. On the public/private ratio in housing construction, see BArch, R 3102/2731, table on German investments 1924–38, 21.3.39.

A4 German Construction Investments

Compared to the figures on total construction output, data on construction investments in different sectors of the economy make it possible to analyse the development in much greater detail. Table A4.1 is an attempt to estimate construction investments in the narrow sense (that is, excluding expenses for repairs and maintenance) for Germany. From 1928 to 1940, the figures refer to Germany with its pre-1938 borders; from 1941 on, they cover Greater Germany with its shifting borders. Figures for the transport sector cover different areas, as explained below. The industry category contains the construction investments of the German industry, the category public utilities covers investments in public electricity, gas and water supply. Transport investments can be split up into investments for railways, the *Autobahn* network, roads, waterways and harbours and the category remaining investments in transport that contains among other things motor traffic, shipping and air traffic. Housing and agriculture form distinct categories, while investments of crafts and trades and remaining parts of the private sector are subsumed under further private sectors. The last category is further public construction, which over the

Table A4.1: Estimate of German construction investment (i.e. excluding repairs and maintenance), 1928/1933–44, million RM, nominal prices.

	1928	1933	1934	1935	1936	1937	1938	1939	1940	1941	1942	1943	1944
Industry	575	123	267	421	651	851	1,145	1,392	1,628	2,002	2,181	1,834	1,903
Public utilities	379	73	107	127	143	157	198	222	185	185	185	185	185
Transport	1,063	637	1,038	1,282	1,532	1,668	2,445	2,726	1,936	2,213	1,624	818	665
of which: railways	*292*	*97*	*194*	*183*	*183*	*212*	*368*	*576*	*555*	*852*	*677*	*584*	*512*
of which: Autobahn	*0*	*3*	*178*	*467*	*692*	*679*	*916*	*1,117*	*505*	*506*	*146*	*108*	*27*
of which: roads, waterways, harbours	*732*	*527*	*651*	*610*	*631*	*746*	*1,115*	*987*	*830*	*809*	*755*	*<80*	*<80*
of which: remaining investment in transport	*39*	*10*	*15*	*22*	*26*	*31*	*46*	*<46*	*<46*	*<46*	*<46*	*<46*	*<46*
Agriculture	95	60	73	80	90	100	125	130	100	100	86	83	48
Housing	2,825	875	1,353	1,563	2,207	2,120	2,150	1,500	650	444	278	222	167
Further private sectors	420	163	175	188	200	200	200	200	75	75	75	75	75
Further public construction	1,833	283	923	932	1,377	1,242	2,125	3,255	2,419	2,555	1,575	<1,613	<1,135
Total investment in construction	**7,190**	**2,214**	**3,936**	**4,593**	**6,200**	**6,338**	**8,388**	**9,425**	**6,993**	**7,574**	**6,004**	**4,830**	**4,178**

course of the 1930s came to contain not only more and more military construction projects like the *Westwall*, but also schools or projects on behalf of the NSDAP. Total investments in those sectors, consisting of machinery, equipment and construction, have recently been reassessed by Jonas Scherner.[18] We can take Scherner's table as a starting point for an estimation of German construction investments.

For some construction investments, we have direct values at hand, others can only be estimated indirectly. The SRA determined the shares of construction investments in total investments for the year 1938.[19] These were 37 percent in public utilities, 10 percent in agriculture, and 25 percent in further private sectors. It has been assumed that these shares remained more or less stable. Furthermore, since investments in those sectors were not very high, potential changes in these shares would only have a very limited effect.[20]

This procedure is not possible in the case of industry, because here, the share of construction investments in total investments increased significantly over the course of time. Scherner estimated industrial construction investments for the years 1936–1943.[21] As the figures show, the share increased from around 25 percent between 1936 and 1938, to almost 26 percent in 1939. During the war, the share quickly rose to over 33 percent in 1943. The reason for this is that the SRA tried to conceal from the public the real, growing extent of German rearmament and therefore omitted considerable investments in armament plants from its 1938 figures. Because the construction ratio in these hidden investments was higher than in the remaining industrial investments, an estimate based on the SRA's 1938 ratio would underrate the industrial construction investments.[22] To estimate construction investments for 1928 and 1933, I employ the SRA's "official" ratio for the mid-1930s (22 percent), because in those years, there were no hidden investments in armament.[23] For 1934–35, I assume that the share of construction investments rose linearly from the 1933 value to the 1936 value calculated by Scherner.[24] For 1944, I assume that the construction share remained on the level of 1943.

Investments in housing and further public construction are directly taken from Scherner, since they consisted exclusively of construction investments.[25]

Construction investments for the *Autobahn* already refer to Greater Germany from 1938 on, because the *Reichsautobahnen* Corporation did not specify which investments were made outside the borders of the *Altreich*, for example in Austria. I apply the

18 Scherner, 'Investment Pattern', 509.
19 BArch, R 3102/2701, fol. 4, confidential report on construction turnover 1929–1940, 10.9.41.
20 Scherner, 'Investment Pattern', 515.
21 Scherner, 'Preparation', 467.
22 Scherner, 455–58.
23 BArch, R 3102/2701, fol. 36.
24 This means: 22.9 percent in 1934 and 23.7 percent in 1935.
25 Scherner, 'Investment Pattern', 509.

method used by the SRA during the 1930s on the figures given by Hartmannsgruber.[26] The SRA excluded expenditure on land acquisition by subtracting it from total expenditure, since it does not constitute construction investment in the narrow sense. Construction investment in railways consists of the National Railway's construction investment and the investment in local lines, underground and tramway networks.[27] To estimate the construction investment in local lines, underground and tramway networks between 1933 and 1938, I applied the construction ratio of the National Railways on the investment figures of the other railways given in the *Statistical Handbook*.[28] I then calculated the construction investment of the other railways until 1944 by assuming that the ratio between the construction investment of the National Railways and the construction investment of the other railways remained stable on its 1938 level.[29] Railway construction investment refers to Greater Germany from 1938 on, too. Construction investment in roads, waterways and harbours can be calculated until 1942 based on data provided by the SRA.[30] They refer to Germany with its pre-1938 borders. For 1943 and 1944, I rely on information by Walter Fey, construction statistician in the Planning Office in the Armaments Ministry.[31] According to Fey, road construction of all kinds, including repairs, amounted to a maximum of 60 million RM per year in 1943 and 1944. This figure is an upper estimate, especially since repairs are included. In a second step, I added 20 million per year for waterways and harbours, because they accounted for approx. 25 percent of the expenditure in roads, waterways and harbours in the second half of the 1930s. In the category remaining investments in transport, construction amounted to 6 percent of the

26 Hartmannsgruber, 'Finanzierung', 675.
27 For the construction investment of the National Railways, see Mierzejewski, *Most Valuable Asset*, 73, 155. Cf. the annual statistical reports of the National Railways in BArch, R5/12222–12232.
28 *Statistisches Handbuch von Deutschland 1928 – 1944*, 606.
29 Note that the investment figures for the other railways in the *Statistical Handbook* refer to Germany with its pre-1938 borders, while the National Railways' investment figures for 1938 already include Austria and the Sudetenland. After the annexation of Austria, 7.5 percent of the National Railways' total track was on former Austrian territory. See BArch, R 5/12227, Statistische Angaben über die Deutsche Reichsbahn im Geschäftsjahr 1938, 23. To calculate the correct ratio, I therefore subtracted 7.5 percent from the National Railways' construction investment of 1938. I did not use this method in the case of the Sudetenland. The National Railways took over the railway network of the Sudetenland in the course of October 1938, and I assume that railway construction activity in this area until the end of the year was not significant compared to the overall figures for 1938, especially since construction activity was low during wintertime.
30 For 1928/1933–38, see *Statistisches Handbuch von Deutschland 1928 – 1944*, 606. For 1939 and 1940, see Scherner, 'Investment Pattern', 514. For 1941 and 1942, I employed the method used by Scherner for 1939–40.
31 *The Effects*, 59.

investments.[32] For 1928–1938, data on remaining investment in transport can be obtained from the *Statistical Handbook*.[33] For the time of the war, no such data are available; I therefore assume that the remaining investments in transport remained constant on their 1938 level. This is certainly an overestimation; however, this construction investment was rather insignificant in absolute numbers and therefore does not distort the overall picture.

32 Scherner, 'Investment Pattern', 515.
33 *Statistisches Handbuch von Deutschland 1928 – 1944*, 606.

A5 Return on Equity of German Construction Joint-Stock Companies

Table A5.1: Return on equity of German construction joint-stock companies as reported in their published commercial balance sheets, 1886–1900, 1907–13, 1924–39, in %.

1886	**1887**	**1888**	**1889**	**1890**	**1891**	**1892**	**1893**	**1894**	**1895**	**1896**	**1897**	**1898**	**1899**	**1900**	
2,44	3,08	4,51	6,92	4,84	7,12	5,42	3,99	3,70	4,02	3,4	3,96	4,36	4,93	4,04	
1907	**1908**	**1909**	**1910**	**1911**	**1912**	**1913**									
8,48	6,85	7,58	9,29	5,91	8,40	4,39									
1924	**1925**	**1926**	**1927**	**1928**	**1929**	**1930**	**1931**	**1932**	**1933**	**1934**	**1935**	**1936**	**1937**	**1938**	**1939**
5,32	2,44	6,31	8,13	8,31	1,72	0,01	-15,56	-12,67	2,31	4,45	5,28	9,08	7,01	10,89	12,12

1886–1900: Eduard Wagon, *Die finanzielle Entwicklung deutscher Aktiengesellschaften von 1870–1900 und die Gesellschaften mit beschränkter Haftung im Jahre 1900*, Sammlung nationalökonomischer und statistischer Abhandlungen des staatswissenschaftlichen Seminars zu Halle a.d.S. 39 (Jena: Gustav Fischer, 1903), 199. 1907–1913: Kaiserliches Statistisches Amt, ed., *Die Geschäftsergebnisse der deutschen Aktiengesellschaften: im Jahre 1907/08 [to] 1913/14*, Vierteljahrshefte zur Statistik des Deutschen Reichs, Ergänzungsheft II (Berlin: Puttkammer u. Mühlbrecht, 1909 to 1915). 1924–1939: *Statistisches Jahrbuch für das Deutsche Reich 1926 [to] 1941/42* (Berlin: Schmidt, 1926 to 1942). Statistisches Reichsamt, ed., *Vierteljahreshefte zur Statistik des Deutschen Reichs 39/II* (Berlin: Schmidt, 1930), 78–94.

A6 Return on Sales of German Construction Firms

As described in the text, the sample of returns on sales of German construction firms is based on 1,094 annual results of 603 firms.[34] This means that results are not available from every firm for all years between 1936 and 1940. The distribution of cases is shown in Table A6.1.

Table A6.1: Distribution of cases in the sample of annual results of 603 German construction firms, grouped according to their annual turnover in 1,000 RM.

	0–100	100–200	200–400	400–800	800–2,000	> 2,000	All firms
1936	8	8	9	22	19	17	83
1937	13	17	20	43	60	42	195
1938	13	32	43	74	107	104	373
1939	8	17	38	51	105	103	322
1940	6	10	25	25	32	23	121
1936–1940	48	84	135	215	323	289	1,094

Table A6.2 shows the average taxable returns on sales that the firms in the sample realised between 1936 and 1940. The results have been grouped according to the firms' annual turnover. The low value in 1939 for firms with an annual turnover of less than 100,000 RM is caused by an outlier, that is, one firm that realised an exceptionally high loss. Altogether only seven of the 1,094 annual results in the sample recorded a loss.

For reasons that have been discussed in chapter III.4.a, the figures in Table A6.2 should be regarded as upper estimates. The same holds true for Table A6.3, which shows that the share of annual results with a RoS of more than 10 percent was higher the smaller the firm was. Maybe the auditors focused primarily on larger firms and only bothered to control small firms when they had strong indications for excess profits.

34 The sample has been compiled from five lists with results of audits by the German Court of Audit and the auditing departments of the *Wehrmacht*. Note that the lists partly overlap and contain cross references. They have been carefully evaluated to avoid double counts: BArch, R 2301/7187, fol. 27–41, list Gesamtverzeichnis (1), 1936–38; ibid., fol. 42–43, list Gesamtverzeichnis (2), August 1937 – February 1939; ibid., fol. 138–48, list I, 1939 – May 1941; ibid., fol. 149–86, second version of list I, 9.11.42; ibid., fol. 187–225, list II, July 1942. The lists are based on the so-called special reports of the German tax authorities. There were at least around 1,200 special reports. Reports no. 500 to no. 1172 can be found in BArch, R 2301/7183–7186 and ibid., R 2301/9950–9951.

Table A6.2: Average taxable return on sales of German construction businesses according to the special reports of German tax authorities, grouped according to their annual turnover in 1,000 RM, 1936–40.

	0–100	100–200	200–400	400–800	800–2000	> 2000	All firms
1936	23,17 %	16,35 %	10,85 %	13,67 %	8,53 %	7,85 %	8,30 %
1937	31,75 %	18,90 %	12,42 %	14,31 %	10,50 %	8,64 %	9,52 %
1938	27,68 %	26,48 %	20,89 %	16,38 %	13,11 %	8,92 %	10,30 %
1939	17,95 %	26,72 %	24,36 %	18,77 %	13,67 %	12,40 %	12,94 %
1940	34,07 %	29,75 %	23,66 %	16,27 %	15,96 %	12,62 %	14,54 %
1936–40 (geometric mean)	26,79 %	23,53 %	18,30 %	15,87 %	12,32 %	10,07 %	11,10 %

Table A6.3: Share of annual results with a RoS of more than 10 %, grouped according to annual turnover in 1,000 RM, 1936–40.

	0–100	100–200	200–400	400–800	800–2000	> 2000	All firms
%	88	83	77	72	59	49	64

A7 The "Purely Norwegian" Construction Volume, 1940–1945

According to the Norwegian Statistical Office, the gross production of Norwegian construction firms amounted to 6.29 billion kroner between 1940 and 1943.[35] As described in the text, I assume that the production decreased between 1943 and 1944 by the same factor as it had between 1942 and 1943, and I assume that there was no Norwegian production in the first months of 1945. Thus, one arrives at a very conservative estimate of 7.34 billion kroner for the occupation years. I deflated this figure with the help of an index constructed by the Norwegian Statistical Office (Table A7.1). As the index only covers the years 1939–1943, I estimated the value for 1944 and put it at the same level as the 1943 value. This seems legitimate as both the Norwegian wholesale price index[36] and Korff's index for wages on *Wehrmacht* construction sites (see below) assume a more or less stable price level in 1943 and 1944, too.

The sums flowing from the occupation account in *Norges Bank* to German construction projects amounted to approximately 4.83 billion kroner in current prices. The sums for the individual years can be calculated based on the information given by Hans Clausen Korff.[37] These numbers have to be deflated, too, albeit with a

35 *NOS X. 102*, 51.
36 *NOS X. 102*, 101–2.
37 Hans Clausen Korff, 'Norwegens Wirtschaft im Mahlstrom der Okkupation' (Unpublished manuscript, 1949), pt. II, 243–252.

Table A7.1: Construction price index by the Norwegian Statistical Office (1944: my own estimate) for the Norwegian part of the sector, 1939–44.*

1939	1940	1941	1942	1943	1944
100	141.4	187.2	185	168.6	168.6

**Statistisk-økonomisk utsyn*, 190.

different index. Price increases on construction sites of the German occupiers were higher than in the Norwegian part of the sector, or as Korff put it: "[. . .] the kroner were worth less in the hands of the *Wehrmacht* than in civilian circulation."[38] Hence, Korff calculated an index for Norwegian wages on German construction sites (Table A7.2).

Table A7.2: Wage index for Norwegians on German construction sites, 1939–44.*

1939	1940	1941	1942	1943	1944
100	160	200	210	200	200

*Korff, pt. II, 368.

After deflation, one arrives at a sum of 1.76 billion kroner that Norwegian construction firms were able to earn on "purely Norwegian" construction sites during the occupation years. The firms realised approximately 41 percent of their wartime turnover on Norwegian construction sites.

This figure can be cross-checked against the results of two Norwegian wartime surveys, which concluded that 33.4 percent and 28 percent of the Norwegian workforce in construction worked on purely Norwegian projects in August 1942 and February 1944, respectively.[39] In 1944, the calculation method used here gives a share of 31.2 percent for purely Norwegian turnover and thus fits very well with the wartime survey of February 1944. For 1942, the method gives a share of 47.2 percent for purely Norwegian turnover, a result that deviates significantly from the wartime survey (33.4 percent of the Norwegian workers on Norwegian-financed construction sites). Put differently: while 66.6 percent of the Norwegian construction workforce worked on German sites in August 1942, these activities resulted only in 52.8 percent of the turnover for the Norwegian part of the construction sector. However, as described in chapter V.2.a, it was precisely in the summer of 1942

38 Korff, pt. II, 367.

39 *Statistisk-økonomisk utsyn*, 221; Tuveng, *Arbeidsløshet*, 161–62.

that EW's projects faced massive start-up difficulties. There was a lack of machinery and workers, and project designs were often still unfinished. It is thus not surprising that productivity was extraordinarily low in the initial phase of EW's activities in August 1942.

A8 Expenditure Einsatzgruppe Wiking

Table A8.1: Expenses of Einsatzgruppe Wiking (without Denmark), accounting years 1942–1944 (1 April 1942 to 31 March 1945) and 1 April to 8 May 1945.*

	1942	1943	1944	1945 (1 April to 8 May)	Total
RM account, Berlin	27,550,194	92,498,780	164,335,801	277,363	**284,662,138**
Converted to Kr.	*48,333,673*	*162,278,561*	*288,308,423*	*486,602*	***499,407,259***
Kr. account in Oslo	242,624,857	358,600,000	472,499,974	21,589,547	**1,095,314,378**
Converted to RM	*138,296,168*	*204,402,000*	*269,324,985*	*12,306,042*	***624,329,195***
Total in RM	165,846,362	296,900,780	433,660,786	12,583,405	**908,991,333**
Total in Kr.	290,958,530	520,878,561	760,808,397	22,076,149	**1,594,721,637**

*For 1942, see RAO, RAFA-2188/1/E8a-L0063, Annual account 1942, 28.3.44. In 1943, EW received 86,750,000 RM in operating funds from the OTZ: ibid., Annual account 1943, 18.9.44. In addition the RM account was debited with another 5,748,780 RM from other sources. See the account statements in RAO, RAFA-2188/1/E8a-L0068. For 1944 and April/May 1945, see RAO, RAFA-2188/1/E8j-L0005, Accounting book for the accounting years 1944 and 1945, 23.10.45.

In the following, EW's expenditure is understood as the sums spent from EW's kroner account in Oslo and from EW's reichsmark account at the OTZ in Berlin. As mentioned in the text, *Wehrmacht* deliveries of construction materials added to EW's economic footprint, but cannot be quantified.

EW's expenditure in its three accounting years in Norway has been reconstructed from EW's annual accounts and bank books.[40] Parts of the data exist in two versions, as both EW in Oslo and the OTZ in Berlin kept their own records on EW's expenditure. The sums registered in Berlin are lower throughout than those registered by EW in Oslo. Apparently, Berlin was not always fully informed about all of EW's transactions.[41] I have thus used the information from EW in Oslo, which can be regarded as more reliable whenever information is conflicting.

40 The German state's accounting year lasted from 1 April until 31 March of the following year. Thus, EW's activities in Norway covered exactly three accounting years plus the weeks from 1 April 1945 to Germany's capitulation on 8 May 1945.

41 See for example RAO, RAFA-2188/1/E8a-L0040, file Schriftwechsel OT-Zentrale betr. Betriebsmittelkonto, EW to OTZ, 26.6.44.

A9 Workforce of Einsatzgruppe Wiking

Historical labour statistics can seek to establish – but seldom achieve – an entirely accurate picture of a historical situation. Furthermore, the statistical overviews on EW's workforce are fragmentary and at times contradictory. Moreover, fluctuations in the labour force could sometimes be considerable within the span of a month. The data that have been gathered in Table A9.1 and displayed in Figure 18 are thus a compromise between best possible accuracy and the desire to give the reader a comprehensive and graphic overview over the phenomenon.

Table A9.1 shows the workforce of EW in occupied Norway (that is, without Denmark) between 1942 and 1945. In the summer of 1944, *Einsatz Finnland* and the construction units of the air force became part of EW. Hence, from August 1944, the figures cover Norway and Finland, including the air force construction sites. In the same summer of 1944, EW also took over the construction sites of the German navy. Their workers were first registered under EW after 1 September 1944. In Table A9.1, they are thus only included in the figures of September 1944 and January 1945.

In the summer and autumn of 1942, 450 Norwegian inmates from the Grini camp as well as approximately 50 Norwegian teachers that had refused to become members of the NS-controlled teacher's union *Norges Lærersamband* worked on OT construction sites in Northern Norway for a limited period of time.[42] These political inmates do not seem to have been registered as "Norwegian inmates", but in the category "Norwegian workers".[43] The number of Norwegians working for the *Vegvesenet* is unknown for October, November and December 1943. Numbers have therefore been estimated by interpolation between the numbers for September 1943 and January 1944. Likewise, the missing value for February 1944 has been estimated as the mean between January and March 1944. Numbers are also missing for September 1944 and January 1945. Here, the value of August 1944 has been used, although it is very likely that this method overestimates the number of Norwegians working for the *Vegvesenet*, particularly in January 1945.

Between October and December 1943, EW did not differentiate between the German and Yugoslavian inmates. Therefore, the number of this merged group has been split up according to the ratio between German and Yugoslavian inmates in September 1943. Again, between February and April 1944, the *Einsatzgruppe* did not give the nationality of the inmates, only an overall number. Therefore, the ratio between German, Yugoslavian and Norwegian inmates in January 1944 has been used to estimate the size of the three groups between February and April.

EW's labour statistics do not include OBL Kirkenes in June and July 1944. However, we know that OBL Kirkenes had 5,208 workers in June and 5,052 in July.

42 Hatlehol, 'Norwegeneinsatz', 186–88.
43 Cf. RAO, RAFA-2188/1/E3c-L0006, Statistics on EW's workers, 15.10.42.

Table A9.1: Workforce of Einsatzgruppe Wiking in Norway (from August 1944 including Finland), 1942–45.*

	German employees and OT staff	Non-German employees and OT staff	German workers	Norwegian workers	Norwegian workers in Vegvesenet	Civilian foreigners	German inmates	Norwegian inmates	Yugoslavian inmates	POWs	German construction soldiers	Sum
Jul 42	1,800		5,720	6,360	1,500	2,650	0	0	1,000	0	0	19,030
Aug 42	2,000		5,452	5,708	2,500	3,883	0	0	869	483	0	20,895
Sep 42	2,339		5,462	7,125	3,500	4,090	0	0	1,436	847	0	24,799
Oct 42	2,550		6,057	7,581	4,450	4,060	2,000	0	1,850	1,214	0	29,762
Nov 42	3,102		6,045	8,392	5,000	3,930	1,982	0	1,840	5,664	0	35,955
Dec 42	3,139		6,172	9,525	4,500	4,160	1,879	0	1,907	6,579	0	37,861
Jan 43	3,157		6,114	9,321	4,300	4,076	1,879	0	1,732	4,615	0	35,194
Feb 43	2,537		6,010	9,796	4,200	4,376	1,518	0	1,432	5,505	0	35,374
Mar 43	2,316		6,026	9,471	4,300	3,968	1,134	0	1,440	5,962	0	34,617
Apr 43	3,064		5,293	11,742	4,399	3,801	1,001	0	1,440	9,609	0	40,349
May 43	3,384		6,586	14,125	4,400	3,758	1,060	0	1,255	11,712	0	46,280
Jun 43	3,745		6,536	14,329	4,200	4,598	1,389	0	1,214	14,354	0	50,365
Jul 43	3,864		6,914	13,948	4,200	5,127	1,310	0	1,395	18,794	0	55,552
Aug 43	3,945		6,603	14,432	3,473	4,923	1,875	0	1,663	22,075	0	58,989

Table A9.1 (continued)

	German employees and OT staff	Non-German employees and OT staff	German workers	Norwegian workers	Norwegian workers in Vegvesenet	Civilian foreigners	German inmates	Norwegian inmates	Yugoslavian inmates	POWs	German construction soldiers	Sum
Sep 43	3,959		6,586	13,845	4,289	4,865	1,582	0	1,895	23,587	0	60,608
Oct 43	4,076		6,489	12,818	*3,975	5,765	*1,551	0	*1,858	26,449	0	62,981
Nov 43	4,157		6,688	12,874	*3,661	5,605	*1,504	0	*1,801	28,431	0	64,721
Dec 43	2,981	1,131	6,620	12,495	*3,347	5,536	*1,559	0	*1,868	28,684	0	64,185
Jan 44	2,918	1,087	6,791	11,807	3,033	5,610	1,396	65	1,564	28,484	0	62,755
Feb 44	2,858	1,127	6,715	11,757	*3,227	5,386	*1,363	*63	*1,527	27,678	0	61,701
Mar 44	2,840	1,069	6,637	11,916	3,421	5,038	*1,359	*63	*1,522	27,657	0	61,522
Apr 44	2,791	910	6,781	12,010	3,607	5,121	*1,285	*60	*1,440	28,005	0	62,010
Jun 44	2,883	925	7,090	11,187	3,854	4,699	1,188	40	34,014		0	71,088
Jul 44	2,947	875	7,535	10,574	3,564	4,725	1,095	641	37,571		0	74,579
Aug 44	3,331	1,185	7,216	20,602	4,887	5,639	1,112	1,175	43,349		2,467	90,963
Sep 44	*3,331	*1,185	7,504	21,828	*4,887	5,412	1,031	1,187	46,519		2,207	95,091
Jan 45	2,804	1,134	7,592	16,673	*4,887	5,710	1,083	1,168	47,969		6,366	95,386

*July 1942–September 1943: RAO, RAFA-2188/1/E3g-L0027, chart on *Einsatzgruppe Wiking's* workforce in Norway, probably from October 1943. October 1943–November 1943: RAO, RAFA-2188/1/E1m-L0047. December 1943 and February – April 1944: BArch, R 50-I/94, fol. 1–4, 49. January 1944 and June 1944–January 1945: Different labour deployment statistics in RAO, RAFA-2188/2/Hh-L0001 and Hh-L0002. Sum for June 1944 includes 5,208 for OBL Kirkenes. Sum for July 1944 includes 5,052 for OBL Kirkenes. Values with * are estimated as described in the text.

Approximately three quarters of them were listed as "prisoners", in all likelihood Soviets. The two values are added on top of the figures for June and July 1944.[44] Furthermore, the category "German inmates" contains two inmates of unknown nationality in June 1944, and 58 in July.

Before December 1944, statistics did not differentiate between German and non-German employees and OT staff members. Furthermore, after spring 1944, the Yugoslavian concentration camp inmates are no longer identifiable as a separate group in EW's labour statistics. Instead, they were counted as prisoners of war. This was a result of the *Wehrmacht* having taken over responsibility for the Yugoslavians in 1943.[45]

Statistics for September 1944 do not give a number either for German nor non-German employees and OT staff. The August values have therefore been used.

44 RAO, RAFA-2188/2/Hh-L0002, file 90-III.
45 Hatlehol, 'Norwegeneinsatz', 284.

List of Tables and Figures

List of Tables

https://doi.org/10.1515/9783110694291-008

List of Figures

Sources

Bundesarchiv Berlin-Lichterfelde (BArch)

R 2, Reichsfinanzministerium
R 3, Reichsministerium für Rüstung und Kriegsproduktion
R 5, Reichsverkehrsministerium
R 13-VIII, Wirtschaftsgruppe Bauindustrie
R 43-I, Reichskanzlei
R 43-II, Reichskanzlei
R 50-I, Organisation Todt
R 2301, Rechnungshof des Deutschen Reiches
R 3101, Reichswirtschaftsministerium
R 3102, Statistisches Reichsamt
R 3903, Reichsanstalt für Arbeitsvermittlung und Arbeitslosenversicherung
R 4601, Generalinspektor für das deutsche Straßenwesen

Bundesarchiv-Militärarchiv, Freiburg im Breisgau (BA-MA)

RH 11-III, General der Pioniere und Festungen im OKH
RH 20–20, Gebirgs-Armeeoberkommando 20
RH 53–12, Wehrkreiskommando XII (Wiesbaden)
RW 39, Wehrmachtbefehlshaber Norwegen

Bundesarchiv Koblenz (BAK)

N 1318, Nachlass Rudolf Wolters
N 1550, Nachlass Hans Clausen Korff

Riksarkivet, Oslo (RAO)

RAFA-2174, Reichskommissariat
RAFA-2188, Organisation Todt, Einsatzgruppe Wiking
S-2079, Industrikomitéen i New York
S-3138, Landssvikarkivet

https://doi.org/10.1515/9783110694291-009

Institutt for historiske studier, NTNU, Trondheim (IHS)

OT-arkiv

Norges Hjemmefrontmuseum, Oslo (NHM)

FO.II/Diverse, Forsvarets Overkommando, diverse dokumenter
FO.II/11.4, Forsvarets Overkommando, Spredte tyske arkivsaker

Unternehmensarchiv der Bilfinger SE, Mannheim

A 11
A 12
A 13
A 14
A 367
A 2797–1
A 4467

Hessisches Hauptstaatsarchiv, Wiesbaden (HHStAW)

Abt. 485, Reichsautobahnen – Oberste Bauleitung Frankfurt am Main

Thüringisches Hauptstaatsarchiv, Weimar (ThHStAW)

Straßenbau AG, Weimar

Archiv des Instituts für Stadtgeschichte, Frankfurt am Main (IfS)

W1-2, Philipp Holzmann AG

Archiv des Instituts für Zeitgeschichte, München (IfZArch)

MA 1555, Nürnberger Kartei (NI-Dokumente)

Stadtarchiv Kaiserslautern

A04/310, Bauunternehmung Gehlen KG

Landesarchiv Berlin (LAB)

A Rep. 250-05-13, Allgemeine Baugesellschaft Lorenz & Co. mbH

Archiv des Deutschen Museums, München (DMM)

FA 010, Dyckerhoff & Widmann AG

Deutsche Zentralbibliothek für Wirtschaftswissenschaften, Leibniz Informationszentrum Wirtschaft (ZBW)

Business reports were collected from the "*Pressemappe 20. Jahrhundert*", the digitalised press archives of the former library of the *Institut für Weltwirtschaft* (IfW) and the *Hamburgisches Welt-Wirtschafts-Archiv* (HWWA):

http://webopac0.hwwa.de/PresseMappe20/index [last accessed: 12 March 2019]

"AHAG" Allgemeine Häuser- und Industriebau-Aktiengesellschaft, Berlin	1928–41
Allgemeine Baugesellschaft – A. Porr AG, Vienna	1939–42
Allgemeine Baugesellschaft Lenz & Co. KG, Berlin	1928–42
Allgemeine Hoch- und Ingenieurbau-Aktiengesellschaft, Düsseldorf	1928–44
Beton- und Monierbau Aktiengesellschaft, Berlin	1928–42
Boswau & Knauer Aktiengesellschaft, Berlin	1928–40
Deutsche Bau-Aktiengesellschaft, Berlin	1937–41
Deutsche Schachtbau Aktiengesellschaft, Nordhausen	1931–38
Friedrich Mehmel Aktiengesellschaft, Hanover	1940
Gebhardt & Koenig – Dt. Schachtbau Aktiengesellschaft, Nordhausen[1]	1939–42
Gebrüder Goedhart Aktiengesellschaft, Düsseldorf	1928–29 & 1931–41
Grün & Bilfinger Aktiengesellschaft, Mannheim	1928–41 & 1943–44
Habermann & Guckes Aktiengesellschaft, Berlin	1928–41
Heilmann & Littmann Bau-Aktiengesellschaft, Munich	1928–42
Hochtief AG für Hoch- und Tiefbauen, vorm. Gebr. Helfmann, Essen	1928–41
HUTA Hoch- und Tiefbau-Aktiengesellschaft, Breslau	1928–42 & 1944

1 Gebhardt & Koenig – Deutsche Schachtbau Aktiengesellschaft, Nordhausen, was established in 1939 as a result of the merger of Tiefbau- und Kälteindustrie AG, vorm. Gebhardt & Koenig, Nordhausen, and Deutsche Schachtbau AG, Nordhausen.

Julius Berger Tiefbau-Aktiengesellschaft, Berlin	1928–43
Karl Kübler AG, Stuttgart	1936–42
Neue Baugesellschaft Wayss & Freytag AG, Frankfurt am Main	1928–42
Philipp Holzmann Aktiengesellschaft, Frankfurt am Main	1928–44
Saar Bauindustrie Aktiengesellschaft, Saarlautern	1938–41
Straßenbau-Aktiengesellschaft, Niederlahnstein	1928–44
Stuag Straßen- und Tiefbau Aktiengesellschaft, Vienna	1938–42
Süddeutsche Held & Francke Bau-Aktiengesellschaft, Munich	1937–42
Thormann & Stiefel Aktiengesellschaft "Thosti", Augsburg	1940–42
Tiefbau- und Kälteindustrie AG, vorm. Gebhardt & Koenig, Nordhausen	1928–38
"Universale" Hoch- und Tiefbau Aktiengesellschaft, Vienna	1936–45

Bibliography

Unpublished manuscripts

Dichgans, Hans. 'Zur Geschichte des Reichskommissars für die Preisbildung'. Unpublished manuscript, 1976. [BArch, R 26-II ANH./1]

Henne, Willi. 'Die Organisation Todt in Norwegen: Aufgabe, Organisation, Art der Baudurchführung mit aufgetretenen Schwierigkeiten, Leistungen'. Unpublished manuscript. Oslo, Akershus fengsel, 1945. [IHS]

Korff, Hans Clausen. 'Norwegens Wirtschaft im Mahlstrom der Okkupation'. Unpublished manuscript, 1949. [BAK, N 1550/1]

Otte, Carlo. 'Die Hauptabteilung Volkswirtschaft der Behörde des Reichskommissars für die besetzten norwegischen Gebiete'. Oslo, 1945. [RAO, Oslo Politikammer, B-sak 3061, box 2, file 3]

Plum, Gustav. 'Die Abschlüsse der Bauindustrie. 1933 bis 1939'. Unpublished manuscript. Bad Homburg, circa 1940. [*ZBW*]

Rentrop, Wilhelm. 'Zu den Materialien zur Geschichte des Reichskommissars für die Preisbildung'. Unpublished manuscript, circa 1976. [BArch, R 26-II ANH./2]

Steffens, Joachim. 'Das Bauwesen – geschichtlich, technisch und wirtschaftlich gesehen'. Unpublished manuscript, circa 1944. [BArch, R 3/Film 77815]

Published sources

Boelcke, Willi A. *Deutschlands Rüstung im Zweiten Weltkrieg. Hitlers Konferenzen mit Albert Speer 1942–1945*. Frankfurt am Main: Akad. Verl.-Ges. Athenaion, 1969.

Das Kabinett von Schleicher. 3. Dezember 1932 Bis 30. Januar 1933. Akten Der Reichskanzlei. Weimarer Republik. Boppard am Rhein: Boldt, 1986.

Daub, Walter. *OT-Verträge. Vertragsrecht der Organisation Todt, Stand 30. Juni 1942*. Wiesbaden: Schellenberg'sche Buchdruckerei, 1942.

Die Regierung Hitler. Band I: 1933/34. Vol. 1. Akten der Reichskanzlei. Regierung Hitler 1933–1945. Boppard am Rhein: Boldt, 1983.

Die Regierung Hitler. Band V: 1938. Akten der Reichskanzlei. Regierung Hitler 1933–1945. München: Oldenbourg Verlag, 2008.

Dittrich, Rudolf. *Vom Werden, Wesen und Wirken der Organisation Todt. Ausarbeitung für die Historical Division / US Army in Europe*. Quellen zur Geschichte der Organisation Todt, 1/2. Osnabrück: Biblio Verlag, 1998.

Domarus, Max, ed. *The Years 1932 to 1934*. Hitler: Speeches and Proclamations 1932–1945. The Chronicle of a Dictatorship 1. London: Tauris, 1990.

Dorsch, Franz Xaver. *Die Organisation Todt*. Quellen zur Geschichte der Organisation Todt, 1/2. 1950. Reprint, Osnabrück: Biblio Verlag, 1998.

Dorsch, Franz Xaver. 'Organization Todt – Operations in the West [MS B-671]'. In *World War II German Military Studies: Part X. Special Topics, Continued*, edited by Donald Detwiler, Vol. 24. World War II German Military Studies: A Collection of 213 Special Reports on the Second World War Prepared by Former Officers of the Wehrmacht for the United States Army. 1947. Reprint, New York, London: Garland Publishing, 1979.

Dorsch, Franz Xaver. 'The Organisation Todt in France and in Germany [MS B-670]'. In *World War II German Military Studies: Part X. Special Topics, Continued*, edited by Donald Detwiler, Vol. 24.

https://doi.org/10.1515/9783110694291-010

World War II German Military Studies: A Collection of 213 Special Reports on the Second World War Prepared by Former Officers of the Wehrmacht for the United States Army. 1946. Reprint, New York, London: Garland Publishing, 1979.

Erstatningsdirektoratet. *Undersøkelser vedrørende norske myndigheters stilling til tyske arbeidsoppdrag den først tid etter 8. april 1940*. Oslo: Erstatningsdirektoratet, 1946.

Generalbevollmächtigter für die Regelung der Bauwirtschaft, ed. *Die Regelung der Bauwirtschaft. Übersicht über die für die Baugenehmigung und Kontingentierung wichtigsten Bestimmungen des GB-Bau*. Berlin: E.S. Mittler & Sohn, 1943.

Handbook of the Organisation Todt by the Supreme Headquarters Allied Expeditionary Force Counter-Intelligence Sub-Division MIRS/MR-OT/5/45. Quellen Zur Geschichte Der Organisation Todt 4. Osnabrück: Biblio Verlag, 1992.

Hauptabteilung IV (Wirtschaft) der Reichsorganisationsleitung der N.S.D.A.P. *Wirtschaftliches Sofortprogramm der N.S.D.A.P.* München: Franz-Eher-Verlag, 1932.

International Military Tribunal. *International Military Tribunal (1945–1946 Nürnberg) Trial of the Major War Criminals before the International Military Tribunal Nuremberg : The Tribunal, 1947–1949*. Vol. 19: Proceedings: 19 July 1946 – 29 July 1946. Nürnberg: The Tribunal, 1948.

Justis- og politidepartementet. *Om landssvikoppgjøret. Innstilling fra et utvalg nedsatt for å skaffe tilveie materiale til en innberetning fra Justisdepartementet til Stortinget. Utvalget oppnevnt 22. desember 1955. Innstilling avgitt 11. januar 1962*. Gjøvik: Mariendals Boktrykkeri, 1962.

Keller, Rolf, and Silke Petry, eds. *Sowjetische Kriegsgefangene im Arbeitseinsatz 1941–1945. Dokumente zu den Lebens- und Arbeitsbedingungen in Norddeutschland*. Schriftenreihe der Stiftung Niedersächsische Gedenkstätten 2. Göttingen: Wallstein Verlag, 2013.

Kotze, Hildegard von, ed. *Heeresadjutant bei Hitler 1938–1943. Aufzeichnungen des Majors Engel*. Stuttgart: Deutsche Verlags-Anstalt, 1974.

Mason, Timothy W. *Arbeiterklasse und Volksgemeinschaft. Dokumente und Materialien zur deutschen Arbeiterpolitik 1936–1939*. Schriften des Zentralinstituts für sozialwissenschaftliche Forschung der Freien Universität Berlin 22. Opladen: Westdeutscher Verlag, 1975.

Moll, Martin, ed. *'Führer-Erlasse' 1939–1945. Edition sämtlicher überlieferter, nicht im Reichsgesetzblatt abgedruckter, von Hitler während des Zweiten Weltkrieges schriftlich erteilter Direktiven aus den Bereichen Staat, Partei, Wirtschaft, Besatzungspolitik und Militärverwaltung*. Stuttgart: Franz Steiner Verlag, 1997.

Office of the Publication Board, Department of Commerce. 'Interrogation of Albert Speer and Members of the Former Reich Ministry of Armaments and War Production [Technical Report No. PB 430]'. Washington, D.C., 1945.

Pollard, Sidney, and Colin Holmes, eds. *The End of the Old Europe 1914–1939*. Documents of European Economic History 3. London: Edward Arnold, 1973.

Schramm, Percy E., ed. *Kriegstagebuch des Oberkommandos der Wehrmacht (Wehrmachtführungsstab) 1940–1945. Band 2: 1. Januar 1942–31. Dezember 1942. Zusammengestellt und erläutert von Andreas Hillgruber. Erster Halbband (= KTB/OKW II/1)*. Kriegstagebuch des Oberkommandos der Wehrmacht (Wehrmachtführungsstab) 1940–1945. Geführt von Helmuth Greiner und Percy Ernst Schramm. Bonn: Bernard & Graefe Verlag, 2000.

Schramm, Percy E. ed. *Kriegstagebuch des Oberkommandos der Wehrmacht (Wehrmachtführungsstab) 1940–1945. Band 2: 1. Januar 1942–31. Dezember 1942. Zusammengestellt und erläutert von Andreas Hillgruber. Zweiter Halbband (= KTB/OKW II/2)*. Kriegstagebuch des Oberkommandos der Wehrmacht (Wehrmachtführungsstab) 1940–1945. Geführt von Helmuth Greiner und Percy Ernst Schramm. Bonn: Bernard & Graefe Verlag, 2000.

Singer, Hedwig. *Entwicklung und Einsatz der Organisation Todt*. Quellen zur Geschichte der Organisation Todt, 1/2. Osnabrück: Biblio Verlag, 1998.

Sozialdemokratische Partei Deutschlands. *Deutschland-Berichte der Sozialdemokratischen Partei Deutschlands (Sopade). Dritter Jahrgang, 1936.* Deutschland-Berichte der Sozialdemokratischen Partei Deutschlands (Sopade) 1934–1940. Salzhausen, Frankfurt am Main: Verlag Petra Nettelbeck, Zweitausendeins, 1980.
Sozialdemokratische Partei Deutschlands. *Deutschland-Berichte der Sozialdemokratischen Partei Deutschlands (Sopade). Vierter Jahrgang, 1937.* Deutschland-Berichte der Sozialdemokratischen Partei Deutschlands (Sopade) 1934–1940. Salzhausen, Frankfurt am Main: Verlag Petra Nettelbeck, Zweitausendeins, 1980.
The Effects of Strategic Bombing on the German War Economy (European Report #3). The United States Strategic Bombing Survey. 1945. Reprint, New York: Garland Publishing, 1976.

Press articles

'1938–1939. Rückblick und Ausblick'. *Die Bauindustrie* 7, no. 1 (1939): 6–10.
Bauch, Botho. 'Die Durchführung der Baupreisverordnung'. *Die Bauindustrie* 8, no. 6 (1940): 121–23.
'Beschaffung von Baumaschinen und Geräten'. *Technische und Wirtschaftliche Rundschau*, Supplement to Der deutsche Baumeister, 37 (1939): 207–8.
'Betrachtungen über die Ausschreibung und Vergebung der Reichsautobahndecken'. *Die Bauindustrie* 3, no. 13 (1935): 174–76.
Daub, Walter. 'Zur Gestaltung von Bauverträgen'. *Die Bauindustrie* 11, no. 5 (1943): 117–21.
'Die arbeits- und sozialrechtlichen Bestimmungen für ausländische Arbeiter und Kriegsgefangene in der Bauwirtschaft nach dem Stande vom 1. Dezember 1942'. *Die Bauindustrie* 10, no. 26 (1942): 690–91.
Dorsch, Franz Xaver. 'Das Wesen der OT'. *Der Deutsche Baumeister* 5, no. 3 (1943): 2–3.
'Dr. Todt gegen die Regiearbeit bei Straßenbauten und Straßenunterhaltung'. *Die Bauindustrie* 1, no. 18 (1933): 233–34.
Enderlein, Kurt. 'Probleme der Preispolitik in der Bauwirtschaft'. *Die Bauindustrie* 11, no. 19 (1943): 509–19.
Fischböck, Hans. 'Sparsame Wirtschaftsführung – Kommerzialisierung der Preisüberwachung – Rede des Reichskommissars für die Preisbildung Staatssekretär Dr. Hans Fischböck auf dem Gefolgschaftsappell des Preiskommissariats am 29. Januar 1943'. *Mitteilungsblatt des Reichskommissars für die Preisbildung Teil 2* 7, no. 3 (1943): 32–37.
'Gegen artfremde Regiearbeiten industrieller Werke'. *Die Bauindustrie* 1, no. 19 (1933): 247.
Glöckner, Werner. 'Die Bauindustrie im Spiegel der Beschäftigung des Jahres 1936 (Fortsetzung)'. *Die Bauindustrie* 5, no. 7 (1937): 109–11.
Henze, Max. 'Die gesetzlichen Grundlagen der Gewinnabführung für 1943'. *Die Bauindustrie* 12, no. 10 (1944): 221–25.
'In einem Monat 15 tödliche Unfälle auf Baustellen der Reichsautobahnen!' *Die Bauindustrie* 3, no. 31 (1935): 463.
Jecht, Reinhard. 'Vereinfachung des Firmeneinsatzes. Zur Anordnung des Reichsministers Speer vom 28. 10.1943'.*Die Bauindustrie* 11, no. 19 (1943): 505–9.
'Keine Benachteiligung wegen Nichtbeteiligung an Ausschreibungen'. *Die Bauindustrie* 3, no. 1 (1935): 8.
'Keine Benachteiligung wegen Nichtbeteiligung an Ausschreibungen'. *Die Bauindustrie* 3, no. 13 (1935): 182–83.
'Keine Regie-Bauarbeiten der Industrie'. *Die Bauindustrie* 1, no. 18 (1933): 234.

'Klare Berichte über die Bauindustrie'. *Die Bauindustrie* 4, no. 6 (1936): 97–98.
Knecht, Karl. 'Vorläufiger Organisationsaufbau der wirtschaftlichen Selbstverantwortung der Bauwirtschaft'. *Die Bauindustrie* 11, no. 14 (1943): 373–77.
Kohary, Theodor von. 'Die Organisation Todt am Jahresbeginn 1941'. *Die Bauindustrie* 9, no. 5 (1941): 1–3.
Krauth, Theodor. 'Die Maschine im Arbeitsbeschaffungsprogramm'. *Die Bauindustrie* 1, no. 21/22 (1933): 271–72.
Krauthause, Hermann. 'Der Lohnfaktor bei der Preissenkung und Gewinnabführung'. *Die Verwaltung* 18, no. 21 (1941): 398–401.
'Leidensweg eines Unternehmers!' *Die Bauindustrie* 3, no. 30 (1935): 445.
Mast, Adolf. 'Das Tiefbaugewerbe an der Jahreswende'. *Deutsche Bauzeitung* 68, no. 1 (1934).
Meierkord, Elisabeth. 'Die Abführung von Übergewinnen für das Wirtschaftsjahr 1941'. *Die Bauindustrie* 10, no. 12 (1942): 312–15.
'Neue Regelung über Kriegslöhne'. *Die Bauindustrie* 7, no. 43 (1939): 1309–11.
Opitz, Gerhard. 'Die Gewinne in der Bauindustrie'. *Die Bauindustrie* 12, no. 7 (1944): 147–50.
Ott, Franz. 'Bauwirtschaftliche Fragen bei der Gewinnabführung 1943'. *Die Bauindustrie* 12, no. 10 (1944): 217–21.
Prange, Herbert. 'Die neuen LSBÖ-Vorschriften'. *Die Bauindustrie* 10, no. 7 (1942): 171–74.
'Preisdisziplin auch für die Bauwirtschaft'. *Neuer Wirtschaftsdienst* 11, no. 7 (1944): 1–2.
Seeger, Anton. 'Kontingentierter Zement'. *Der deutsche Baumeister* 1, no. 11 (1939): 17–21.
Speer, Albert. 'Die Grundlagen der deutschen Rüstung. Reichsminister Speer über Bewährung und Aufgaben der Selbstverantwortung'. *Berliner Börsen-Zeitung*, 18 June 1944.
Speer, Albert. 'Selbstverantwortung in der Rüstungsindustrie'. *Der Vierjahresplan*, no. 7 (1943): 242–46.
Steffens, Joachim. 'Die Bauwirtschaft im Kriege'. *Der deutsche Baumeister* 1, no. 11 (1939): 9–11.
Steffens, Joachim. 'Die deutsche Bauwirtschaft im Kriege'. *Die Bauindustrie* 8, no. 6 (1940): 118–20.
Struck, Josef. 'Eisen und Stahl für die Bauwirtschaft im Kriege. Entwicklung der Lenkung des Eiseneinsatzes in der Bauwirtschaft von 1936 bis heute.' *Der deutsche Baumeister* 1, no. 12 (1939): 14–16.
Todt, Fritz. 'Die Neuordnung des deutschen Bauwesens'. *Der deutsche Baumeister* 1, no. 1 (1939): 5–9.
Todt, Fritz. 'Regelung der Bauwirtschaft'. *Der Vierjahresplan* 3, no. 12 (1939): 762–64.
'Umschulungskursus für Straßenbau'. *Die Bauindustrie* 1, no. 19 (1933): 249.
'Verlängerung der Lieferschlußfrist für Inanspruchnahme der Steuerfreiheit für Ersatzbeschaffungen'. *Die Bauindustrie* 3, no. 4 (1935): 57.
Vögler, Eugen. 'Die Lage der Bauwirtschaft. Rede auf der Tagung der Wirtschaftsgruppe Bauindustrie 1936 in Essen-Ruhr'. *Die Bauindustrie* 4, no. 12 (1936): 202–10.
'Was bringen die neuen kriegswirtschaftlichen Verordnungen für die Bauwirtschaft? Eine Übersicht, abgeschlossen am 10. September 1939'. *Die Bauindustrie* 7, no. 37 (1939): 1198–1200.

Literature

Abelshauser, Werner, Jan-Otmar Hesse, and Werner Plumpe. 'Wirtschaftsordnung und Unternehmen im Nationalsozialismus: Neuere Forschungen zur Wirtschaftsgeschichte des Nationalsozialismus'. In *Wirtschaftsordnung, Staat und Unternehmen: Neue Forschungen zur Wirtschaftsgeschichte des Nationalsozialismus*, edited by Werner Abelshauser, Jan-Otmar Hesse, and Werner Plumpe. Essen: Klartext Verlag, 2003.

Absolon, Rudolf. *Die Wehrmacht im Dritten Reich*. Vol. V. Boppard am Rhein: Harald Boldt Verlag, 1988.

Absolon, Rudolf. *Die Wehrmacht im Dritten Reich*. Vol. VI. Boppard am Rhein: Harald Boldt Verlag, 1995.

Adamek, Robert, and Friedrich Saake, eds. *Die Straßenkosten und ihre Finanzierung. Hauptband*. Forschungsarbeiten aus dem Strassenwesen 8. Bielefeld: Kirschbaum-Verlag, 1952.

Andenæs, Johs. *Det vanskelige oppgjøret: rettsoppgjøret etter okkupasjonen*. 1979. Reprint, Oslo: Tano Aschehoug, 1998.

Andersen, Ketil G. *Flaggskip i fremmed eie: Hydro 1905–1945*. Oslo: Pax Forlag, 2005.

Andersen, Ketil G. '"Teknisk æresoppdrag av høyeste orden". Organisation Todt og byggingen av Hitlers polarjernbane'. *Historisk Tidsskrift* 97, no. 3 (2018): 206–23.

Andersen, Steen. 'Between Imperative and Risk. The Case of Christiani & Nielsen's Market Entry in Norway 1941–45'. *Scandinavian Economic History Review* 59, no. 1 (2011): 3–28.

Andersen, Steen. *De gjorde Danmark større . . . De multinationale danske entreprenørfirmaer i krise og krig 1919–1947*. København: Lindhardt og Ringhof, 2005.

Andersen, Steen. 'The French Opportunity. A Danish Construction Company Working for the Germans in France, 1940–1944'. In *Economies under Occupation. The Hegemony of Nazi Germany and Imperial Japan in World War II*, edited by Marcel Boldorf and Tetsuji Okazaki. London, New York: Routledge, 2015.

Aukrust, Odd, and Petter Jakob Bjerve. *Hva krigen kostet Norge*. Oslo: Dreyers Forlag, 1945.

Bader, Uwe, and Beate Welter. 'Das SS-Sonderlager/KZ Hinzert'. In *Hinzert, Auschwitz, Neuengamme*. Der Ort des Terrors. Geschichte der nationalsozialistischen Konzentrationslager 5. München: C.H. Beck, 2007.

Bajari, Patrick, and Steven Tadelis. 'Incentives versus Transaction Costs: A Theory of Procurement Contracts'. *RAND Journal of Economics* 32, no. 3 (2001): 387–407.

Banken, Ralf. *Hitlers Steuerstaat: Die Steuerpolitik im Dritten Reich*. Das Reichsfinanzministerium im Nationalsozialismus 2. München, Wien: De Gruyter Oldenbourg, 2018.

Barkai, Avraham. *Nazi Economics. Ideology, Theory, and Policy*. New Haven, London: Yale University Press, 1990.

Baumeister, Ludwig. *Preisermittlung und Veranschlagen von Hoch-, Tief- und Eisenbetonbauten*. 8th ed. Berlin, Heidelberg: Springer-Verlag, 1941.

Bassett, Edward O. *Industrial Sales, Output and Productivity. Pre-War Area of Germany 1939–44 (Special Paper No. 8)*. United States Strategic Bombing Survey, 1946.

Bera, Matt. *Lobbying Hitler. Industrial Associations between Democracy and Dictatorship*. New York: Berghahn Books, 2016.

Bergh, Trond, Harald Espeli, and Knut Sogner. *Brytningstider: storselskapet Orkla 1654–2004*. Oslo: Orion, 2004.

Bergmann, Armin. 'Organisation Todt (OT)'. In *Enzyklopädie des Nationalsozialismus*, edited by Wolfgang Benz, Hermann Graml, and Hermann Weiß. München: Deutscher Taschenbuch Verlag, 1997.

Beton- und Tiefbau Mast Aktiengesellschaft. *Von einem, der auszog das Bauen zu lernen: 50 Jahre Beton- und Tiefbau Mast Aktiengesellschaft*. Berlin: Mast, 1955.

Bettinger, Dieter, and Martin Büren. *Der Westwall. Die Geschichte Der Deutschen Westbefestigungen Im Dritten Reich*. Vol. 1. 2 vols. Osnabrück: Biblio Verlag, 1990.

Bjerkholt, Olav. *Kunnskapens krav: om opprettelsen av Forskningsavdelingen i Statistisk sentralbyrå*. Oslo, Kongsvinger: Statistisk Sentralbyrå, 2000.

Boelcke, Willi A. *Die Kosten von Hitlers Krieg. Kriegsfinanzierung und finanzielles Kriegserbe in Deutschland 1933–1948*. Paderborn: Ferdinand Schöningh, 1985.

Bohn, Robert. *Reichskommissariat Norwegen. 'Nationalsozialistische Neuordnung' und Kriegswirtschaft*. München: Oldenbourg, 2000.
Bopp, Wolfgang. 'The Evolution of the Pricing Policy for Public Orders During the Third Reich'. In *After the Slump. Industry and Politics in 1930s Britain and Germany*, edited by Christoph Buchheim and Redvers Garside, 149–60. Frankfurt am Main: Peter Lang, 2000.
Botzet, Christiane. 'Ministeramt, Sondergewalten und Privatwirtschaft. Der Generalbevollmächtigte für die Regelung der Bauwirtschaft'. In *Hitlers Kommissare: Sondergewalten in der nationalsozialistischen Diktatur*, edited by Rüdiger Hachtmann and Winfried Süß. Göttingen: Wallstein Verlag, 2006.
Bracher, Karl Dietrich. *The German Dictatorship. The Origins, Structure, and Consequences of National Socialism*. 1969. Reprint, London: Penguin Books, 1973.
Brechtken, Magnus. *Albert Speer. Eine deutsche Karriere*. München: Siedler, 2017.
Breloer, Heinrich. *Unterwegs zur Familie Speer: Begegnungen, Gespräche, Interviews*. Berlin: Propyläen, 2005.
Broszat, Martin. *Der Staat Hitlers. Grundlegung und Entwicklung seiner inneren Verfassung*. München: Deutscher Taschenbuch Verlag, 2000.
Brünger, Sebastian. *Geschichte und Gewinn. Der Umgang deutscher Konzerne mit ihrer NS-Vergangenheit*. Göttingen: Wallstein Verlag, 2017.
Bry, Gerhard. *Wages in Germany 1871–1945*. Princeton: Princeton University Press, 1960.
Buchheim, Christoph. 'Das NS-Regime und die Überwindung der Weltwirtschaftskrise in Deutschland'. *Vierteljahreshefte für Zeitgeschichte* 56, no. 3 (2008): 381–414.
Buchheim, Christoph. 'Der Mythos vom "Wohlleben". Der Lebensstandard der deutschen Zivilbevölkerung im Zweiten Weltkrieg'. *Vierteljahreshefte für Zeitgeschichte* 58, no. 3 (2010): 299–328.
Buchheim, Christoph. 'Die Wirtschaftsentwicklung im Dritten Reich. Mehr Desaster als Wunder. Eine Erwiderung auf Werner Abelshauser'. *Vierteljahreshefte für Zeitgeschichte* 49, no. 4 (2001): 653–64.
Buchheim, Christoph. 'Unternehmen in Deutschland und NS-Regime: Versuch einer Synthese'. *Historische Zeitschrift* 282, no. 2 (2006): 351–90.
Buchheim, Christoph, and Jonas Scherner. 'Anmerkungen zum Wirtschaftssystem des "Dritten Reichs"'. In *Wirtschaftsordnung, Staat und Unternehmen: Neue Forschungen zur Wirtschaftsgeschichte des Nationalsozialismus*, edited by Werner Abelshauser, Jan-Otmar Hesse, and Werner Plumpe. Essen: Klartext Verlag, 2003.
Buchheim, Christoph, and Jonas Scherner. 'The Role of Private Property in the Nazi Economy: The Case of Industry'. *The Journal of Economic History* 66, no. 2 (2006): 390–416.
Budraß, Lutz, Jonas Scherner, and Jochen Streb. 'Fixed-Price Contracts, Learning and Outsourcing: Explaining the Continuous Growth of Output and Labour Productivity in the German Aircraft Industry during the Second World War'. *Economic History Review* 63, no. 1 (2010): 107–36.
Buggeln, Marc. *Das System der KZ-Außenlager: Krieg, Sklavenarbeit und Massengewalt*. Bonn: Friedrich-Ebert-Stiftung, 2012.
Buggeln, Marc. *Der U-Boot-Bunker 'Valentin': Marinerüstung, Zwangsarbeit und Erinnerung*. Bremen: Edition Temmen, 2017.
Buggeln, Marc. '"Menschenhandel" als Vorwurf im Nationalsozialismus. Der Streit um den Gewinn aus den militärischen Großbaustellen am Kriegsende (1944/45)'. In *Rüstung, Kriegswirtschaft und Zwangsarbeit im 'Dritten Reich'*, edited by Andreas Heusler, Mark Spoerer, and Helmuth Trischler, 199–218. München: Oldenbourg Verlag, 2010.
Buggeln, Marc. '"Menschenhandel" als Vorwurf im Nationalsozialismus. Der Streit um den Gewinn aus den militärischen Großbaustellen am Kriegsende (1944/45)'. In *Rüstung, Kriegswirtschaft*

und Zwangsarbeit im 'Dritten Reich', edited by Andreas Heusler, Mark Spoerer, and Helmuth Trischler. München: Oldenbourg, 2010.
Bundeskriminalamt Wiesbaden, ed. *Wirtschaftsdelikte (einschließlich der Korruption). Arbeitstagung im Bundeskriminalamt Wiesbaden vom 8. April bis 13. April 1957 über Bekämpfung der Wirtschaftsdelikte einschließlich der Korruption*. Wiesbaden: Bundeskriminalamt, 1957.
Bunting, Madeleine. *The Model Occupation: The Channel Islands under German Rule 1940–1945*. London: HarperCollins, 1995.
Christensen, Sverre A., and Harald Rinde. *Nasjonale utlendinger: ABB i Norge 1880–2010*. Oslo: Gyldendal akademisk, 2009.
Christopher, John, ed. *Organisation Todt: From Autobahns to the Atlantic Wall*. Stroud: Amberley, 2014.
Cohen, Laurie R. *Smolensk under the Nazis: Everyday Life in Occupied Russia*. Rochester: University of Rochester Press, 2013.
Custodis, Johann. 'Employing the Enemy: The Economic Exploitation of POW and Foreign Labor from Occupied Territories by Nazi Germany'. In *Paying for Hitler's War: The Consequences of Nazi Hegemony for Europe*, edited by Jonas Scherner and Eugene N. White, 67–100. New York: Cambridge University Press, 2015.
Dahl, Hans Fredrik, Guri Hjeltnes, Berit Nøkleby, Nils Johan Ringdal, and Øystein Sørensen, eds. *Norsk krigsleksikon 1940–1945*. Oslo: J. W. Cappelens Forlag, 1995.
Denkiewicz-Szczepaniak, Emilia. 'Polske OT-tvangsarbeidere og krigsfanger i Norge under annen verdenskrig'. *Historisk Tidsskrift* 76, no. 2 (1997): 268–81.
Deutsche Bau- und Bodenbank. 'Die Entwicklung der deutschen Bauwirtschaft im Jahre 1937', 1938.
Dickert, Elena. *Die 'Nutzbarmachung' des Produktionspotentials besetzter Gebiete durch Auftragsverlagerungen im Zweiten Weltkrieg: Staatliche Regulierung und Verlagerungsverhalten von Maschinenbau- und Automobilunternehmen*. Trondheim: NTNU, 2014.
Didriksen, Jan. *Industrien under hakekorset*. Oslo: Universitetsforlaget, 1987.
Die Wirtschaftslage des In- und Auslands im Frühjahr 1938 – im Herbst 1938. Halbjahresberichte zur Wirtschaftslage. 13. Jahrgang 1938/39. Hamburg: Hanseatische Verlagsanstalt, 1938.
Diehl, Markus Albert. *Von der Marktwirtschaft zur nationalsozialistischen Kriegswirtschaft. Die Transformation der deutschen Wirtschaftsordnung 1933–1945*. Stuttgart: Franz Steiner Verlag, 2005.
Donges, Alexander. *Die Vereinigte Stahlwerke AG im Nationalsozialismus. Konzernpolitik zwischen Marktwirtschaft und Staatswirtschaft*. Paderborn: Ferdinand Schöningh, 2014.
Dreier, Horst. 'Rechtszerfall und Kontinuität. Zur asynchronen Entwicklung von Staatsrecht und Wirtschaftssytem in der Zeit des Nationalsozialismus'. In *Wirtschaftskontrolle und Recht in der nationalsozialistischen Wirtschaft*, edited by Dieter Gosewinkel. Frankfurt am Main: Vittorio Klostermann, 2005.
Eastwood, James. *The Maginot and Siegfried Lines: Walls of Death*. London: Pallas Publishing, 1939.
Ed. Züblin AG, ed. *75 Jahre Züblin-Bau, 1898–1973*. Stuttgart: Karl Krämer Verlag, 1973.
Ehrler, Karl. *Die statistische Erfassung der Bauwirtschaft. Methode und Auswertung*. Dresden: M. Dittert & Co, 1940.
Eichholtz, Dietrich. *Geschichte der deutschen Kriegswirtschaft 1939–1945. Band I: 1939–1941*. 1996. Reprint, München: K. G. Saur, 2003.
Eichholtz, Dietrich. *Geschichte der deutschen Kriegswirtschaft 1939–1945. Band II/2: 1941–1943*. 1985. Reprint, München: K. G. Saur, 2003.
Eichholtz, Dietrich. *Geschichte der deutschen Kriegswirtschaft 1939–1945. Band III/1: 1943–1945*. 1996. Reprint, München: K. G. Saur, 2003.
Ellingsen, Dag. *Krigsprofitørene og rettsoppgjøret*. Oslo: Gyldendal, 1993.

Eng, Vidar. 'Hvorfor arbeidet gode nordmenn på flyplassene i Oslo og Trondheim i april 1940?' *Historisk Tidsskrift* 89, no. 3 (2010): 425–46.

Erbe, René. *Die nationalsozialistische Wirtschaftspolitik 1933–1939 im Lichte der modernen Theorie*. Zürich: Polygraphischer Verlag, 1958.

Espeli, Harald. '"Cooperation on a Purely Matter-of-Fact Basis": The Norwegian Central Bank and Its Relationship to the German Supervisory Authority during the Occupation, 1940–1945'. *Scandinavian Economic History Review* 62, no. 2 (2014): 188–212.

Espeli, Harald. '"Det gavner ingenting å gjøre store vanskeligheter i små saker. Dette er ikke store saker". Norges Bank, Administrasjonsrådet og etableringen av okkupasjonskontoen i 1940'. *Historisk Tidsskrift* 90, no. 4 (2011): 559–84.

Espeli, Harald. 'Det økonomiske forholdet mellom Tyskland og Norge 1940–45'. In *Danske tilstander – norske tilstander. Forskjeller og likheter under tysk okkupasjon 1940–45*, edited by Hans Fredrik Dahl, Hans Kirchhoff, Joachim Lund, and Lars-Erik Vaale. Oslo: Forlaget Press, 2010.

Espeli, Harald. 'Economic Consequences of the German Occupation of Norway, 1940–1945'. *Scandinavian Journal of History* 38, no. 4 (2013): 502–24.

Espeli, Harald. 'Incentive Structures and State Regulations'. In *Industrial Collaboration in Nazi-Occupied Europe: Norway in Context*, edited by Hans Otto Frøland, Mats Ingulstad, and Jonas Scherner, 245–71. London: Palgrave Macmillan, 2016.

Evans, Richard J. *The Third Reich in Power 1933–1939*. London: Penguin Books, 2006.

Feinstein, Charles H., Peter Temin, and Gianni Toniolo. *The European Economy Between the Wars*. Oxford: Oxford University Press, 1997.

Feldenkirchen, Wilfried. *Siemens 1918–1945*. München, Zürich: Piper, 1995.

Fings, Karola. *Krieg, Gesellschaft und KZ: Himmlers SS-Baubrigaden*. Paderborn: Ferdinand Schöningh, 2005.

Fischer, Curt. *Die Gesellschaft mit beschränkter Haftung*. Berlin, Göttingen, Heidelberg: Springer-Verlag, 1948.

Fredriksen, Gaute. '" . . . og det er en annen historie": Den økonomiske historieskrivingen om årene 1940–1945 i Norges historie. En historiografi.' Universitetet i Oslo, 2011.

Frei, Norbert, Ralf Ahrens, Jörg Osterloh, and Tim Schanetzky. *Flick: Der Konzern, die Familie, die Macht*. München: Karl Blessing Verlag, 2009.

Fremdling, Rainer, and Reiner Stäglin. 'Work Creation and Rearmament in Germany 1933–1938 – A Revisionist Assessment of NS-Economic Policy Based on Input-Output Analysis'. *DIW Berlin Discussion Paper No. 1473*, 2015.

Fried, John H. E. *The Exploitation of Foreign Labour by Germany*. Montreal: International Labour Office, 1945.

Frøland, Hans Otto. 'Facing Disincentives? Norwegian Aluminium Companies Working for the German Aircraft Industry'. In *Industrial Collaboration in Nazi-Occupied Europe: Norway in Context*, edited by Hans Otto Frøland, Mats Ingulstad, and Jonas Scherner, 331–57. London: Palgrave Macmillan, 2016.

Frøland, Hans Otto. 'Nazi Planning and the Aluminium Industry'. In Alan S. *Milward and a Century of European Change*, edited by Fernando Guirao, Frances Lynch, and Sigfrido Ramirez Perez. New York: Routledge, 2012.

Frøland, Hans Otto. 'Organisation Todt som byggherre i Norge'. *Historisk Tidsskrift* 97, no. 3 (2018): 167–88.

Frøland, Hans Otto, and Gunnar D. Hatlehol. 'Organisation Todt and Forced Labour in Norway during the Nazi Occupation: Preliminary Remarks from an Ongoing Research Project'. In *Painful Heritage. Studies in the Cultural Landscape of the Second World War*, edited by Marek E. Jasinski and Leiv Sem. Trondheim: DKNVS, 2015.

Frøland, Hans Otto, Gunnar D. Hatlehol, and Mats Ingulstad. 'Regimenting Labour in Norway during Nazi Germany's Occupation'. *Working Papers of the Independent Commission of Historians Investigating the History of the Reich Ministry of Labour (Reichsarbeitsministerium) in the National Socialist Period No. 12*, 2017.
Frøland, Hans Otto, Mats Ingulstad, and Jonas Scherner, eds. *Industrial Collaboration in Nazi-Occupied Europe: Norway in Context*. London: Palgrave Macmillan, 2016.
Frøland, Hans Otto, and Jan Thomas Kobberrød. 'The Norwegian Contribution to Göring's Megalomania. Norway's Aluminium Industry during World War II'. *Cahiers d'histoire de l'aluminium* 42–43 (2009): 137–53.
Frøland, Hans Otto, and Anders Lervold. 'Forced Labour in Norway during the German Occupation: French and Soviet Workers in the Light Metals Program'. *Revue d'histoire Nordique* 17 (2014): 71–100.
Fuchs, Karlheinz, and Senta Everts-Grigat. *Züblin. 100 Jahre Bautechnik, 1898–1998*. Stuttgart: Ed. Züblin AG, 1998.
Fygle, Svein, Svein Lundestad, and Inge Strand. *Banken, folket og fylket. Nordlandsbanken og Nordlands næringsliv gjennom 100 år: 1893–1993*. Bodø: Nordlandsbanken, 1993.
Gemzell, Carl-Axel. *Raeder, Hitler und Skandinavien. Der Kampf für einen maritimen Operationsplan*. Lund: C.W.K. Gleerup, 1965.
Gerlach, Christian. *Kalkulierte Morde. Die deutsche Wirtschafts- und Vernichtungspolitik in Weißrußland 1941 bis 1944*. Hamburg: Hamburger Edition, 1999.
Geyer, Michael. 'Rüstungsbeschleunigung und Inflation. Zur Inflationsdenkschrift des Oberkommandos der Wehrmacht vom November 1938'. *Militärgeschichtliche Mitteilungen* 30, no. 2 (1981): 121–86.
Gilles, Franz-Otto. *Effizienz ohne Ethik? Zur Praxis der Verwaltungs- und Finanzkontrolle in den während des Zweiten Weltkrieges von Deutschland annektierten bzw. besetzten Gebieten*. Berliner Arbeitshefte und Berichte zur sozialwissenschaftlichen Forschung 78. Berlin, 1992.
Gilles, Franz-Otto, and Gerhard Otto. 'Ordnungsgemäße Beuteverwaltung? Zur Finanz-, Verwaltungs- und Wirtschaftskontrolle in den während des Zweiten Weltkrieges von Deutschland besetzten Gebieten'. In *Autonomie und Kontrolle: Beiträge zur Soziologie des Finanz- und Steuerstaates*, edited by Theo Pirker. Berlin: Schelzky u. Jeep, 1989.
Gilles, Franz-Otto, and Gerhard Otto, eds. *Verwalteter Beutepartikularismus. Finanz-, Verwaltungs- und Wirtschaftskontrolle und nationalsozialistische Besatzungspolitik in den von Deutschland besetzten Gebieten: Ein Tagungsbericht*. Berlin: Freie Universität Berlin, 1991.
Glauning, Christine. *Entgrenzung und KZ-System. Das Unternehmen „Wüste" und das Konzentrationslager in Bisingen 1944/45*. Geschichte der Konzentrationslager 1933–1945 7. Berlin: Metropol Verlag, 2006.
Goldhagen, Erich. 'Albert Speer, Himmler and the Secrecy of the Final Solution'. *Midstream* 17 (1971): 43–50.
Gorgass, Paul. *Das Baugewerbe und seine Eingliederung in die Organisation der gewerblichen Wirtschaft: Eine Untersuchung der Gliederung und betrieblichen Struktur des Baugewerbes*. Berlin: Triltsch & Huther, 1937.
Gossweiler, Kurt. 'Der Übergang von der Weltwirtschaftskrise zur Rüstungskonjunktur in Deutschland 1933 bis 1934. Ein historischer Beitrag zur Problematik staatsmonopolistischer "Krisenüberwindung"'. *Jahrbuch für Wirtschaftsgeschichte* 9, no. 2 (1968): 55–116.
Grebler, Leo. 'Work Creation Policy in Germany, 1932–1935'. *International Labour Review* 35 (1937): 328–51, 505–27.
Gregor, Neil. *Daimler-Benz in the Third Reich*. New Haven, London: Yale University Press, 1998.

Gregor, Neil. 'Wissenschaft, Politik, Hegemonie. Zum Boom der NS-Unternehmensgeschichte'. In *Unternehmen im Nationalsozialismus: Zur Historisierung einer Forschungskonjunktur*, edited by Norbert Frei and Tim Schanetzky, 79–93. Wallstein Verlag, 2010.

Grimnes, Ole Kristian. 'Kollaborasjon og oppgjør'. In *I krigens kjølvann. Nye sider ved norsk krigshistorie og etterkrigstid*, edited by Stein Ugelvik Larsen, 47–57. Oslo: Universitetsforlaget, 1999.

Gross, Manfred. *Der Westwall zwischen Niederrhein und Schnee-Eifel*. Köln: Rheinland-Verlag, 1982.

Grötecke, Johannes. *Edertalsperre, Wiederaufbau nach der Zerstörung: 1943–1945*. Marburg: Jonas Verlag, 1996.

Grotkopp, Wilhelm. *Die grosse Krise. Lehren aus der Überwindung der Wirtschaftskrise 1929/32*. Düsseldorf: Econ Verlag, 1954.

Gruner, Wolf. 'Juden bauen die "Straßen des Führers". Zwangsarbeit und Zwangsarbeitslager für nichtdeutsche Juden im Altreich 1940 bis 1943/44'. *Zeitschrift für Geschichtswissenschaft* 44 (1996): 789–808.

Guttermann, Bella. 'Jews in the Service of Organisation Todt in the Occupied Soviet Territories, October 1941–March 1942'. *Yad Vashem Studies* 29 (2001): 65–107.

Hachtmann, Rüdiger. *Industriearbeit im 'Dritten Reich'. Untersuchungen zu den Lohn- und Arbeitsbedingungen in Deutschland 1933–1945*. Göttingen: Vandenhoeck & Ruprecht, 1989.

Hachtmann, Rüdiger, and Winfried Süß, eds. *Hitlers Kommissare: Sondergewalten in der nationalsozialistischen Diktatur*. Göttingen: Wallstein Verlag, 2006.

Handbuch der deutschen Aktiengesellschaften 1943: Das Spezial-Archiv der deutschen Wirtschaft. Vol. 48. Berlin: Hoppenstedt, 1943.

Hanisch, Tore Jørgen, and Even Lange. *Veien til velstand: industriens utvikling i Norge gjennom 50 år*. Oslo: Universitetsforlaget, 1986.

Harlander, Tilman. *Zwischen Heimstätte und Wohnmaschine. Wohnungsbau und Wohnungspolitik in der Zeit des Nationalsozialismus*. Basel, Berlin, Boston: Birkhäuser, 1995.

Hartmannsgruber, Friedrich. '" . . . ungeachtet der noch ungeklärten Finanzierung". Finanzplanung und Kapitalbeschaffung für den Bau der Reichsautobahnen 1933–1945'. *Historische Zeitschrift* 278, no. 3 (2004): 625–81.

Hatlehol, Gunnar D. 'Einsatzgruppe Wiking og kampen om arbeidskraften 1942–1945'. *Historisk Tidsskrift* 97, no. 3 (2018): 240–55.

Hatlehol, Gunnar D. '"Norwegeneinsatz" 1940–1945. Organisation Todts arbeidere i Norge og gradene av tvang'. NTNU Trondheim, 2015.

Hatlehol, Gunnar D. 'Tvangsstyringen av arbeidslivet under hakekorset 1940–1945: Diktat og kollaborasjon'. *Arbeiderhistorie* 22, no. 1 (2018): 49–71.

Häusler, Hermann. 'Von der Wehrgeologie in Norwegen 1940–45 zum „Salzburger Kreis" der Geomechanik'. In *14. Wissenschaftshistorisches Symposium 'Geologie und Medizin', 4. Dezember 2015*, 56–84. Berichte der Geologischen Bundesanstalt 113. Wien: Geologische Bundesanstalt, 2015.

Hayes, Peter. 'Corporate Freedom of Action in Nazi Germany'. *Bulletin of the German Historical Institute* 45 (2009): 29–42.

Hayes, Peter. *From Cooperation to Complicity: Degussa in the Third Reich*. Cambridge: Cambridge University Press, 2004.

Hayes, Peter. *Industry and Ideology. IG Farben in the Nazi Era*. 1987. Reprint, Cambridge: Cambridge University Press, 2001.

Henning, Hansjoachim. 'Kraftfahrzeugindustrie und Autobahnbau in der Wirtschaftspolitik des Nationalsozialismus 1933 bis 1936'. *Vierteljahrschrift für Sozial- und Wirtschaftsgeschichte* 65, no. 2 (1978): 217–42.

Hensler, Ulrich. *Die Stahlkontingentierung im Dritten Reich*. Stuttgart: Franz Steiner Verlag, 2008.

Hensler, Ulrich. 'Iron and Steel Rationing During the Third Reich'. In *German Industry in the Nazi Period*, edited by Christoph Buchheim. Stuttgart: Franz Steiner Verlag, 2008.
Herbert, Ulrich. *Fremdarbeiter: Politik und Praxis des 'Ausländer-Einsatzes' in der Kriegswirtschaft des Dritten Reiches*. 1985. Reprint, Bonn: Dietz, 1999.
Herbert, Ulrich. 'Labour and Extermination: Economic Interest and the Primacy of Weltanschauung in National Socialism'. *Past & Present* 138 (1993): 144–95.
Hildebrand, Klaus. *The Third Reich*. London, New York: Routledge, 1984.
Hjeltnes, Guri. *Hverdagsliv*. Norge i krig. Fremmedåk og frihetskamp 1940–1945 5. Oslo: Aschehoug, 1986.
Hodne, Fritz. *Norges økonomiske historie 1815–1970*. Oslo: J. W. Cappelens Forlag, 1981.
Hodne, Fritz. *The Norwegian Economy 1920–1980*. New York: St. Martin's Press, 1983.
Hodne, Fritz, and Ola Grytten. *Norsk økonomi 1900–1990*. Oslo: Tano, 1992.
Hodne, Fritz, and Ola Grytten. *Norsk økonomi i det tyvende århundre*. Bergen: Fagbokforlaget, 2002.
Hopmann, Barbara, Mark Spoerer, Birgit Weitz, and Beate Brüninghaus. *Zwangsarbeit bei Daimler-Benz*. Stuttgart: Franz Steiner Verlag, 1994.
Höschle, Gerd. *Die deutsche Textilindustrie zwischen 1933 und 1939. Staatsinterventionismus und ökonomische Rationalität*. Stuttgart: Franz Steiner Verlag, 2004.
Humann, Detlev. *'Arbeitsschlacht'. Arbeitsbeschaffung und Propaganda in der NS-Zeit 1933–1939*. Göttingen: Wallstein Verlag, 2011.
Ingulstad, Mats. 'I nyordningens tegn? Organisation Todt og reguleringen av den norske byggebransjen'. *Historisk Tidsskrift* 97, no. 3 (2018): 256–76.
James, Harold. *Die Deutsche Bank im Dritten Reich*. München: C.H. Beck, 2003.
James, Harold. *Krupp. A History of the Legendary German Firm*. Princeton and Oxford: Princeton University Press, 2012.
James, Harold. *The German Slump. Politics and Economics 1924–1936*. Oxford: Clarendon Press, 1986.
Kaftan, Kurt. *Der Kampf um die Autobahnen: Geschichte und Entwicklung des Autobahngedankens in Deutschland von 1907–1935 unter Berücksichtigung ähnlicher Pläne und Bestrebungen im übrigen Europa*. Berlin: H. Wigankow Verlag, 1955.
Kaienburg, Hermann. 'Jüdische Arbeitslager an der "Straße der SS"'. *1999. Zeitschrift für Sozialgeschichte des 20. und 21. Jahrhunderts* 11, no. 1 (1996): 13–39.
Kaiserliches Statistisches Amt, ed. *Die Geschäftsergebnisse der deutschen Aktiengesellschaften: im Jahre 1907/08 – 1913/14*. Vierteljahrshefte zur Statistik des Deutschen Reichs, Ergänzungsheft II. Berlin: Puttkammer u. Mühlbrecht, 1909.
Kalmbach, Peter. *Wehrmachtjustiz*. Berlin: Metropol Verlag, 2012.
Karlsch, Rainer, and Raymond G. Stokes. *'Faktor Öl': Die Mineralölwirtschaft in Deutschland, 1859–1974*. München: C.H. Beck, 2003.
Karlsen, Ola. *Profitørene: De ukjente landssvikerne*. Oslo: Gyldendal Norsk Forlag, 2017.
Kershaw, Ian. *Das Ende: Kampf bis in den Untergang, NS-Deutschland 1944/45*. München: Deutsche Verlags-Anstalt, 2011.
Kershaw, Ian. *The Nazi Dictatorship: Problems and Perspectives of Interpretation*. 3rd ed. 1985. Reprint, London: Arnold, 1993.
Kilian, Jürgen. *Krieg auf Kosten anderer: Das Reichsministerium der Finanzen und die wirtschaftliche Mobilisierung Europas für Hitlers Krieg*. Berlin, Boston: De Gruyter Oldenbourg, 2017.
Kirkhusmo, Anders. 'Kommentar til Vidar Engs artikkel: Hvorfor arbeidet gode nordmenn på flyplassene i Oslo og Trondheim i april 1940?' *Historisk Tidsskrift* 90, no. 4 (2011): 587–91.
Klass, Gerd v. *Weit spannt sich der Bogen. 1865 – 1965. Die Geschichte der Bauunternehmung Dyckerhoff & Widmann*. Wiesbaden: Verlag für Wirtschaftspublizistik, 1965.

Klein, Burton H. *Germany's Economic Preparations for War*. Cambridge/Mass.: Harvard University Press, 1959.
Klemann, Hein, and Sergei Kudryashov. *Occupied Economies. An Economic History of Nazi-Occupied Europe, 1939–1945*. London, New York: Berg, 2012.
Knechtel, Erhard F. *Arbeitseinsatz im Baugewerbe im Deutschen Reich 1939–1945*. Berlin: Hauptverband der Deutschen Bauindustrie, 2001.
Knechtel, Erhard F. *Auf Dialog gebaut: 50 Jahre Hauptverband der Deutschen Bauindustrie; Eine Chronik*. Wiesbaden, Berlin: Bauverlag, 1998.
Koch, Birgit. 'De sovjetiske, polske og jugoslaviske krigsfanger i tysk fangenskap i Norge 1941–1945'. Universitetet i Oslo, 1988.
Kranig, Andreas. *Lockung und Zwang. Zur Arbeitsverfassung im Dritten Reich*. Stuttgart: Deutsche Verlags-Anstalt, 1983.
Krengel, Rolf. 'Die langfristige Entwicklung der Brutto-Anlage-Investitionen der westdeutschen Industrie von 1924 bis 1955/56'. *Vierteljahrshefte zur Wirtschaftsforschung*, no. 2 (1957): 168–84.
Kroener, Bernhard R. 'Management of Human Resources, Deployment of the Population, and Manning the Armed Forces in the Second Half of the War (1942–1944)'. In *Organization and Mobilization of the German Sphere of Power*, edited by Militärgeschichtliches Forschungsamt. Germany and the Second World War, V/II. Oxford: Clarendon Press, 2015.
Kruszyński, Piotr. 'Die Ausbeutung der Häftlingsarbeit im Komplex Riese im Eulengebirge durch die Organisation TODT und mitarbeitende Firmen'. In *Die Ausnutzung der Zwangsarbeit der Häftlinge des KL Groß-Rosen durch das Dritte Reich*, 40–54. 1999. Reprint, Wałbrzych: Muzeum Gross-Rosen, 2004.
Kugler, Anita. 'Airplanes for the Führer. Adam Opel AG as Enemy Property, Model War Operation, and General Motors Subsidiary, 1939–1945'. In *Working for the Enemy: Ford, General Motors, and Forced Labor in Germany during the Second World War*, edited by Reinhold Billstein, Karola Fings, Anita Kugler, and Nicholas Levis, 33–81. New York, Oxford: Berghahn Books, 2000.
Kunze, Thomas, and Rainer Stommer. 'Geschichte der Reichsautobahn'. In *Reichsautobahn: Pyramiden des Dritten Reiches. Analysen zur Ästhetik eines unbewältigten Mythos*, edited by Rainer Stommer, 22–48. Marburg: Jonas Verlag, 1982.
Küppers, Hans, and Rudolf Bannier. *Einsatzbedingungen der Ostarbeiter sowie der sowjetischen Kriegsgefangenen*. Sonderveröffentlichung des Reichsarbeitsblattes. Berlin: Geschäftsstelle des Reichsarbeitsblattes, 1943.
Kwiet, Konrad. 'Judenmord als Amtsanmaßung. Das Feldurteil vom 12. März 1943 gegen Johann Meißlein'. *Dachauer Hefte* 16 (2000): 125–35.
Laffont, Jean-Jacques, and Jean Tirole. *A Theory of Incentives in Procurement and Regulation*. Cambridge/Mass.: The MIT Press, 2002.
Landmann, Elisabeth. 'Der Einfluss des Krieges auf den Baumarkt'. Friedrich-Wilhelms-Universität Berlin, 1925.
Lange, Even. *Samling om felles mål: 1935–1970*. Aschehougs norgeshistorie 11. Oslo: Aschehoug, 1998.
Lärmer, Karl. *Autobahnbau in Deutschland 1933 bis 1945. Zu den Hintergründen*. Berlin (Ost): Akademie Verlag, 1975.
Laube, Stefan. 'Technik, Arbeit und Zerstörung. Die Organisation Todt in Prag (1944–1945)'. *Bohemia* 40, no. 2 (1999): 387–416.
Lemmes, Fabian. 'Arbeiten für das Reich: die Organisation Todt in Frankreich und Italien, 1940–1945'. European University Institute Firenze, 2009.

Lemmes, Fabian. 'Arbeiten für den Besatzer: Lockung und Zwang bei der Organisation Todt in Frankreich und Italien 1940–1945'. In *Zwangsarbeit in Hitlers Europa. Besatzung – Arbeit – Folgen*, edited by Dieter Pohl and Tanja Sebta. Berlin: Metropol Verlag, 2013.

Lemmes, Fabian. 'Zwangsarbeit im besetzten Europa. Die Organisation Todt in Frankreich und Italien, 1940–1945'. In *Rüstung, Kriegswirtschaft und Zwangsarbeit im 'Dritten Reich'*, edited by Andreas Heusler, Mark Spoerer, and Helmuth Trischler. München: Oldenbourg Verlag, 2010.

Leonhardt, Fritz. *Baumeister in einer umwälzenden Zeit: Erinnerungen*. Stuttgart: Deutsche Verlags-Anstalt, 1984.

Lie, Einar. *Ambisjon og tradisjon: Finansdepartementet 1945–1965*. Oslo: Universitetsforlaget, 1995.

Lofti, Gabriele. 'SS-Sonderlager im nationalsozialistischen Terrorsystem'. In *Ausbeutung, Vernichtung, Öffentlichkeit: neue Studien zur nationalsozialistischen Lagerpolitik*, edited by Norbert Frei, Sybille Steinbacher, and Bernd C. Wagner, 209–29. Darstellungen und Quellen zur Geschichte von Auschwitz 4. München: K. G. Saur, 2000.

Longerich, Peter. *Holocaust: The Nazi Persecution and Murder of the Jews*. Oxford: Oxford University Press, 2010.

Lorentz, Bernhard, and Paul Erker. *Chemie und Politik. Die Geschichte der Chemischen Werke Hüls 1938 bis 1979. Eine Studie zum Problem der Corporate Governance*. München: C.H. Beck, 2003.

Ludwig, Karl-Heinz. *Technik und Ingenieure im Dritten Reich*. Düsseldorf: Droste Verlag, 1974.

Lütgenau, Stefan, and Alexander Schröck. *Zwangsarbeit in der österreichischen Bauindustrie. Die Teerag-Asdag AG 1938–1945*. Innsbruck, Wien, München: Studien-Verlag, 2001.

Marcon, Helmut. *Arbeitsbeschaffungspolitik der Regierungen Papen und Schleicher. Grundsteinlegung für die Beschäftigungspolitik im Dritten Reich*. Frankfurt am Main: Peter Lang, 1974.

Mason, Timothy W. *Social Policy in the Third Reich. The Working Class and the 'National Community'*. Providence and Oxford: Berg, 1993.

Mast, Adolf. *Unternehmer aus Berufung: Adolf Mast erzählt aus seinem Wirken als Bauingenieur und im öffentlichen Leben*. Wiesbaden, Berlin: Bauverlag, 1964.

Matis, Herbert, and Dieter Stiefel. *'Mit der vereinigten Kraft des Capitals, des Credits und der Technik . . .': Die Geschichte der österreichischen Bauwesens am Beispiel der Allgemeinen Baugesellschaft – A. Porr Aktiengesellschaft*. Wien, Köln, Weimar: Böhlau, 1994.

Meyer-Heinrich, Hans. *Philipp Holzmann Aktiengesellschaft im Wandel von hundert Jahren*. Frankfurt am Main: Umschau Verlag, 1949.

Middlebrook, Martin, and Chris Everitt. *The Bomber Command War Diaries. An Operational Reference Book 1939–1945*. Barnsley: Pen & Sword Aviation, 2014.

Mierzejewski, Alfred C. *The Most Valuable Asset of the Reich. A History of the German National Railway. Volume 2, 1933–1945*. Chapel Hill, London: The University of North Carolina Press, 2000.

Milward, Alan S. *The Fascist Economy in Norway*. Oxford: Clarendon Press, 1972.

Mommsen, Hans, and Manfred Grieger. *Das Volkswagenwerk und seine Arbeiter im Dritten Reich*. Düsseldorf: Econ Verlag, 1996.

Müller, Rolf-Dieter. 'Albert Speer and Armaments Policy in Total War'. In *Organization and Mobilization of the German Sphere of Power*, edited by Militärgeschichtliches Forschungsamt. Germany and the Second World War, V/II. Oxford: Clarendon Press, 2015.

Müller, Rolf-Dieter. 'Todt Organization'. In *The Oxford Companion to World War II*, edited by Ian Dear, 870. Oxford: Oxford University Press, 2001.

Myrvang, Christine. *Troskap og flid: Kongsberg Våpenfabrikks historie 1814–1945*. Oslo: Pax Forlag, 2014.

Naasner, Walter. *Neue Machtzentren in der deutschen Kriegswirtschaft 1942–1945*. Boppard am Rhein: Harald Boldt Verlag, 1994.

Neitzel, Sönke. *Die deutschen Ubootbunker und Bunkerwerften. Bau, Verwendung und Bedeutung verbunkerter Ubootstützpunkte in beiden Weltkriegen*. Koblenz: Bernard & Graefe Verlag, 1991.

Nilssen, Trond Risto. 'Jugoslaviske fanger i Norge under andre verdenskrig'. In *Sotavangit ja internoidut. Prisoners of War and Internees. A Book of Articles by the National Archives*, edited by Lars Westerlund, 166–81. Helsinki: Kansallisarkisto, 2008.

Nøkleby, Berit. *Hitlers Norge: Okkupasjonsmakten 1940–1945*. Oslo: Cappelen Damm, 2016.

Nolte, Ernst. 'Big Business and German Politics: A Comment'. *The American Historical Review* 75, no. 1 (1969): 71–78.

Norges Bank. *Norges Bank under okkupasjonen*. Oslo: Norges Bank, 1945.

NOS X. 102. Nasjonalinntekten i Norge 1935–1943. Realkapitalen 1939 og kapitalreduksjonen under krigen. Okkupasjonskostnadene. Oslo: Statistisk Sentralbyrå, 1946.

NOS XII. 291. Historisk Statistikk 1978. Oslo: Statistisk Sentralbyrå, 1978.

Olsen, Kåre. *Innledning til katalog for arkivet etter Organisation Todt Einsatzgruppe Wiking 1940 (–42) – 1945*. Oslo: Riksarkivet Oslo, 2012.

Ottmer, Hans-Martin. 'Skandinavien in der deutschen marinestrategischen Planung der Reichs- bzw. Kriegsmarine'. In *Neutralität und totalitäre Aggression: Nordeuropa und die Großmächte im Zweiten Weltkrieg*, edited by Robert Bohn, Jürgen Elvert, Hain Rebas, and Michael Salewski, 49–72. Stuttgart: Franz Steiner Verlag, 1991.

Ottnad, Bernd, ed. *Baden-Württembergische Biographien 1*. Stuttgart: Kohlhammer, 1994.

Overy, Richard. *The Nazi Economic Recovery 1932–1938*. London: Macmillan, 1982.

Overy, Richard. 'Unemployment in the Third Reich'. In *War and Economy in the Third Reich*, edited by Richard Overy, 37–67. Oxford: Clarendon Press, 1994.

Overy, Richard. *War and Economy in the Third Reich*. Oxford: Clarendon Press, 1994.

Partridge, Colin. *Hitler's Atlantic Wall*. Guernsey: D.I. Publications, 1976.

Paulsen, Helge. 'Norwegen'. In *Verwalteter Beutepartikularismus. Finanz-, Verwaltungs- und Wirtschaftskontrolle und nationalsozialistische Besatzungspolitik in den von Deutschland besetzten Gebieten: Ein Tagungsbericht*, edited by Franz-Otto Gilles and Gerhard Otto, 37–44. Berlin: Freie Universität Berlin, 1991.

Paulsen, Helge. 'Reichskommissariat og "motytelsene" under riksrådsforhandlingene'. In *Norge og den 2. verdenskrig: 1940 – fra nøytralt til okkupert*, edited by Samtidshistorisk Forskningsgruppe, 285–356. Oslo: Universitetsforlaget, 1969.

Pedersen, Per Helge, and Øyvind Steen. *Skaperevne og initiativ. Byggenæringen i 100 år, 1905–2005*. Oslo: Byggeindustrien / Bygg og Anlegg Media AS, 2005.

Perz, Bertrand. 'Die "Neue Österreichische Tunnelbaumethode" und ihre weniger bekannten Bezüge zum Nationalsozialismus'. In *Wissenschaft, Technologie und industrielle Entwicklung in Zentraleuropa im Kalten Krieg*, edited by Wolfgang L. Reiter, Juliane Mikoletzky, Herbert Matis, and Mitchell G. Ash, 245–64. Berlin, Münster, Wien, Zürich, London: LIT Verlag, 2017.

Petersen, Erling. 'Økonomiske forhold'. In *Norges krig 1940–1945: 3*, by Sverre Steen, 495–568. Oslo: Gyldendal, 1950.

Petersen, Joachim. *Hitlers Polareisenbahnpläne 1940 bis 1945 in Nordnorwegen*. Friedberg: Podzun-Pallas-Verlag, 1992.

Petzina, Dietmar. *Autarkiepolitik im Dritten Reich. Der nationalsozialistische Vierjahresplan*. Stuttgart: Deutsche Verlags-Anstalt, 1968.

Petzina, Dietmar. 'Hauptprobleme der deutschen Wirtschaftspolitik 1932/33'. *Vierteljahreshefte für Zeitgeschichte* 15, no. 1 (1967): 18–55.

Pfahlmann, Hans. *Fremdarbeiter und Kriegsgefangene in der deutschen Kriegswirtschaft 1939–1945*. Darmstadt: Wehr und Wissen, 1968.

Pohl, Manfred. *Die Strabag: 1923 bis 1998*. München: Piper, 1998.

Pohl, Manfred. *Philipp Holzmann: Geschichte eines Bauunternehmens 1849–1999*. München: C.H. Beck, 1999.
Pohl, Manfred, and Birgit Siekmann. *HOCHTIEF und seine Geschichte: Von den Brüdern Helfmann bis ins 21. Jahrhundert*. München, Zürich: Piper, 2000.
Pritzkoleit, Kurt. *Gott erhält die Mächtigen: Rück- und Rundblick auf den deutschen Wohlstand*. Düsseldorf: Karl Rauch Verlag, 1963.
Prollius, Michael von. *Das Wirtschaftssystem der Nationalsozialisten 1933–1939. Steuerung durch emergente Organisation und Politische Prozesse*. Paderborn: Ferdinand Schöningh, 2003.
Raim, Edith. *Die Dachauer KZ-Außenkommandos Kaufering und Mühldorf. Rüstungsbauten im letzten Kriegsjahr 1944/45*. Landsberg am Lech: Landsberger Verlagsanstalt Martin Neumeyer, 1992.
Rauh-Kühne, Cornelia. 'Hitlers Hehler? Unternehmerprofite und Zwangsarbeiterlöhne'. *Historische Zeitschrift* 275, no. 1 (2002): 1–55.
Recker, Marie-Luise. *Nationalsozialistische Sozialpolitik im Zweiten Weltkrieg*. München: Oldenbourg Verlag, 1985.
Reichs-Kredit-Gesellschaft. *Deutschlands wirtschaftliche Lage in der Jahresmitte 1939*. Berlin: Reichs-Kredit-Gesellschaft, 1939.
Reichs-Verdingungs-Ausschuß. *Verdingungsordnung für Bauleistungen (VOB)*. Berlin: Bauwelt-Verlag, 1926.
Reichsversicherungsamt. *Die deutsche Sozialversicherung 1937 – mit einem Blick auf das Jahr 1938*. Sonderveröffentlichung des Reichsarbeitsblatts und der Amtlichen Nachrichten für Reichsversicherung. Berlin: Reichsversicherungsamt, 1938.
Renn, Walter Frederik. 'Hitler's West Wall: Strategy in Concrete and Steel 1938–1945'. The Florida State University, 1970.
Reuss, Erich. *Die Verbände der Bauindustrie*. Wiesbaden: Bauverlag, 1966.
Richter, Rudolf, and Eirik G. Furubotn. *Institutions and Economic Theory. The Contribution of the New Institutional Economics*. 2nd ed. Ann Arbor: The University of Michigan Press, 2010.
Rinde, Harald. *Det moderne fylket*. Nordlands historie 3. Oslo: Fagbokforlaget, 2015.
Ritschl, Albrecht. 'Hat das Dritte Reich wirklich eine ordentliche Beschäftigungspolitik betrieben?' *Jahrbuch für Wirtschaftsgeschichte* 44, no. 1 (2003): 125–40.
Roseman, Mark. *The Villa, the Lake, the Meeting: Wannsee and the Final Solution*. London: Penguin Books, 2002.
Rutar, Sabine. 'Arbeit und Überleben in Serbien. Das Kupfererzbergwerk Bor im Zweiten Weltkrieg'. *Geschichte und Gesellschaft* 31, no. 1 (2005): 101–34.
Sæveraas, Torgeir Ekerholt. '"Beton macht Geschichte" – Organisation Todt og utbyggingen av Festung Norwegen'. NTNU Trondheim, 2016.
Sæveraas, Torgeir Ekerholt. 'OT, Wehrmacht og byggingen av Festung Norwegen'. *Historisk Tidsskrift* 97, no. 3 (2018): 189–205.
Sager & Woerner AG. *Sager & Woerner 1864–1964*. Vol. 2: Sammlung-Einsatz-Krise-Erstarkung, 1920–1945. München: Sager & Woerner AG, 1969.
Saldern, Adelheid von. *Mittelstand im 'Dritten Reich': Handwerker – Einzelhändler – Bauern*. Frankfurt am Main, New York: Campus Verlag, 1979.
Sandvik, Pål Thonstad. *Falconbridge Nikkelverk 1910 – 1929 – 2004: Et internasjonalt selskap i Norge*. Kristiansand: Falconbridge Nikkelverk, 2004.
Sappington, David E. M. 'Incentives in Principal-Agent Relationships'. *The Journal of Economic Perspectives* 5, no. 2 (1991): 45–66.
Sarholz, Thomas. 'Die Auswirkungen der Kontingentierung von Eisen und Stahl auf die Aufrüstung der Wehrmacht von 1936 bis 1939'. TU Darmstadt, 1983.

Schalm, Sabine. *Überleben durch Arbeit? Außenkommandos und Außenlager des KZ Dachau 1933–1945*. Geschichte der Konzentrationslager 1933–1945 10. Berlin: Metropol Verlag, 2009.
Scherner, Jonas. 'Anreiz statt Zwang. Wirtschaftsordnung und Kriegswirtschaft im "Dritten Reich"'. In *Unternehmen im Nationalsozialismus: Zur Historisierung einer Forschungskonjunktur*, edited by Norbert Frei and Tim Schanetzky, 140–55. Göttingen: Wallstein Verlag, 2010.
Scherner, Jonas. '"Armament in Depth" or "Armament in Breadth"? German Investment Pattern and Rearmament during the Nazi Period'. *Economic History Review* 66, no. 2 (2013): 497–517.
Scherner, Jonas. 'Bericht zur deutschen Wirtschaftslage 1943/44. Eine Bilanz des Reichsministeriums für Rüstung und Kriegsproduktion über die Entwicklung der deutschen Kriegswirtschaft bis Sommer 1944'. *Vierteljahreshefte für Zeitgeschichte* 55, no. 3 (2007): 499–546.
Scherner, Jonas. 'Das Verhältnis zwischen NS-Regime und Industrieunternehmen – Zwang oder Kooperation?' *Zeitschrift für Unternehmensgeschichte / Journal of Business History* 51, no. 2 (2006): 166–90.
Scherner, Jonas. 'Der deutsche Importboom während des Zweiten Weltkriegs. Neue Ergebnisse zur Struktur der Ausbeutung des besetzten Europas auf der Grundlage einer Neuschätzung der deutschen Handelsbilanz'. *Historische Zeitschrift* 294, no. 1 (2012): 79–113.
Scherner, Jonas. *Die Logik der Industriepolitik im Dritten Reich – Die Investitionen in die Autarkie- und Rüstungsindustrie und ihre staatliche Förderung*. Stuttgart: Franz Steiner Verlag, 2008.
Scherner, Jonas. 'Keine primäre moralische Verantwortung? Zum Umfang und zu Grenzen unternehmerischer Handlungsspielräume in der NS-Zeit'. In *Ökonomie und Ethik. Beiträge aus Wirtschaft und Geschichte*, edited by Gert Kollmer-von Oheimb-Loup, Sibylle Lehmann-Hasemeyer, and Stefanie van de Kerkhof, 149–87. Ostfildern: Jan Thorbecke Verlag, 2017.
Scherner, Jonas. 'Nazi Germany's Preparation for War: Evidence from Revised Industrial Investment Series'. *European Review of Economic History* 14 (2010): 433–68.
Scherner, Jonas. 'The Institutional Architecture of Financing German Exploitation: Principles, Conflicts, and Results'. In *Paying for Hitler's War: The Consequences of Nazi Hegemony for Europe*, edited by Jonas Scherner and Eugene N. White, 43–66. New York: Cambridge University Press, 2015.
Scherner, Jonas, and Jochen Streb. 'Das Ende eines Mythos? Albert Speer und das sogenannte Rüstungswunder'. *Vierteljahresschrift für Sozial- und Wirtschaftsgeschichte* 93, no. 2 (2006): 172–96.
Schiller, Karl. *Arbeitsbeschaffung und Finanzordnung in Deutschland*. Berlin: Junker und Dünnhaupt, 1936.
Schlimper, Joachim. 'Die Preisermittlung im Hochbaugewerbe'. Handels-Hochschule Leipzig, 1940.
Schmidt, Matthias. *Albert Speer. Das Ende eines Mythos*. 1982. Reprint, Berlin: NZ Netzeitung, 2005.
Schneider, Michael. 'The Development of State Work-Creation Policy in Germany, 1930–1933'. In *Unemployment and the Great Depression in Weimar Germany*, edited by Peter D. Stachura, 163–86. New York: St. Martin's Press, 1986.
Schneider, Michael. *Unterm Hakenkreuz: Arbeiter und Arbeiterbewegung 1933 bis 1939*. Bonn: Dietz, 1999.
Schneider, Michael C. *Unternehmensstrategien zwischen Weltwirtschaftskrise und Kriegswirtschaft. Chemnitzer Maschinenbauindustrie in der NS-Zeit 1933–1945*. Essen: Klartext Verlag, 2005.
Schulze-Fielitz, Günther. *Die Bauwirtschaft im Kriege*. Berlin: Junker und Dünnhaupt, 1941.
Schütz, Erhard, and Eckhard Gruber. *Mythos Reichsautobahn. Bau und Inszenierung der 'Straßen des Führers' 1933–1941*. Berlin: Christoph Links Verlag, 1996.
Schweitzer, Arthur. 'Plans and Markets: Nazi Style'. *Kyklos* 30, no. 1 (1977): 88–115.

Seidler, Franz W. 'Das Nationalsozialistische Kraftfahrkorps und die Organisation Todt im Zweiten Weltkrieg. Die Entwicklung des NSKK bis 1939'. *Vierteljahreshefte für Zeitgeschichte* 32, no. 4 (1984): 625–36.
Seidler, Franz W. *Die Organisation Todt. Bauen für Staat und Wehrmacht 1938–1945.* 2nd ed. 1987. Reprint, Koblenz: Bernard & Graefe Verlag, 1998.
Seidler, Franz W. *Fritz Todt. Baumeister des Dritten Reiches.* 1986. Reprint, München: F. A. Herbig, 1988.
Seidler, Franz W. 'L'Organisation Todt'. *Revue d'histoire de la Deuxième Guerre mondiale et des conflits contemporains* 34, no. 134 (1984): 33–58.
Sereny, Gitta. *Das Ringen mit der Wahrheit: Albert Speer und das deutsche Trauma.* München: Kindler, 1995.
Shand, James D. 'The Reichsautobahn: Symbol for the Third Reich'. *Journal of Contemporary History* 19, no. 2 (1984): 189–200.
Shirer, William L. *The Rise and Fall of the Third Reich: A History of Nazi Germany.* London: Secker and Warburg, 1961.
Siedbürger, Hermann. *Abgabe der Gewinnerklärung im Handwerk: Gewinnprüfung, Preissenkung und Gewinnabführung.* Wiesbaden: Betriebswirtschaftlicher Verlag Dr. Th. Gabler, 1942.
Siegel, Tilla. *Leistung und Lohn in der nationalsozialistischen 'Ordnung der Arbeit'.* Opladen: Westdeutscher Verlag, 1989.
Silverman, Dan P. *Hitler's Economy: Nazi Work Creation Programs, 1933–1936.* Cambridge/Mass., London: Harvard University Press, 1998.
Simonsen, Kjetil Braut. 'Nazifisering, kollaborasjon, motstand. En analyse av Politidepartementet og Forsyningsdepartementet (Næringsdepartementet), 25. september 1940 – 8. mai 1945'. Universitetet i Oslo, 2016.
Skodvin, Magne. *Norsk historie 1939–1945: Krig og okkupasjon.* Oslo: Det Norske Samlaget, 1991.
Solbrekken, Ingeborg. *Galskap og rettergang: landssviksaken mot Kirsten Flagstad og Henry Johansen.* Oslo: Transit, 2007.
Soleim, Marianne Neerland, ed. *Prisoners of War and Forced Labour: Histories of War and Occupation.* Newcastle: Cambridge Scholars Publishing, 2010.
Soleim, Marianne. *Sovjetiske krigsfanger i Norge 1941–1945. Antall, organisering og repatriering.* Tromsø: Universitetet i Tromsø, 2004.
Soleim, Marianne. *Sovjetiske krigsfanger i Norge 1941–1945. Antall, organisering og repatriering.* Oslo: Spartacus Forlag, 2009.
Soo, Scott. 'Ambiguities at Work: Spanish Republican Exiles and the Organisation Todt in Occupied Bordeaux'. *Modern & Contemporary France* 15, no. 4 (2007): 457–77.
Speer, Albert. *Der Sklavenstaat. Meine Auseinandersetzung mit der SS.* Berlin: Ullstein, 1981.
Speer, Albert. *Erinnerungen.* Frankfurt am Main, Berlin: Ullstein, 1969.
Speer, Albert. *Infiltration: How Heinrich Himmler Schemed to Build an SS Industrial Empire.* New York: Macmillan, 1981.
Speer, Albert. *Inside the Third Reich.* New York: Macmillan, 1970.
Speer, Albert. *Spandau: The Secret Diaries.* New York: Macmillan, 1976.
Speer, Albert. *Spandauer Tagebücher.* Berlin: Propyläen, 1975.
Spoerer, Mark. 'Industrial Profitability in the Nazi Economy'. In *After the Slump. Industry and Politics in 1930s Britain and Germany*, edited by Christoph Buchheim and Redvers Garside, 53–80. Frankfurt am Main: Peter Lang, 2000.
Spoerer, Mark. 'Moralische Geste oder Angst vor Boykott? Welche Großunternehmen beteiligten sich aus welchen Gründen an der Entschädigung ehemaliger NS-Zwangsarbeiter?' *Perspektiven der Wirtschaftspolitik* 3, no. 1 (2002): 37–48.

Spoerer, Mark. 'Profitierten Unternehmen von KZ-Arbeit? Eine kritische Analyse der Literatur'. *Historische Zeitschrift* 268, no. 1 (1999): 61–95.
Spoerer, Mark. *Von Scheingewinnen zum Rüstungsboom: Die Eigenkapitalrentabilität der deutschen Industrieaktiengesellschaften 1925–1941*. Stuttgart: Franz Steiner Verlag, 1996.
Spoerer, Mark. *Zwangsarbeit unter dem Hakenkreuz. Ausländische Zivilarbeiter, Kriegsgefangene und Häftlinge im Deutschen Reich und im besetzten Europa 1939–1945*. Stuttgart, München: Deutsche Verlags-Anstalt, 2001.
Spoerer, Mark, and Jochen Streb. *Neue deutsche Wirtschaftsgeschichte des 20. Jahrhunderts*. München: Oldenbourg Verlag, 2013.
Statistisches Handbuch von Deutschland 1928 – 1944. München: Franz Ehrenwirth-Verlag, 1949.
Statistisches Jahrbuch für das Deutsche Reich 1926 [to] 1941/42. Berlin: Schmidt, 1926.
Statistisches Jahrbuch für das Deutsche Reich 1936. Berlin: Schmidt, 1936.
Statistisches Jahrbuch für das Deutsche Reich 1937. Berlin: Schmidt, 1937.
Statistisches Jahrbuch für das Deutsche Reich 1938. Berlin: Schmidt, 1938.
Statistisches Jahrbuch für das Deutsche Reich 1939/40. Berlin: Schmidt, 1940.
Statistisches Jahrbuch für das Deutsche Reich 1941/42. Berlin: Schmidt, 1942.
Statistisches Reichsamt. *Die berufliche und soziale Gliederung der Bevölkerung des Deutschen Reichs*. Vol. 2. Statistik des Deutschen Reichs 453. Berlin: Verlag für Sozialpolitik, Wirtschaft und Statistik, 1936.
Statistisches Reichsamt, ed. *Vierteljahreshefte zur Statistik des Deutschen Reichs 39/II*. Berlin: Schmidt, 1930.
Statistisk-økonomisk utsyn over krigsårene. Oslo: Statistisk Sentralbyrå, 1945.
Steen, Øyvind. *I første rekke. Ingeniør F. Selmer A/S 1906–1981*. Oslo: Ingeniør F. Selmer A/S, 1981.
Steffenak, Einar Kristian. *Russerfangene*. Oslo: Humanist Forlag, 2008.
Steiner, André. 'Der Reichskommissar für die Preisbildung – "eine Art wirtschaftlicher Reichskanzler"?' In *Hitlers Kommissare: Sondergewalten in der nationalsozialistischen Diktatur*, edited by Rüdiger Hachtmann and Winfried Süß, 93–114. Göttingen: Wallstein Verlag, 2006.
Steiner, André. 'Industry and Administrative Price Regulation 1933–1938/39'. In *German Industry in the Nazi Period*, edited by Christoph Buchheim. Stuttgart: Franz Steiner Verlag, 2008.
Stier, Bernhard, and Martin Krauß. *Drei Wurzeln – ein Unternehmen. 125 Jahre Bilfinger Berger AG*. Heidelberg, Ubstadt-Weiher, Basel: Ifu im Verl. Regionalkultur, 2005.
Stokke, Michael. 'Sovjetiske og franske sivile tvangsarbeidere i Norge 1942–1945: en sammenligning av arbeids- og leveforhold'. Universitetet i Bergen, 2008.
Stommer, Rainer, ed. *Reichsautobahn: Pyramiden des Dritten Reiches. Analysen zur Ästhetik eines unbewältigten Mythos*. Marburg: Jonas Verlag, 1982.
Storeide, Anette. *Norske krigsprofitører – Nazi-Tysklands velvillige medløpere*. Oslo: Gyldendal, 2014.
Strasser, Gregor. *Das Wirtschaftliche Aufbauprogramm Der NSDAP*. Berlin: Hempel, 1932.
Streb, Jochen. 'Das nationalsozialistische Wirtschaftssystem: Indirekter Sozialismus, gelenkte Marktwirtschaft oder vorgezogene Kriegswirtschaft?' In *Der Staat und die Ordnung der Wirtschaft: Vom Kaiserreich bis zur Berliner Republik*, edited by Werner Plumpe and Joachim Scholtyseck, 61–83. Stuttgart: Franz Steiner Verlag, 2012.
Streb, Jochen. 'Das Scheitern der staatlichen Preisregulierung in der nationalsozialistischen Bauwirtschaft'. *Jahrbuch für Wirtschaftsgeschichte*, no. 1 (2003): 27–48.
Streb, Jochen. 'Negotiating Contract Types and Contract Clauses in the German Construction Industry during the Third Reich'. *The RAND Journal of Economics* 40, no. 2 (2009): 364–79.
Streb, Jochen, and Sabine Streb. 'Optimale Beschaffungsverträge bei asymmetrischer Informationsverteilung: Zur Erklärung des nationalsozialistischen "Rüstungswunders"

während des Zweiten Weltkriegs'. *Zeitschrift für Wirtschafts- und Sozialwissenschaften* 118 (1998): 275–97.
Streit, Christian. *Keine Kameraden. Die Wehrmacht und die sowjetischen Kriegsgefangenen 1941–1945*. Stuttgart: Deutsche Verlags-Anstalt, 1978.
Svendsen, Gun Helena. *Anleggsprat. Jubileumsnummer 1990: 50 år*. Trondheim: A/S Anlegg, 1991.
Tarifordnung über den Leistungslohn im Baugewerbe nebst den bisher erschienenen Anhängen 1 bis 13. Düsseldorf: Droste Verlag, 1953.
Taylor, Telford. *Sword and Swastika: The Wehrmacht in the Third Reich*. London: Victor Gollancz Ltd., 1953.
Temin, Peter. 'Soviet and Nazi Economic Planning in the 1930s'. *Economic History Review* 44, no. 4 (1991): 573–93.
Theis, Kerstin. *Wehrmachtjustiz an der 'Heimatfront': Die Militärgerichte des Ersatzheeres im Zweiten Weltkrieg*. Berlin, Boston: De Gruyter Oldenbourg, 2016.
Thormann-und-Stiefel-AG Augsburg. *75 Jahre 'Thosti', Thormann & Stiefel Aktiengesellschaft Augsburg, 1876–1951*. Darmstadt: Hoppenstedt, 1951.
Tjelmeland, Hallvard. 'Kva gjorde krigen med Norge?' *Nytt Norsk Tidsskrift* 13, no. 1 (1996): 28–42.
Tooze, Adam. 'No Room for Miracles: German Industrial Output in World War II Reassessed'. *Geschichte Und Gesellschaft* 31, no. 3 (2005): 439–64.
Tooze, Adam. *The Wages of Destruction: The Making and Breaking of the Nazi Economy*. London: Penguin Books, 2007.
Trommer, Isabell. *Rechtfertigung und Entlastung: Albert Speer in der Bundesrepublik*. Frankfurt am Main: Campus Verlag, 2016.
Turner, Henry A. 'Unternehmen unter dem Hakenkreuz'. In *Unternehmen im Nationalsozialismus*, edited by Lothar Gall and Manfred Pohl, 15–23. München: C.H. Beck, 1998.
Tuveng, Morten. *Arbeidsløshet og beskjeftigelse i Norge før og under krigen. En statistisk analyse med henblikk på beskjeftigelsesspørsmålet etter krigen*. Bergen: John Griegs Forlag, 1946.
Umbreit, Hans. 'German Rule in the Occupied Territories 1942–1945'. In *Organization and Mobilization of the German Sphere of Power*, edited by Militärgeschichtliches Forschungsamt. Germany and the Second World War, V/II. Oxford: Clarendon Press, 2015.
Valen, Terje. *De tjente på krigen: hjemmefronten og kapitalen*. Oslo: Oktober, 1978.
Vat, Dan van der. *The Good Nazi. The Life & Lies of Albert Speer*. Boston, New York: Houghton Mifflin Company, 1997.
Vogt, Johan. *Vår økonomiske stilling: En kritisk vurdering av gjenreisningsperioden*. Oslo: Aschehoug, 1950.
Vogt, Lorentz. *Det organiserte, offisielle, økonomiske samarbeide med okkupantene*. Oslo, 1947.
Volkmann, Hans-Erich. 'Politik, Wirtschaft und Aufrüstung unter dem Nationalsozialismus'. In *Hitler, Deutschland und die Mächte: Materialien zur Außenpolitik des Dritten Reiches*, edited by Manfred Funke, 269–91. Düsseldorf: Droste Verlag, 1976.
Volkmann, Hans-Erich. 'The National Socialist Economy in Preparation for War'. In *The Build-up of German Aggression*, edited by Militärgeschichtliches Forschungsamt. Germany and the Second World War, I. Oxford: Clarendon Press, 2015.
Wagemann, Ernst. *Konjunkturstatistisches Handbuch 1936*. Hamburg: Hanseatische Verlagsanstalt, 1935.
Wagenführ, Rolf. *Die deutsche Industrie im Kriege 1939–1945*. Berlin: Duncker & Humblot, 1954.
Wagner, Bernd C. *IG Auschwitz. Zwangsarbeit und Vernichtung von Häftlingen des Lagers Monowitz 1941–1945*. München: K. G. Saur, 2000.
Wagon, Eduard. *Die finanzielle Entwicklung deutscher Aktiengesellschaften von 1870–1900 und die Gesellschaften mit beschränkter Haftung im Jahre 1900*. Sammlung nationalökonomischer und

statistischer Abhandlungen des staatswissenschaftlichen Seminars zu Halle a.d.S. 39. Jena: Gustav Fischer, 1903.
Wasberg, Gunnar, and Arnljot Svendsen. *Industriens historie i Norge*. Oslo: Norges Industriforbund, 1969.
Weber, Konrad. 'Baugeräte'. In *Gelenkte Bauwirtschaft*, edited by Eitelfritz Kühne. Berlin: Otto Elsner, 1941.
Werner, Wolfgang Franz. *'Bleib übrig!' – Deutsche Arbeiter in der nationalsozialistischen Kriegswirtschaft*. Düsseldorf: Schwann, 1983.
Westlie, Bjørn. *Fangene som forsvant: NSB og slavearbeiderne på Nordlandsbanen*. Oslo: Spartacus Forlag, 2015.
Westlie, Bjørn. 'NSB, krigsfangene og andre verdenskrig'. Universitetet i Oslo, 2018.
Wicken, Olav. 'Industrial Change in Norway during the Second World War: Electrification and Electrical Engineering'. *Scandinavian Journal of History* 8 (1983): 119–50.
Willems, Susanne. *Der entsiedelte Jude: Albert Speers Wohnungsmarktpolitik für den Berliner Hauptstadtbau*. Berlin: Edition Hentrich, 2002.
Williamson, Oliver E. *The Economic Institutions of Capitalism. Firms, Markets, Relational Contracting*. New York: The Free Press, 1985.
Willmot, Louise. 'The Goodness of Strangers: Help to Escaped Russian Slave Labourers in Occupied Jersey, 1942–1945'. *Contemporary European History* 11, no. 2 (2002): 211–27.
Wulff, Birgit. *Arbeitslosigkeit und Arbeitsbeschaffungsmaßnahmen in Hamburg 1933–1939. Eine Untersuchung zur nationalsozialistischen Wirtschafts- und Sozialpolitik*. Frankfurt am Main: Peter Lang, 1987.
Ziegler, Dieter. '"A Regulated Market Economy": New Perspectives on the Nature of the Economic Order of the Third Reich, 1933–1939'. In *Business in the Age of Extremes. Essays in Modern German and Austrian Economic History*, edited by Hartmut Berghoff, Jürgen Kocka, and Dieter Ziegler, 139–52. Cambridge: Cambridge University Press, 2013.

Index

https://doi.org/10.1515/9783110694291-011

www.ingramcontent.com/pod-product-compliance
Ingram Content Group UK Ltd.
Pitfield, Milton Keynes, MK11 3LW, UK
UKHW051029041026
15470UKWH00016B/1810

9 783110 694109